NATO ASI Series

Advanced Science Institutes Series

A series presenting the results of activities sponsored by the NATO Science Committee, which aims at the dissemination of advanced scientific and technological knowledge, with a view to strengthening links between scientific communities.

The Series is published by an international board of publishers in conjunction with the NATO Scientific Affairs Division

A Life Sciences	Plenum Publishing Corporation
B Physics	London and New York
C Mathematical and Physical Sciences	Kluwer Academic Publishers Dordrecht, Boston and London
D Behavioural and Social Sciences	
E Applied Sciences	
F Computer and Systems Sciences	Springer-Verlag Berlin Heidelberg New York
G Ecological Sciences	London Paris Tokyo
H Cell Biology	

The ASI Series Books Published as a Result of
Activities of the Special Programme on
SENSORY SYSTEMS FOR ROBOTIC CONTROL

This book contains the proceedings of a NATO Advanced Research Workshop held within the activities of the NATO Special Programme on Sensory Systems for Robotic Control, running from 1983 to 1988 under the auspices of the NATO Science Committee.

The books published so far as a result of the activities of the Special Programme are:

Vol. F25: Pyramidal Systems for Computer Vision. Edited by V. Cantoni and S. Levialdi. 1986.

Vol. F29: Languages for Sensor-Based Control in Robotics. Edited by U. Rembold and K. Hörmann. 1987.

Vol. F33: Machine Intelligence and Knowledge Engineering for Robotic Applications. Edited by A.K.C. Wong and A. Pugh. 1987.

Vol. F42: Real-Time Object Measurement and Classification. Edited by A.K. Jain. 1988.

Vol. F43: Sensors and Sensory Systems for Advanced Robots. Edited by P. Dario. 1988.

Vol. F44: Signal Processing and Pattern Recognition in Nondestructive Evaluation of Materials. Edited by C.H. Chen. 1988.

Vol. F45: Syntactic and Structural Pattern Recognition. Edited by G. Ferraté, T. Pavlidis, A. Sanfeliu and H. Bunke. 1988.

Vol. F50: CAD Based Programming for Sensory Robots. Edited by B. Ravani. 1988.

Vol. F52: Sensor Devices and Systems for Robotics. Edited by Alícia Casals. 1989.

Sensor Devices and Systems for Robotics

Edited by
Alícia Casals
Facultat d'Informàtica de Barcelona (U.P.C.)
c/ Pau Gargallo, 5, 08028 Barcelona (Spain)

Springer-Verlag Berlin Heidelberg GmbH

Proceedings of the NATO Advanced Research Workshop on Sensor Devices and Systems for Robotics held in Catalonia, Spain, October 13–16, 1987.

Library of Congress Cataloging-in-Publication Data. NATO Advanced Research Workshop on Sensor Devices and Systems for Robotics (1987: Catalonia, Spain) Sensor devices and systems for robotics/ edited by Alícia Casals. p. cm.—(NATO ASI series. Series F, Computer and systems sciences; vol. 52) "Proceedings of the NATO Advanced Research Workshop on Sensor Devices and Systems for Robotics, held in Catalonia, Spain, October 13–16, 1987"—T.p. verso. "Published in cooperation with NATO Scientific Affairs Division."

ISBN 978-3-642-74569-0   ISBN 978-3-642-74567-6 (eBook)
DOI 10.1007/978-3-642-74567-6

1. Robotics—Congresses. 2. Detectors—Congresses. I. Casals, Alícia, 1955-. II. North Atlantic Treaty Organization. Scientific Affairs Division. III. Title. IV. Series: NATO ASI series. Series F, Computer and systems sciences; vol. 52.
TJ210.3.N3755 1987   629.8'92—dc 19   89-5995

© Springer-Verlag Berlin Heidelberg 1989
Originally published by Springer-Verlag Berlin Heidelberg New York in 1989
Softcover reprint of the hardcover 1st edition 1989

2145/3140-543210 – Printed on acid-free paper

PREFACE

As robots improve in efficiency and intelligence, there is a growing need to develop more efficient, accurate and powerful sensors in accordance with the tasks to be robotized. This has led to a great increase in the study and development of different kinds of sensor devices and perception systems over the last ten years. Applications that differ from the industrial ones are often more demanding in sensorics since the environment is not usually so well structured. Spatial and agricultural applications are examples of situations where the environment is unknown or variable. Therefore, the work to be done by a robot cannot be strictly programmed and there must be an interactive communication with the environment. It cannot be denied that evolution and development in robotics are closely related to the advances made in sensorics.

The first vision and force sensors utilizing discrete components resulted in a very low resolution and poor accuracy. However, progress in VLSI, imaging devices and other technologies have led to the development of more efficient sensor and perception systems which are able to supply the necessary data to robots.

This book is a collection of papers presented at the NATO Advanced Research Workshop (ARW) "Sensor Devices and Systems for Robotics" held in October 1987 at the Costa Brava, Catalonia, Spain. About 40 researchers from the academic community and from industry from all over the world were invited and 22 of these presented papers. Many of the participants are well-known researchers in the field of Robotics and Sensorics.

The workshop was divided into sessions dealing with the different kinds of sensors and related systems for data interpretation with special emphasis on their application to Robotics. The sessions were in the form of lectures and panel discussions. In addition to the discussion panels the debates that arose from the lectures contributed greatly to the success of the workshop.

Robotics and Sensorics are subjects of considerable interest today and the participants recognized the need to promote contacts among researchers in order to encourage further development and progress. In this connection, the question of whether to include the teaching of Robotics in Electrical and Electronic studies or in Mechanical Engineering studies was discussed. However, the field is so wide that it would not be feasible to restrict Robotics to one or other of these fields. Nevertheless, agreement was reached on the need to set up new centers.

This book includes lectures concerning the sensors, their use in perception systems and their applications. It is divided into parts dealing with the different kinds of parameters to be measured in accordance with each application. Different aspects of force and torque sensors, which are of special interest in assembly tasks, are treated: making models of the interaction between the robot and its environment so as not to damage the manipulated objects and obtaining information from the robot for calibration. A number of punctual force sensors may be used to build a tactile surface. Tactile sensors are analyzed in two ways: using sensitive materials (carbon and piezoelectric film) and evaluating their own geometry and influence on the sensor performance.

Although acoustic sensors have a poorer resolution than vision sensors, acoustic sensors can measure distances more easily than vision systems. Acoustic sensors are used in robotics to provide feedback for robot motion control. Acoustic signals are used in range finder systems to control a robot task. Acoustic sensors can also be used to obtain a 3D map of the robot environment.

Optical sensors and vision systems are treated in different papers. The first group of papers concerns the analysis of the optical sensors and imaging devices. The second group deals with the use of these devices to build vision systems, a combined laser/camera vision system for calibration. The last two papers are devoted to parameter estimation in digital signal processing and object shape estimation from partial information. The last section on sensors ends with communications on other kinds of sensors such as dynamic weighing, position deviation and multisensorial fusion.

Finally, some applications arising form the use of force or vision sensors are described.

This book is, in short, an overview of the current state of robot sensors and their application.

ACKNOWLEDGEMENTS

The editor would like to acknowledge the support and encouragement provided by the vision group of the Computer's Technology Department, especially that of J. Amat and V. Llario.

The helpful suggestions of P. Dario and H. van Brussel are greatly appreciated. The editor is also indebted to Mr. Blackbourn of the U.S. Navy for his interest in the workshop and to Mr. L.V. DaCunha and Mr. G.A. Venturi, Program Directors, for their support in the organization of the workshop.

The grant received from the CIRIT (Research Agency of the Catalan Government) is gratefully appreciated.

The editor would also like to thank all the lecturers and participants whose active participation throughout helped to make the workshop fruitful and enjoyable.

Barcelona, December 1988   Alícia Casals

TABLE OF CONTENTS

IV. Optical sensors

V. Other kinds of sensors

VI. Applications

I. FORCE AND TORQUE SENSORS

JOINT FORCE SENSING FOR
UNIFIED MOTOR LEARNING

A. Mukerjee
Department of Computer Science
Texas A&M University
College Station, Texas 77843-3112 (USA)

ABSTRACT

Motor learning consists of using in-built sensors to learn more about one's own motion behavior. In this paper, we present an approach for motor learning based on a perturbed parameter scheme. A technique is developed for determining the link inertias based on joint reaction data, obtained through force sensors. Due to inexactness of the model, the parameters thus estimated are likely to differ from their true values. This perturbed parameter set can be thought of as a "learned" model of the executed motion. Running the dynamics procedure with these altered parameters results in a more accurate prediction of the control torques needed for the desired motion.

1. Introduction

In the manipulator task execution domain, motions are usually executed with all degrees of freedom under direct control and without regard to the high-level representation for the motion. The trajectory is reduced to a joint motion history and this is implemented through a suitable control scheme. It is well recognized that there are several interrelated effects which make the computations inadequate. The dynamics is highly coupled and non-linear. Also the model usually ignores significant effects like joint compliance, loading effects and coulomb friction. Finally, there are uncertainties in the model parameters such as inertias and geometrical terms. These inaccuracies, as well the computation-intensive nature of the algorithms have caused many robot controller designs to forego any realistic consideration of the dynamics.

NATO ASI Series, Vol. F52
Sensor Devices and Systems for Robotics
Edited by A. Casals
© Springer-Verlag Berlin Heidelberg 1989

An alternative to this approach is *motor learning*, which can be understood as a scheme for refining the motion control by executing many motions, and efficiently storing the execution data for future recall. With such a scheme, the desired motion is compared to known motions which have been "learned" through prior training, so that all control computations do not have to be performed during every execution.

In this paper we develop a motor learning scheme which is based on joint force sensing. The first part looks at motor learning schemes in man, and compares these with some of the attempts for motor learning in robots. The second part develops a method for obtaining link inertia estimates from joint torques, and the final part shows how this method may be applied to obtain motor learning.

2. Motor Learning in Man

Motor learning is an integral part of the human motor system, one of the most versatile manipulation systems known. Motor learning in man consists of two phases:

a) Skill acquisition or motion refinement through repeated trials.
b) An efficient storage and recall mechanism for the learned motions.

Acquired skill in motor control is sometimes explained by the theory of the *motor program*, whereas organization of learned knowledge at the task level is explained in terms of the *schema*. The schema generates flexible motor programs which are capable of adapting to different parameters within the same task framework. Another theory, first proposed by Adams, is the *open-loop* theory, and is analogous to the computed-torque control for robot manipulators.

2.1. Motor Program

The motor program theory postulates that most skilled motions are executed from memory without feedback [Lashley 1951]. There are two phases - in the *learning* phase (mostly during infancy), various motions are attempted and slowly improved. Sensory feedback is widely used at this stage. The *perceptual trace* corresponding to successful executions are gradually reinforced, and it is this trace which is then stored as the motor program. Eventually, the motion can even be executed even in the absence of sensory feedback.*

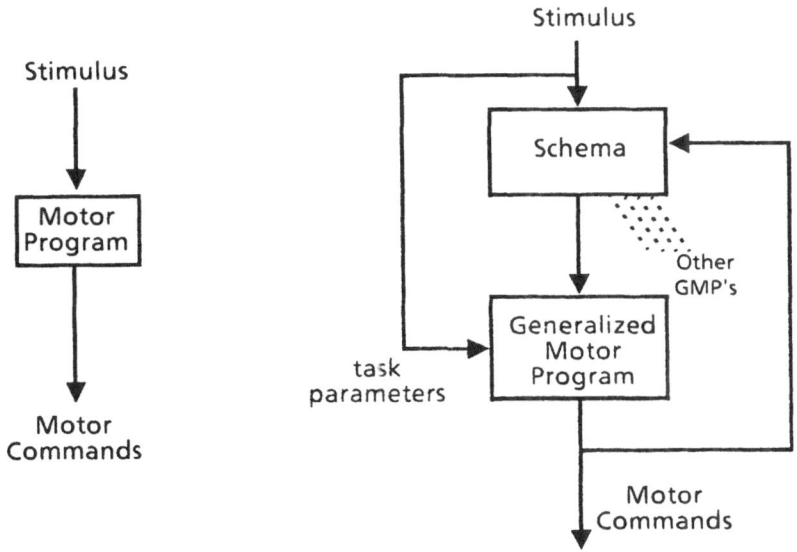

Figure 1. Theories of Sensory-motor control.

The *Motor Program* is purely open-loop and consists of a large number of stored muscle commands, each corresponding to a motion. The appropriate signal is chosen by comparing the stimulus with an "efferent map", which is a representation of stored sensory experiences.

The *Schema* is a high-level task organizer which generates motor commands via a "generalized motor program" which takes into account some of the task parameters at a very low-level. This can explain activities, e.g. in sports, where each motion is novel in response to a slightly different set of parameters.

That some form of motor trace retention exists cannot be doubted, for many of our everyday motions (e.g., speech) are executed with such rapidity as to defy a computational control explanation. What has given rise to extensive debate is the exact structure of the perceptual trace. If one thinks

---

* The original idea for this theory was developed by Karl Lashley in 1917 after experiments on a nerve-damaged gunshot patient indicated that many normal leg motions can be executed even if the sensory pathways from the leg have been destroyed.

of the motor program as analogous to the joint torques required to execute a manipulator motion, then the perceptual trace is nothing but the command history - the sequence of motor torques necessary to execute the motion.

Several aspects of motor behavior have been hard to explain using this theory alone. Separate motor programs are required for all discrete movements. The number of such programs (100,000 for speech alone) would be very large. While not enough is known to determine whether so many programs can actually be stored in the human cortex, it appears to be rather unlikely.

In addition to the storage problem, state-space models for learning face several other drawbacks. The state-space relationships are extremely non-linear and ill-behaved, and interpolation between grid points is inexact. It is always possible to overshoot the state space using sudden jerky movements. Also, the large tables are not easy to fill, and iterations of the same order as table size are required for motor learning. Furthermore, skilled movements, e.g. in sports, are never truly duplicated, each motion having been performed under a novel set of parameters. Motor program theory also has difficulty explaining error correction.

2.2. Schema

Recent psychological experiments seem to point to a feedback assisted theory for motor learning and control. One of the more likely explanations involve the concept of schema, which is a high level construct for recognizing and recalling motor tasks [Schmidt 1976]. Motion execution is achieved through *generalized* motor programs, called *schema*, which may require event-specific details for execution (Figure 1). This structure is viewed as a template for a general class of motions; substituting the variable values result in a specific motor program. *Schema* can be thought of as analogous to hierarchical task planning. So far there appear to be no developments in robot learning analogous to the Schema.

3. Learning models in robot control

3.1. State-space Control

One of the more direct methods for representing the motor program is by storing the actuator motion history itself. In terms of a manipulator, this constitutes a relation between the task and the joint torque history. It is extremely difficult to characterize the task space with sufficient clarity to allow storing the motor program in the task space directly. A somewhat easier alternative is to embed the joint torques not in the task space but in the manipulator state space, typically consisting of the joint variables and their derivatives. This approach is usually called *state-space control.*

[Albus 1975] proposed such a scheme by mapping the joint torque τ as a function of the joint variable θ, velocity $\dot{\theta}$, and acceleration $\ddot{\theta}$:

$$\tau = f (\theta, \dot{\theta}, \ddot{\theta})$$

The state space for this model, consisting of the parameters $(\theta, \dot{\theta}, \ddot{\theta})$ for each of the n joints of the manipulator, has 3n dimensions. Thus, even if we consider as coarse a tessellation as 10 data points for each variable, the number of entries needed, 10^{3n}, is too large to be realistically used.

Subsequently [Raibert 1977] proposed a smaller model of the form

$$\tau = J(\theta) \ddot{\theta} + G (\theta, \dot{\theta})$$

where $J(\theta)$ represented the total robot inertia, which is a function of the configuration θ and can be maintained as a separate table, and $G (\theta, \dot{\theta})$ represented the effects of the Corioli's force, centrifugal forces and gravity. This formulation resulted in a 2n-dimensional state-space, which was tested on a 3 DOF arm. Even then, extremely coarse tessellations had to be used. Some performance improvement could be observed but convergence was slow. In [Raibert and Horn 1977] the state-space for this model was further reduced to tables of only θ, which is the configuration space since there is no velocity dependence. However, with these reductions in memory, computational demands increased considerably.

3.2. Iterative Control Optimization

Over the last few years several researchers have developed "learning control" based algorithms for improving the performance of a manipulator over many repetitions of the identical motion.

In this method, the robot and controller are assumed to be a single unit, and the desired trajectory is input to the controller. Depending on the output error, the input is adjusted using one of several algorithms ([Arimoto et al 1984], [Furuta and Yamakita 1986]) until satisfactory accuracy has been achieved. This method is equivalent to determining the inverse plant through an iterative method. Convergence can be guaranteed for certain models of robot behavior by suitably choosing the input adjustment method.

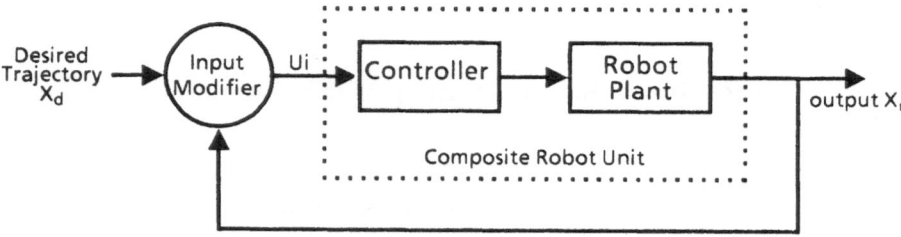

Figure 2. Inverse Plant model for robot learning: If the inverse plant can be identified as $[P^{-1}]$ then the input u required to generate the desired output X_d can be computed as $u = [P^{-1}] X_d$. However, $[P^{-1}]$ cannot be computed directly, and we can achieve the same effect by an iterative procedure:

$$u_0 = X_d$$
$$u_{i+1} = u_i + H(e_i)$$

where e_i is the trajectory error and $H(e_i)$ is a suitable adaptation function. The iteration is continued until $|X_i - X_d| < \epsilon_0$. Based on an assumed general model of robot behavior, convergence can be guaranteed if the function $H(e_i)$ is bounded.

This method restricts the learning to repeatedly executed motions, and is in essence an optimization technique for a given trajectory. The motion needs to be executed in a highly structured environment and even the smallest variations may lead to instability of convergence. Each motion is then stored as a sequence of altered trajectory point specifications, the physical significance of which is not easily determined.

4. Perturbed Parameter models

In this paper we propose a new learning scheme based on direct parameter estimation. This makes the learning sensitive to the dynamic model, and the results have direct physical significance. Moreover, each motion can be stored very compactly, and the learning can be transferred to "proximate" motions.

The basic idea of a perturbed model is very simple. Since an exact model for the system behavior would be very complex, the behavior of the system is approximated by taking a simpler model and tailoring the parameters of this model to suit the observed behavior. These perturbed parameters can then be used to predict the behavior of the system in this region of the operating space. A simple example of this is where the excitation energy of a Helium electron can be determined using the Hydrogen atom model but with incorrect parameters (the nucleus is increased). Unfortunately, the effect of such perturbations on a complex non-linear system like robot control raises many issues such as stability and convergence, which have not been adequately answered.

Which of the model parameters are to be perturbed? This can be best ascertained by performing a sensitivity analysis. There are three kinds of parameters in the dynamic equations:

    (i)   forces and torques,

    (ii) inertial and geometric parameters, and

    (iii) joint displacements and velocities.

While the groups (i) and (iii) are variable during motion, parameters in (ii) are treated as relatively slow to change. These three kinds of terms are related by the dynamic equations (Figure 3). This grouping motivates three kinds of solution techniques:

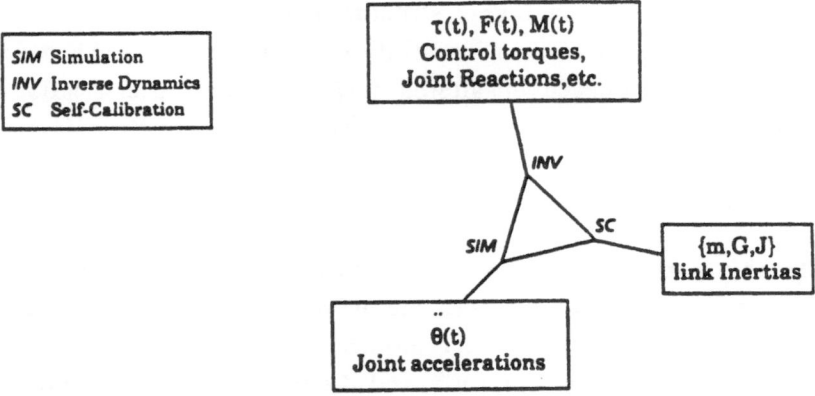

Figure 3. Dynamic parameter groups and processes.

The parameters of the manipulator dynamics can be grouped in three categories, any two of which can be used to solve for the third group. Calibration is concerned with obtaining the inertias if the torques and joint motions are known.

a) Simulation or Forward dynamics: Motor torques and inertial parameters are known, and used to solve for the motion. These are the computations executed by a simulator.

b) Inverse dynamics: Desired joint trajectories and link inertias are known, and used to determine the control torques. These computations are used in real-time control.

The remaining method, which is the calibration approach, is:

c) Self Calibration: Joint motions and reactions/torques are known, and used to solve for the inertial parameters.

An analogous trichotomy can be made for the geometric parameters. The relevant problems are *forward kinematics, inverse kinematics*, and *geometric calibration*.

Since our interest is in the dynamics, we direct our efforts towards the calibration of the inertias rather than the geometric or kinematic parameters.

4.1. Inertia Estimation

As a simple example, let us consider the free body motion of a rigid mass m. Let a force of magnitude F and moment of magnitude M_C be acting at the center of gravity C. The equations of motion of this body are given by

$$F = ma_C, \qquad (1)$$

and

$$M_C = J_C \omega, \qquad (2)$$

where a_C is the translational acceleration at C and ω the angular acceleration. From equation (1), we can solve for the mass m if we know F and a_C. Similarly, from equation (2), we can solve for the mass moment of inertia component (about the spin axis) J_C, if we know M_C and ω.

In this example the inertial parameters (m and J_C) linearly relate acceleration (a_C and ω) to loads (F and M_C). This is a feature which is preserved in more complex systems but in a more intricate form.

4.1.1. Free-Body Dynamics Equations

For general articulated kinematic chains, there are ten inertia parameters associated with each link. These are: the mass, the three center of gravity coordinates, and the six independent components of the mass moment of inertia. In order to determine these parameters (or at least those that are relevant to the motion), we will need to measure joint forces and moments, as well as link velocities and accelerations.

We derive the following expressions for the free-body force and moment balance the ith link in terms of the instantaneous velocities (V, ω) and accelerations measured in the inertial frame that coincides with the link$_i$ frame.

$$F = m\dot{V} + \dot{\omega} \times mG + m\omega \times V + \omega \times (\omega \times mG)$$

$$M = [J]\dot{\omega} + mG \times \dot{V} + \omega \times [J]\omega + mG \times (\omega \times V)$$

where [J] is the inertia tensor, G the vector from the origin of the link coordinate frame to the link center of gravity, and m the mass of the ith link.

4.1.2. Linear Regression Model

Although the above equation is nonlinear in the velocity and kinematic terms, we note that all the inertia terms $\{m, mG, [J]\}$ are linear in this equation. This forms the basis for our subsequent linearized regression model. Written in matrix form,

$$\mathbf{F} = m\dot{\mathbf{V}} + [\mathbf{S}]^T \dot{\boldsymbol{\omega}} + m[\boldsymbol{\omega}]\mathbf{V} + [\boldsymbol{\omega}][\mathbf{S}]^T \boldsymbol{\omega}$$

$$\mathbf{M} = [\mathbf{J}]\dot{\boldsymbol{\omega}} + [\mathbf{S}]\dot{\mathbf{V}} + [\boldsymbol{\omega}][\mathbf{J}]\boldsymbol{\omega} + [\mathbf{S}][\boldsymbol{\omega}]\mathbf{V}$$

where $[\omega]$ is the cross product matrix for ω, and $[S]$ represents the cross product matrix for the mass moment vector $mG = S$.

Expanding this equation into all the component terms and reorganizing according to the inertia parameters, we get the linear regression equations, which are of the form

$$\{\mathbf{F}^T, \mathbf{M}^T\}^T = [\text{Kin}]\{m, \mathbf{S}^T, \mathbf{J}^T\}^T$$

where [Kin] is the matrix containing all kinematic information, $\mathbf{S} = mG$, and $\mathbf{J} = (J_{11}, J_{22}, J_{33}, J_{23}, J_{31}, J_{12})^T$. This equation represents the fact that the forces are the product of the kinematics and the inertias, the same structure as $\mathbf{F} = m\mathbf{a}_c$.

4.2. Perturbed Inertias model

For a system with unmodeled dynamic effects such as friction, the parameter set determined by this approach may bear no relation with the actual values, but this set of calibrated parameters represents a better model of the system behavior in the neighborhood of the given trajectory. For example if it is determined that the system is behaving as if $link_n$ is 20% heavier then a perturbed model with a 20% higher $link_n$ mass will predict the system dynamics much more accurately in the vicinity of the current motion.

On the other hand, the degree of adaptation using this method is limited by the extent to which the inertia perturbations will be able to model the path. The predictive power may be improved by splitting the trajectory into smaller segments.

4.3. Sensors and Instrumentation

Any calibration system is only as good as its sensors. In the context of articulated chains, the parameters to be measured are primarily position, velocity, and force. Reliable position sensors are now widely implemented in industrial robots in the form of optical encoders or resolvers. Determination of velocity and acceleration is more difficult but good progress has been made towards relatively compact tachometers and accelerometers. It is possible to obtain acceleration and velocity information by differentiating the position data, but this process is inherently too noisy to be meaningful in the self-calibration context [An, Atkeson and Hollerbach 1985]. There are a number of other sensors under development, e.g., visual position sensors, which may be advantageously used in the analysis of spatial systems. Significant performance improvements may be achievable by using independent position sensors, which eliminate the accumulation of error typical of multi-joint measurements.

The most significant problem with force sensors is mounting it on the joint. Since force/torque sensors need to be very flexible in order to be sensitive, the overall joint becomes much more flexible than the drive shaft, and this causes a loss of controllability at the joint. Implementing such sensors for all six joints would be very ill-advised. Another alternative attempted at the University of Edinburgh was to extend the drive shaft and mount a dedicated design of strain gauges on the shaft itself.

Other inertia estimation algorithms exist that do not require the installation of force/torque sensors at all joints. An, Atkeson and Hollerbach [1985] have shown that all the inertias that affect the motion can be estimated by measuring the joint torques alone. In this method, each link is not considered as a free-body but the entire system dynamics is considered as a whole. This requires the kinematic terms to be propagated across the link chain, which further magnifies the errors, especially in the acceleration term. The mapping matrix is singular, but by adding small values to the diagonal term a pseudo-inverse can be found which is used to obtain estimates of inertia values. This makes the algorithm further sensitive to propagated errors. Also, some of the parameters remain coupled in linear combinations.

5. Conclusion

Here we have presented a simple idea for implementing a motor learning system in articulated chain manipulators. One of the fundamental limitations of this approach is that due to the nature of the unmodeled effects, the degree of accuracy that can be achieved by varying only the inertias is likely to be limited. This is especially true for longer motions, but can be compensated for by breaking up the trajectory into several smaller portions, and incorporating a parameter switch at the transitions points between these regions.

Another issue that remains open is that of convergence. Nonconservative effects such as friction and backlash render the results of the model erratic, and convergence is likely to be slow at best. The same effects were observed in the Inverse plant and State-space control implementations.

One advantage of the perturbed parameters approach is that some rudimentary knowledge can be transferred between proximal trajectories [Mukerjee 1987]. We hope that the concepts developed in this paper will instigate some of the future work that needs to be done to focus on this extremely difficult problem.

REFERENCES

Adams, J.A., 1976, Issues for a closed-loop theory of motor learning, in "Motor Control: Issues and Trends," ed. G.E. Stelmach, Academic Press, NY.

Albus, J.S., 1975, A new approach to manipulator control: the Cerebellar Model Articulation Controller (CMAC), *J. Dynamic Systems, Measurement, and Control, Trans. ASME(G)*, v.97(3), pp.220-227.

An, C.H., C.H. Atkeson, and J.M. Hollerbach, 1985, Estimation of inertial parameters of rigid body loads for manipulators, Proc. 24th IEEE Conference on Decision and Control, Fort Lauderdale, December 1985.

Arimoto, Suguru, Sadao Kawamura, and Fumio Miyazaki, 1984, "Bettering operation of robots by learning", *J. of Robotic Systems*, v.1(2):123-140.

Furuta, K., and M. Yamakita, 1986, "Iterative generation of optimal input of a manipulator", *Proc. IEEE International Conference on Robotics and Automation*, San Francisco, April 1986, pp.579-584.

Lashley, K.S., 1951, "The problem of serial order in behavior", in *Cerebral Mechanisms in Behavior*, ed. Jeffries, L.A., Wiley 1951, pp.112-136.

Mukerjee, A., 1987. "Robot learning: Transferring execution knowledge between trajectories", Texas A&M University Department of Computer Science, TR 87-003.

Mukerjee, A., 1985. "Self-calibration strategies for robot manipulators", Univ. of Rochester Department of Computer Science, TR 193.

Raibert, M.H., 1977, "Motor control and learning by the state-space model", Ph.D. Thesis, MIT Dept. of Psychology, September 1977, AI-TR-439.

Raibert, M.H., and B.K.P. Horn, 1978, "Manipulator control using the configuration space method", *The Industrial Robot*, June 1978, pp. 69-73.

Schmidt, Richard A., 1976, "The Schema as a solution to some persistent problems in motor learning theory", in *Motor Control: Issues and Trends*, ed. George E. Stelmach, Academic Press, 1976.

MODELLING THE INTERACTION BETWEEN ROBOT AND ENVIRONMENT

A.A. Goldenberg
Robotics and Automation Laboratory
Department of Mechanical Engineering
University of Toronto   (Canada)

ABSTRACT

The contact between robot and environment generates forces of interaction which need to be controlled. In addition, the motion of the robot interacting with the environment must also be controlled. The unification of these two objectives is usually titled 'force and position control'. In the recent past this issue has been extensively addressed in the research, and several basic approaches to force control have emerged. The synthesis of controllers for force and position control has also been addressed. The paper reviews some of the relevant work in this area in particular the techniques used to generate models of the interaction between robot and environment. The paper also addresses conceptually the implementation of impedance control and presents an application of this technique as well as of the descriptor system to control of interaction forces in a dexterous multifingered hand.

NATO ASI Series, Vol. F52
Sensor Devices and Systems for Robotics
Edited by A. Casals
© Springer-Verlag Berlin Heidelberg 1989

1. Introduction

The interaction between robot and environment generates forces which must be controlled to preserve the integrity of the product being manipulated. This interaction, and the associated forces generated as a result of its occurance, may either be undesirable, hence subject to rejection or compensation, or may be desired for example when the robot performs work on the environment. In either case the force of interaction is a non-exogenous signal which must be regulated to a desired level (zero when the interaction is undesired and non-zero in the case of work done on the environment).

The regulation of the force is inherently dependent on the motion of the robot and object particularly of the point (area) of interaction. In principle if this relationship is known (for ex. Hooke's law) then the force can be regulated either in a force control loop (force input/force output) or a motion control loop (force input ~ motion input/motion output). If we consider the simple one DOF system composed of a spring with stiffness K, then clearly control of force can be performed, but controlling both, force and position is impossible without certain refinement of the control law.

The force regulator problem has been addressed by numerous authors. The most notable contribution is that of [Raibert and Craig, 1981] who present the concept of hybrid control. Hybrid control amounts to controlling the mutually dependent variables, position and force, along mutually orthogonal axes. This concept can be used for any system of any number of degrees of freedom (higher than 1). The methods of controlling either force or position are routine classical or modern control approaches. The literature has presented so far numerous cases of force control in the context of hybrid control. The linear system case is dealt with quite conventionally, whereas the nonlinear case requires certain refinements such as feedback linearization or computed torque technique, and various schemes using nonlinear feedback have been proposed. In principle, these schemes always translate into an application of modern regulator theory to the specific case of force control.

In parallel to hybrid control strategy it is possible to demand that force and motion be controlled along the same axis. In this case the stiffness needs to be controlled as performed by [Salisbury and Craig, 1982] and later in the context of impedance control by [Hogan, 1985] and [Kazerooni, 1986].

In more recent work by [Goldenberg, 1987] it has been proposed to combine both force and impedance control in order to unify the control objectives when interaction between the robot and environment occurs. In this paper this problem is addressed with a particular emphasize on modelling and control of interaction forces, for example in grasping and manipulation with multifingered hands. The paper has five sections. Section 2 presents an implementation of force and impedance control which is later extended to manipulation with dexterous hands (Section 3). Section 4 addresses the modelling of robot/environment interactions using the concept of descriptor system. Section 5 presents an application of this modelling technique to manipulation with dexterous hands. Section 6 includes conclusions.

It should be emphasized that the two basic models used here (compliant and singular) allow both control of force and motion by way of controlling the relationship between these two variables. At this stage of research either model is characteristic and suitable for interaction studies such as grasping and manipulation. The important feature of such models is that they bring into consideration characteristics of the environment which are ignored in standard 'force control' algorithms where the measurement of force is the only information available about the interaction.

2. Force and Impedance Control Using Compliant Models

The objectives of this control are: i) regulation of force exerted on the environment; ii) generation of a desired (target) relationship between the force and the relative location of the point of interaction (ex. tool) with respect to the commanded location.

According to [Hogan, 1985], impedance control is a general approach to control in which the robot behaves as a mass-spring-dashpot system whose parameters (inertia-stiffness-damping) can be specified arbitrarily. This methodology, which is based on replacing the robot with a target impedance, can be realized by feedback and feedforward compensation of robot control variables including force and position sensing. In principle the target impedance can be implemented in two modes: flow or force.

In order to develop conceptually these two modes let us consider the well known dynamics model of an n-DOF rigid manipulator [Greenwood, 1977].

$$\Phi(q)\ddot{q} + H(\dot{q},q) + G(q) = \tau \tag{1}$$

where $\Phi(q)$ is the mass inertia matrix, $H(q,\dot{q})$ represents the rate and configuration dependent terms of the moment equation, $G(q)$ is the gravity (conservative force), τ is the actuator input and q is the vector of generalized coordinates. Assuming an equilibrium configuration we can analyse small motions about an equilibrium point by reducing equation (1) to

$$M\ddot{X} + C\dot{X} + KX = F \tag{2}$$

where X, $\dot{X}$, $\ddot{X}$ are deviations of the tip (point of interaction with the environment) from an equilibrium point and F is the deviation from the nominal torque (corresponding to the equilibrium configuration). A similar model could be derived for the environment [Hogan, 1987].

Let us now define X_o to be the equilibrium location ($\dot{X}_o = 0$, $\ddot{X}_o = 0$), X_1 the location (of the robot) command with respect to X_o, and X_2 the actual location (of the robot tip) in contact with the environment. We also define Z_r as the robot impedance, Z_2 the environmental impedance, F_e the force impressed on the environment and F_{in} the applied force to the robot. Consequently the following relationships always hold

$$F_{in} = Z_r V_c + F_e \qquad (3)$$

$$F_e = Z_2 V_e \qquad (4)$$

where V_c is the commanded motion of the robot and V_e is the motion of the point (node) of interaction. These equations form the basic robot/environment system compliant model.

The basic methodology of control is to generate a target impedance or admittance in lieu of the robot's nominal one in order to obtain a desired closed-loop behaviour, in particular in order to "shape up" the transfer function relating the force exerted and the relative motion of the point of interaction with respect to the commanded robot motion. The main concern, subsequent to the definition of a model for a given environment, is to define the target impedance. The generation of specifications for the target impedance has been dealt with extensively in terms of frequency domain in the papers by [Kazerooni et al. 1986]. [Anderson et al. 1987] also addresses the issue in a more direct way using simple 1st order models. We reformulate Anderson's approach using a simple "duality criterion" which is common in electrical network design. In principle, this criterion states that, if the load is inductive, then the source impedance (or admittance) must be capacitive and vice versa a capacitive load would demand an inductive "internal" impedance of the source. This criterion when applied to mechanical systems reveals an intuitive design approach of adding a mass to a spring-damper load and a damper to a mass-spring load. In other words, an "inductive" environment should be connected to "capacitive" robot and vice versa. For example, let us consider that the two subsystems, robot and environment are interfaced in a force mode (voltage source). The schematic diagram is shown in Fig. 1. In this diagram

$$\frac{F_e}{F_{in}} = \frac{v}{v_{in}} = \frac{Z_e}{Z_e + Z_r} \qquad (5)$$

and

$$\frac{V}{F_{in}} = \frac{i}{v_{in}} = \frac{1}{Z_r + Z_e}. \qquad (6)$$

The above diagram is shown schematically using mass-damper-spring elements in Fig. 2.

In order to implement the target impedance the strategy is to replace Z_r by a target impedance (task dependent) Z_d such that

$$Z_d = Z_r \qquad (7)$$

In the force mode (5), (7) implies that

$$\frac{F_e}{F_{in}} = \frac{Z_e}{Z_e + Z_d} \qquad (8)$$

and furthermore that

$$\frac{F_e}{F_{in}-F_e} = \frac{Z_e}{Z_d} \tag{9}$$

or, by definition (4), that

$$\frac{V_e}{\Delta F} = \frac{1}{Z_d} \tag{10}$$

where V_e denotes the motion of the point of interaction between the robot and environment and ΔF is the difference between F_{in} and F_e (exerted on the environment). Alternatively from (9), using definitions (3) and (4) the following is obtained

$$\frac{F_e}{V_{in}-V_e} = \frac{F_e}{\Delta V} = Z_d \tag{11}$$

In a flow mode the following can be obtained

$$\frac{V_e}{V_{in}-V_e} = \frac{Z_d}{Z_e} \tag{12}$$

Hence

$$\frac{F_e}{\Delta V} = Z_d \tag{13}$$

which is similar to (10).

The expression (10) (or (13)) is the transfer function relevant to the impedance control strategy. This strategy relates the difference between the applied and exerted (impressed) motion of the robot (on the environment) and the reaction force resulting from the interaction (with the environment). This relationship is of a primary concern in terms of the need to maintain "control" over the interaction force which may exceed desirable upper or lower levels generated by unwanted or uncertain motion, causing improper execution of tasks. For each task a desired (target) impedance must be defined to ensure the successful execution of the task when uncertainty or unwanted motion of the robot and environment occurs. Let us now consider the implementation of impedance control. This strategy can be implemented in flow or force mode.

The flow (motion) control mode is shown schematically in Fig. 3 where V_c and V_e denote the commanded and actual robot tip motion respectively, Z_m denotes the feedforward controller and Z_c denotes the feedback controller. Using elementary algebra we can obtain the following transfer functions

motion control

$$\frac{V_e}{V_c} = \frac{Z_c+Z_m}{Z_r+Z_e+Z_c} \tag{14}$$

force control

$$\frac{F_e}{V_c} = \frac{(Z_c+Z_m)Z_e}{Z_r+Z_e+Z_c}$$ (15)

impedance control

$$\frac{F_e}{\Delta V} = \frac{Z_e(Z_c+Z_m)}{Z_e+Z_r-Z_m} = Z_d$$ (16)

If we choose

$$Z_m = Z_r + Z_e$$ (17)

then $\frac{V_e}{V_c} = 1$, $\frac{F_e}{V_c} = Z_e$, $\frac{F_e}{\Delta V} \to \infty$ and Z_c is selected to "shape up" the transient response. But Z_e may be unknown, or it can not be known exactly, hence this alternative of modelling Z_m is not considered. If we choose Z_c very large it can be shown that $\frac{V_e}{V_c} \to 1$, $\frac{F_e}{V_c} \to Z_e$ and $\frac{F_e}{\Delta V} \to \frac{Z_e Z_c}{Z_r+Z_e-Z_m}$. This particular choice of the compensator Z_c does not lead to a convenient and practical design. In order to control the impedance, i.e. $\frac{F_e}{\Delta V} = Z_d$, one can choose $Z_m = 0$ which leads to $Z_c = Z_d \frac{Z_r+Z_e}{Z_e}$. This choice is not considered also since it requires knowledge of Z_e.

Alternatively, let us consider

$$Z_m = Z_r$$ (18)

Then, in order to implement impedance control the following must hold

$$Z_c = Z_d - Z_m$$ (19)

In this case the motion control is obtained as

$$\frac{V_e}{V_c} = \frac{Z_d}{Z_d+Z_e}$$ (20)

and the force control is

$$\frac{F_e}{V_c} = \frac{Z_d Z_e}{Z_d+Z_e}$$ (21)

Subsequently $\frac{F_e}{\Delta V} = Z_d$ and schematically the robot/environment system is shown in Fig. 4.

If the model of the robot is not available precisely then

$$Z_m = Z_r + \delta Z_r$$ (22)

where δZ_r is the uncertainty in robot modelling. Hence, on the basis of the above discussion (19)

$$Z_c = Z_d - Z_r - \delta Z_r \qquad (23)$$

and

$$\frac{F_e}{\Delta V} = \frac{Z_e Z_d}{Z_e + \delta Z_r} \qquad (24)$$

which indicates the effect of the robot model uncertainty in implementing impedance control in motion control mode.

Let us now consider the implementation of both force and impedance control using the proposed mode of operation. We shall consider that an exact model of the robot exists, i.e. (18) holds, hence (19) can be implemented. The proposed implementation (Figure 5) generates the relationships (20), (21) and the desired impedance $Z_d = \dfrac{F_e}{\Delta V}$. In this diagram the input/output 'perfect control' relationship is also satisfied, i.e. $\dfrac{F_e}{F_c} = 1$.

In this implementation the reference (desired force F_c) is used to generate the desired relative motion of the point of interaction with respect to the commanded motion which is denoted ΔV_c^o. The error ΔF generates a correction δV which is added to ΔV_c^o to form the commanded ΔV_c. The relationship $F_e/\Delta V$ is implemented using the motion control mode diagram of Fig. 3. Clearly, the correction δV may seem to be unnecessary if the model of the robot is perfect. However if we consider the case of the uncertain model (22) the following is obtained from (21).

$$\frac{F_e}{V_c} = \frac{Z_d Z_e}{Z_d + Z_e + \delta Z_r} \qquad (25)$$

Furthermore

$$\frac{F_e}{F_c} = \frac{(Z_c + Z_d) Z_e}{(Z_c + Z_d) Z_e + Z_c \delta Z_r} \qquad (26)$$

Clearly for $\delta Z_r = 0$ the previous relationships are obtained. In order to compensate for the effect of the uncertainty, a robust compensation must be added to cancel the effects of δZr.

The above analysis can also be carried out using non-linear robot models. As in the linear case, here we assume that an exact model of the robot exists; i.e. $Z_m = Z_r$. For simplicity we shall first consider the implementation of the motion loop with a non-linear model of the robot (see Fig. 3). The equation of motion of the robot is

$$\Phi(V_e)\ddot{V}_e + H(V_e, \dot{V}_e) + G(V_e) = F_{in} + Z_c(V_c - V_c) - F_e , \qquad (27)$$

where F_{in} is generated by the following model equation:

$$\Phi(V_c)\ddot{V}_c + H(V_c, \dot{V}_c) + G(V_c) = F_{in} \tag{28}$$

In arriving at Eq. (28) we implicitly assumed that the functional forms of $\Phi(\cdot)$, $H(\cdot)$, and $G(V)$ have not been modified. This stresses the fact that an exact model of the robot exists as assumed earlier, and the uncertainty of the robot model is neglected. For a realistic model this assumption must be removed, but for a conceptual treatment we shall accept it. We assume that the deviation from the commanded motion V_c is small; i.e.

$$|| V_c - V_e || = || \Delta V_c || < \varepsilon \quad ; \quad \varepsilon > 0 \tag{29}$$

This assumption is crucial for the validation of the following analysis and, therefore, must be justified on experimental ground. Eq. (27) then can be expanded and simplified using Eq. (28) as follows:

$$\left[\Phi(V_c) - \frac{\partial \Phi}{\partial V} \bigg|_{V_c} \Delta V_c \right] \Delta \ddot{V}_c + \left[\frac{\partial H}{\partial \dot{V}} \bigg|_{V_c} \right] \Delta \dot{V}_c + \frac{\partial}{\partial V} [\Phi(V)\underline{\ddot{V}} + H(V, \dot{V}) + G(V)]_{V_c} = -Z_c \Delta \underline{V}_c + F_e \tag{30}$$

Let us define $Z_c = Z_d - \hat{Z}$ where

$$\hat{Z} = \left[\Phi(V_c) - \frac{\partial \Phi}{\partial V} \bigg|_{V_c} \Delta V_c \right] \frac{d^2}{dt^2} + \left[\frac{\partial H}{\partial \dot{V}} \bigg|_{V_c} \right] \frac{d}{dt} + \frac{\partial}{\partial V} [\Phi(V)\ddot{V} + H(V, \dot{V}) + G(V)]_{V_c} \tag{31}$$

With this definition Eq. (30) becomes

$$Z_d \, \Delta V_c - F_e = 0 \tag{32}$$

Hence

$$F_e / \Delta V_c = Z_d \tag{33}$$

Eq. (33) defines the desired impedance as in the linear model. The same analysis can be applied to the force mode. It is found that, as in the linear case, the perfect control law $F_c / F_e = 1$ can be achieved provided an exact model exists, i.e. $G_m = G$.

When both the force and motion loops are implemented as shown in Fig. 5, the motion loop generates the desired impedance $Z_d = F_e / \Delta V_c$ and the force loop gives $F_c / F_e = 1$. A profound difference between the linear and nonlinear implementations of force and impedance control is reflected in the definition of Z_c. In the linear case $Z_c = Z_d - Z_r$, where Z_r is the impedance of the robot, whereas in the latter case $Z_c = Z_d - \hat{Z}$, for which $\hat{Z}$ is defined by (31). This demonstrates that nonlinear implementation of force and impedance control is feasible provided on-line computations like those given in Eq. (31) are available at a suitable rate.

3.   **Application of Compliant Models to Grasping & Manipulation**

One of the relevant application area for the concepts described in the previous section is dexterous mechanical hands. This area of research has applications in flexible manufacturing systems and biomechanical engineering. In flexible manufacturing it can increase productivity by introducing versatility in robotics applications and, in biomechanical engineering it may allow the disabled to manipulate the environment with more dexterity. In this context we address the topic of grasping and manipulation of objects. In order to clarify this topic the following definitions [Payandeh et al., 1987] are introduced.

Grasping: grasping is defined as the coordinated movement of fingers in order to establish contact (point, line, surface) with an object and exert forces (grasping force) on the object such that, when object is held between fingers, the grasp is stable.

Manipulation: manipulation is defined as a displacement with respect to fingers' base (palm) of a grasped object such that the stability of grasp is maintained.

Stable grasp: a grasp is defined to be stable when: a) the object is in equilibrium; i.e. there is no net force and moment; b) all forces acting on the object (grasping force) must be within the cone of friction; c) the grasped object returns to its original position when it is displaced by an arbitrary small amount.

The models of objects can be in the most general case, represented using the inertia and/or restoring and/or dissipative representation of its structural and/or material properties. A model of the object at contact areas can be obtained by serial and/or parallel combination of these representations. These models may vary as the location of the contact areas on the object change.

In a dexterous mechanical hand, each finger is some collection of a physical structure, sensors and actuators (hardware) combined with controlling algorithms (software). The dynamics model of this combination of hardware and software (closed-loop dynamics), as it is apparent at the finger contact area, can be written using a constitutive equation that in general expresses the dynamics of a mechanical system (Newton's equation as shown in equation 2). Notice that contact areas of each finger can have various geometries (e.g. point, line, surface, curved).

The equation (2) expresses the closed-loop dynamics of each finger where X is the location (position & orientation) of the finger contact geometry expressed with respect to the commanded location and F is the force acting on the finger contact area. In this equation M represents the apparent inertia, C represents the apparent dissipative parameter and K is the apparent restoring parameter of the closed-loop dynamics in the direction of finger motion.

Although equation (2) represents a simple way of expressing the closed-loop dynamics of fingers, the selection of the parameters of the closed-loop dynamics is a difficult undertaking which will remain the designer's choice for each desired task.

Let us consider that equation (2) represents the closed-loop dynamics of a finger contact geometry before a contact with an object is made ($F = 0$), where the parameters of the equation are selected such that the closed-loop dynamics is stable and certain performance specifications are satisfied. As soon as the finger establishes a contact with the object, the stability and performance are affected since the closed-loop dynamics is now dependent on the object's dynamics model parameters. However, one can control the apparent dynamics parameters of the closed-loop model of the finger, through the choice of the target impedance, such that the stability is preserved and the performance specifications are satisfied.

The closed-loop (targeted) dynamics of each finger can be implemented using various approaches, for example as shown in Fig. 5.

In Fig. 6 the problem above mentioned for the case of two fingers making contact with a compliant object is shown [Payandeh et al., 1987]. The object is modelled as a capacitive load in a force mode loop (6a) and flow mode (6b) respectively for an inductive load. This approach allows the synthesis of control strategies for any number of fingers making contact with an object.

4. Force Control Using Singular Models

Trajectory control during constrained manipulation, resulting from contact of the robot with the environment represents an important class of control problems. In this context both contact force exerted by the manipulator and the position of the robot while in contact with the surface must be controlled. The proposed model exploits the structure of the constrained dynamic formulation of the equations of motion [Greenwood, 1987]. When the end-effector is in contact with a rigid surface, a kinematic constraint is imposed on the manipulator motion, which corresponds to an algebraic constraint amongst the manipulator state variables. This algebraic constraint give rise to a system of differential algebraic equations which are characterized by a singular matrix premultiplying the vector of state derivatives. [McClamroch and Huang, 1985] and [McClamroch, 1986] present singular dynamic models which represent manipulator dynamics under several different circumstances when subject to kinematic constraints. With this formulation, the control problem is naturally expressed in terms of variables of a state vector.

Singular systems of differential equations are of a fundamentally different nature than nonsingular or state variable systems, which describe unconstrained manipulator dynamics. A significant difference is the introduction of impulsive behaviour of the system, which depends on the system initial conditions [Cobb, 1983].

Let us consider the dynamic equations of motion of a rigid, n degree of freedom manipulator in contact with a rigid, frictionless constraint surface. It is assumed that contact with the environment is made by the end-effector and occurs at a point. Let $p \in R^3$ denote a position vector from a fixed reference frame to the constraint surface. The constraint surface is assumed to satisfy the following scalar relation

$$\phi(p) = 0 \tag{34}$$

where: $\phi(\cdot) = fat\nabla \phi : \mathbf{R}^3 \rightarrow \mathbf{R}$, is a given scalar function with continuous gradient.

Let us define the manipulator forward kinematics as follows,

$$z = g(q) \tag{35}$$

where $g(\cdot)$ is a map from joint space to task space and q is the n-tuple of manipulator generalized coordinates.

As noted earlier, the manipulator is in contact with a rigid constraint surface hence the position vector p is defined as a function of the joint coordinates as follows:

$$p = L(q) \tag{36}$$

A contact force exists and the contact or *workless* force is given by the following

$$f = D^T(p)\lambda \tag{37}$$

where:

$$D(p) = \frac{\partial\phi(p)}{\partial p} \tag{38}$$

and λ is the Lagrange multiplier. Thus, based on the Lagrange formulation, the equations of motion of a manipulator, constrained by one point in contact with a rigid frictionless surface are given by

$$\Phi(q)\ddot{q} + H(\dot{q}, q) + G(q) = \tau + J^T(q)D^T(p)\lambda \tag{39}$$

where: $J(q)$ is the manipulator Jacobian, $J(\cdot) \in \mathbf{R}^{3 \times n}$, $J = \frac{\partial L(q)}{\partial q}$.

Equation (39) can be rewritten as in [McClamroch, 1986] to give,

$$\begin{bmatrix} I & 0 & 0 \\ 0 & \Phi(q) & 0 \\ 0 & 0 & 0 \end{bmatrix} \begin{bmatrix} \dot{q} \\ \ddot{q} \\ \dot{\lambda} \end{bmatrix} = \begin{bmatrix} 0 & I & 0 \\ 0 & 0 & J^T D^T \\ 0 & 0 & 0 \end{bmatrix} \begin{bmatrix} q \\ \dot{q} \\ \lambda \end{bmatrix}$$

$$+ \begin{bmatrix} 0 \\ -H(\dot{q}, q) - G(q) \\ \phi(p) \end{bmatrix} + \begin{bmatrix} 0 \\ I \\ 0 \end{bmatrix} \tau, \tag{40}$$

where all terms have been previously defined.

We assume that the application of a force at a point on a constraint surface is such that the amplitude of variations in joint position, velocity and acceleration about some nominal value remains small. This permits the manipulator dynamics to be represented by a perturbation model about a nominal state given by

$$(q_o, \dot{q}_o, \lambda_o)^T \tag{41}$$

where q_o is the nominal manipulator joint configuration, $\dot{q}_o$ is the nominal joint velocity λ_o is the nominal value of the Lagrange multiplier.

We wish to linearize the nonlinear dynamic equations, given by (39), about the nominal state (41) [Mills et al., 1987]. At equilibrium, the manipulator is at rest hence $\dot{q}_o = 0$ and $\ddot{q}_o = 0$, resulting in simplification of the linearized equations of motion. Writing (39) in the form

$$\tau = \tau(q, \dot{q}, \ddot{q}, \lambda) \tag{49}$$

and expanding about the nominal state (41) using a multivariable Taylor series with the definitions $\delta q \triangleq q - q_o$, $\delta \dot{q} \triangleq \dot{q} - \dot{q}_o$, $\delta \ddot{q} \triangleq \ddot{q} - \ddot{q}_o$, $\delta \tau \triangleq \tau - \tau_o$ and $\delta \lambda \triangleq \lambda - \lambda_o$ yields, where higher order terms have been dropped,

$$\delta \tau = \Phi(q_o)\delta \ddot{q} + \frac{\partial}{\partial q}(G - J^T D^T \lambda) \mid_o \delta q - J^T D^T \mid_o \delta \lambda \tag{50}$$

where we have made use of the fact that $H(\dot{q}, q) \mid_o = 0$ and $\frac{\partial}{\partial q}H(\dot{q}, q) \mid_o = 0$ due to the fact that $H(\dot{q}, q)$ is homogeneous in $\dot{q}$, as in [Kazerooni et al., 1986]. Note that τ_o and λ_o are as yet undefined. We must also linearize the constraint equation given by (34). Differentiating (34) with respect to q and evaluating the result at q_o gives

$$\frac{\partial \phi(p)}{\partial p} \frac{\partial p}{\partial q} \mid_o \delta q = 0 \tag{51}$$

$$\text{or} \quad D(p)J(q) \mid_o \delta q = 0. \tag{52}$$

Using (50) and (52), the linearized dynamic equations are written in a form similar to (40) as follows:

$$\begin{bmatrix} I & 0 & 0 \\ 0 & \Phi(q_o) & 0 \\ 0 & 0 & 0 \end{bmatrix} \begin{bmatrix} \delta \dot{q} \\ \delta \ddot{q} \\ \delta \dot{\lambda} \end{bmatrix} = \begin{bmatrix} 0 & I & 0 \\ -\frac{\partial}{\partial q}(G - J^T D^T \lambda) \mid_o & 0 & J^T D^T \mid_o \\ DJ \mid_o & 0 & 0 \end{bmatrix}$$

$$\begin{bmatrix} \delta q \\ \delta \dot{q} \\ \delta \lambda \end{bmatrix} + \begin{bmatrix} 0 \\ I \\ 0 \end{bmatrix} \delta \tau \tag{53}$$

Once again, equation (53) represents a singular system of differential algebraic equations, due to the singularity of the matrix on the left hand side of (53). All coefficients of (53) are time invariant due to the linearization of the nonlinear system (40) about a nominal state given by (41).

In order to determine the nominal applied torque τ_o corresponding to the nominal state (41), it is only necessary to evaluate (40) at $q = q_o$, $\dot{q} = 0$, $\ddot{q} = 0$, $\lambda = \lambda_o$ to yield the following

$$\tau_o = G(q_o) - J^T(q_o)D^T(L(q_o))\lambda_o \tag{54}$$

We note that the *workless* force is to be controlled, thus at equilibrium

$$D^T(L(q_o))\lambda_o = f^{ref} \tag{55}$$

and using (36) and (37), τ_o becomes

$$\tau_o = G(q_o) - J^T(q_o)f^{ref} \tag{56}$$

where all quantities on the right hand side of (55) are known. The nominal Lagrange multiplier λ_o is easily determined from equation (54) as follows

$$\lambda_o = (DD^T)^{-1}Df^{ref} \tag{57}$$

where the scalar (DD^T) is always invertible.

The linearized equations of motion developed for a robotic manipulator constrained to be in contact, at a point, with a rigid frictionless constraint surface can be used to describe manipulation of rigid objects with dexterous hands.

5. Modelling of Manipulation of Rigid Objects as a Singular System

Most of the literature on dexterous robotic hands is related to hand designs [Salisbury et al., 1982] and grasping strategies [Fearing, 1986]. Manipulation has only been lightly discussed [Kerr et al., 1986], [Kobayashi, 1985], [Okada, 1982]. A singular system model of manipulation of rigid objects is presented below. The subject here is the manipulation of rigid objects with a dexterous robotic hand. The problem is modelled as a singular system. Such a system is subject to an algebraic constraint which can be holonomic or non-holonomic. The system, together with the constraint, forms the following state equation (see equation 53):

$$E\dot{x} = Ax + Bu \tag{58}$$

where x is the state vector and u is the input, and A, B, E are system matrices and E is singular.

Manipulation is defined here [Hui et al., 1987] to be the deliberate and finite displacement of a grasped object, with respect to the fingers holding it, from an initial stable grasp to a final stable grasp. This represents a different type of constrained manipulator task where each finger on the robotic hand is a constrained manipulator. The object forms the constraint surface and it is necessary to control its position as well as the contact forces. Therefore, the singular system approach is intrinsic and necessary. Since a stable grasp can only be maintained with friction forces, they have to be considered as contact forces as well.

A model of rolling manipulation of rigid objects is presented here. The manipulation surface - the portion of the finger surface which actually comes into contact with the object during the manipulation - is also assumed to be rigid.

Point contact is represented by the following equations (Fig. 7).

$$b + C_{lb}\,{}^b c = t + C_{lt}\,{}^t c \tag{59a}$$

$$C_{lt}\,{}^t n = C_{lb}\,{}^b n \tag{59b}$$

where b is the position vector of object center of gravity (c.g.) with respect to (w.r.t.) inertial frame origin of object frame; t is the position vector of reference point on finger w.r.t. inertial frame, origin of finger frame; ${}^b c$ is the position of contact point w.r.t. object frame; ${}^t c$ is the position of contact point w.r.t. finger frame; ${}^b n$ is the surface normal vector from object surface w.r.t. object frame; ${}^t n$ is the surface normal vector from finger surface w.r.t. finger frame; C_{lb} is the transformation matrix from object frame to inertial frame; C_{lt} is the transformation matrix from finger frame to inertial frame

Eq. (59b) describes the surface normals as being aligned but opposite in direction. This is natural at a point contact between two surfaces. The model is based on the assumption that two surfaces in point contact are rolling with respect to each other if a *particle* attached to one surface at the point of contact has exactly the same velocity as another *particle* attached to the other surface at the same contact point. This is mathematically described by the following non-holonomic constraint.

$$\dot{b} + \omega_b^x (C_{lb}\,{}^b c) = \dot{t} + \omega_t^x (C_{lt}\,{}^t c) \tag{60}$$

where ω_b is the angular velocity of the object c.g. with respect to the inertial frame, ω_t is the angular velocity of the finger reference point with respect to the inertial frame and ω^x is the skew symmetric matrix corresponding to vector ω.

As pointed out by [Kerr and Roth, 1986], since t, $\dot{t}$, C_{lt}, ω_t and ${}^t c$ are functions of q and $\dot{q}$ and C_{lb} and ${}^b c$ are functions of b and $\dot{b}$, the above constraints can be expressed in the following matrix equation.

$$D\dot{q} = \begin{bmatrix} N_1 & N_2 \end{bmatrix} \begin{bmatrix} \dot{b} \\ \omega_b \end{bmatrix} \tag{61}$$

where D, N_1 and N_2 are matrix functions of q, $\dot{q}$ and b.

Two fundamental examples of manipulation are considered. The first scenario is the rolling of a round object by a robotic finger against a fixed surface (Fig. 6). The second example is the rolling of a round object between two fingers.

In the first case, since the object is round, the surface can be defined by

$$\Phi(b) = 0 \tag{62}$$

where $\Phi(b)$ is a scalar differentiable function.

The dynamics of a round object can be represented by

$$m\ddot{b} = mg + f_1 + f_2 \tag{63}$$

$$I_b\dot{\omega}_b = (C_{lb}\,{}^bc_1)^x f_1 + (C_{lb}\,{}^bc_2)^x f_2 \tag{64}$$

where f_1 is the contact force vector at the contact point between the finger and the object, w.r.t. inertial frame; f_2 is the contact force vector at the contact point between the object and the surface, w.r.t. inertial frame; ${}^bc_1$ is the position vector of contact point between finger and object, w.r.t. object frame; ${}^bc_2$ is the position vector of contact point between surface and object, w.r.t. object frame.

Together with the dynamics equation of the finger, which is similar to that of a standard manipulator, the following singular system is obtained.

$$\begin{bmatrix} M(q) & 0 & 0 & 0 & 0 \\ 0 & mI & 0 & 0 & 0 \\ 0 & 0 & I_b & 0 & 0 \\ 0 & 0 & 0 & 0 & 0 \\ 0 & 0 & 0 & 0 & 0 \end{bmatrix} \begin{bmatrix} \ddot{q} \\ \ddot{b} \\ \dot{\omega}_b \\ \dot{f}_1 \\ \dot{f}_2 \end{bmatrix} = \begin{bmatrix} 0 & 0 & 0 & J^T & 0 \\ 0 & 0 & 0 & I & I \\ 0 & 0 & 0 & (C_{lb}\,{}^bc_1)^x & (C_{lb}\,{}^bc_2)^x \\ D & -N_1 & -N_2 & 0 & 0 \\ 0 & 0 & 0 & 0 & 0 \end{bmatrix} \tag{65}$$

$$\begin{bmatrix} \tau \\ 0 \\ 0 \\ 0 \\ 0 \end{bmatrix} + \begin{bmatrix} -C(q,\dot{q}) & -G(q) \\ mg \\ 0 \\ 0 \\ \phi(b) \end{bmatrix}$$

The system equation for the scenario of rolling a round object between two fingers is quite similar. First, the constraint in eq. (61) has to be repeated for the second finger as shown below.

$$D_2\dot{q}_2 = \begin{bmatrix} L_1 & L_2 \end{bmatrix} \begin{bmatrix} \dot{b} \\ \omega_b \end{bmatrix} \tag{66}$$

where D_2, L_1 and L_2 are matrix functions of q_2, the joint angle vector of the second finger.

The system equation can now be written as:

$$
\begin{bmatrix}
M_1(q_1) & 0 & 0 & 0 & 0 & 0 \\
0 & M_2(q_2) & 0 & 0 & 0 & 0 \\
0 & 0 & mI & 0 & 0 & 0 \\
0 & 0 & 0 & I_b & 0 & 0 \\
0 & 0 & 0 & 0 & 0 & 0 \\
0 & 0 & 0 & 0 & 0 & 0
\end{bmatrix}
\begin{bmatrix}
\ddot{q}_1 \\
\ddot{q}_2 \\
\dot{b} \\
\dot{\omega}_b \\
\dot{f}_1 \\
\dot{f}_2
\end{bmatrix}
\tag{67}
$$

$$
=
\begin{bmatrix}
0 & 0 & 0 & 0 & J_1^T & 0 \\
0 & 0 & 0 & 0 & 0 & J_2^T \\
0 & 0 & 0 & 0 & I & I \\
0 & 0 & 0 & 0 & (C_{lb}{}^b c_1)^x & (C_{lb}{}^b c_2)^x \\
D_1 & 0 & -N_1 & -N_2 & 0 & 0 \\
0 & D_2 & -L_1 & -L_2 & 0 & 0
\end{bmatrix}
$$

$$
+
\begin{bmatrix}
\tau_1 \\
\tau_2 \\
0 \\
0 \\
0 \\
0
\end{bmatrix}
-
\begin{bmatrix}
C_1(q_1,\dot{q}_1) + G_1(q_1) \\
C_2(q_2,\dot{q}_2) + G_2(q_2) \\
0 \\
0 \\
0 \\
0
\end{bmatrix}
$$

where subscript 1 denotes variables associated with the first finger and subscript 2 is associated with the second finger.

6.   CONCLUSIONS

Interaction between robot and environment occurs often and generates effects which must be controlled. Controlling force and motion at the point of interaction in multi DOF systems has generated two basic approaches: force regulation in the context of hybrid control and impedance and force control in the context of force and motion regulation. The later approach is addressed in the paper and models suitable for control synthesis are presented. Also, models based on compliant and constraint motion are discussed and shown to be useful to the same class of problems, control of dexterous hands. These models underline a special feature which is the inclusion of environmental parameters in the model definition. This particular issue is critical to obtaining desirable performance and robustness. Current research is conducted to investigate the two types of models for real-time control in the context of grasping and manipulation.

7. REFERENCES

[1] Raibert, M.H., Craig, J.J., "Hybrid Position/Force Control of Manipulators", ASME Journal of Dynamic Systems, Measurement and Control, Vol. 102, June 1981, pp. 126-133.

[2] Salisbury, J.K. and Craig, J.J., "Articulated hands: Force control and kinematic issues", The Int. Journal of Robotics Research, Vol. 1, No. 1, 1982.

[3] Hogan, N., "Impedance Control: An Approach to Manipulation: Part I – Theory", pp. 1-7. "Part II – Implementation", pp. 8-16, "Part III – Applications", pp. 17-24, ASME Journal of Dynamic Systems, Measurement and Control, Nov. 107, 1985.

[4] Kazerooni, H., Sheridan, T.B., Houpt, P.K., "Robust Compliant Motion for Manipulators: Part I – The Fundamental Concepts of Compliant Motion", pp. 83-92, "Part II – Design Method", pp. 93-105, IEEE Journal of Robotics and Automation, Vol. RA-2, No. 2, 1986.

[5] Goldenberg, A.A., "Force and Impedance Control of Robot Manipulators", to appear in Proceedings of ASME Winter Annual Meeting, 1987.

[6] Greenwood, D.T. "Classical Dynamics", Prentice Hall Inc., 1977.

[7] Hogan, N., "On the Stability of Manipulators Performing Contact Tasks", submitted for publication IEEE Journal of Robotics and Automation, 1987.

[8] Anderson, R.J., Spong, M.W., "Hybrid Impedance Control of Robotic Manipulators", Proceedings of 1987 IEEE Conference on Robotics and Automation, pp. 1073-1080.

[9] Payandeh, S., Goldenberg, A.A., "Closed Loop Dynamics of Dexterous Fingers (Hand) for Control of Grasping and Manipulation", RAL Technical Report, 1987.

[10] Hui, R., Goldenberg, A.A., "Modelling of Manipulation of Rigid Objects as a Singular System", RAL Technical Report, 1987.

[11] McClamroch, N.H. and Huang, H.P., "Dynamics of a closed chain manipulator", American Control Conf., Boston, 1985.

[12] McClamroch, N.H., "Singular Systems of Differential Equations as Dynamic Models for Constrained Robot Systems", Proceedings of 1986 IEEE Conference on Robotics and Automation, pp. 21-28.

[13] Cobb, D., "Descriptor variable systems and optimal state regulation", IEEE Trans. on Aut. Control, Vo. AC-28, No. 5, 1983.

[14] Goldenberg, A.A. and Mills, J.K., "Force and position control of constrained manipulators", Proc. of IEEE Int. Symp. on Intelligent Control, Philadelphia, 1987, pp. 294-301.

[15] Fearing, R.S., "Simplified grasping and manipulation with dextrous robot hands", IEEE Journal of Robotics and Automation, Vol. RA-2, No.4, 1986.

[16] Kerr, J. and Roth B., "Analysis of multifingered hands", The Int. Journal of Robotics Research, Vol. 4, No. 4, 1986.

[17] Kobayashi, H., "Control and geometrical considerations for an articulated robot hand", The International Journal of Robotics Research, Vol. 4, No. 1, 1985.

[18] Okada, T., "Computer control of multijointed finger system for precise object-handling", IEEE Trans. System Man and Cybernetics, Vol. 12, No. 3, 1982.

34

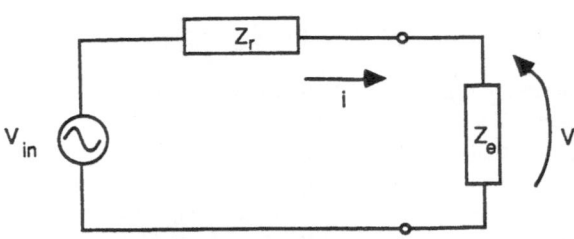

Fig. 1: Diagram of force mode

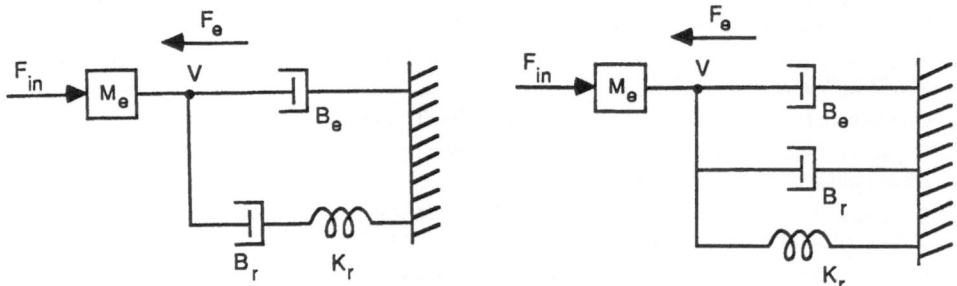

Fig. 2: Schematic diagrams of force mode

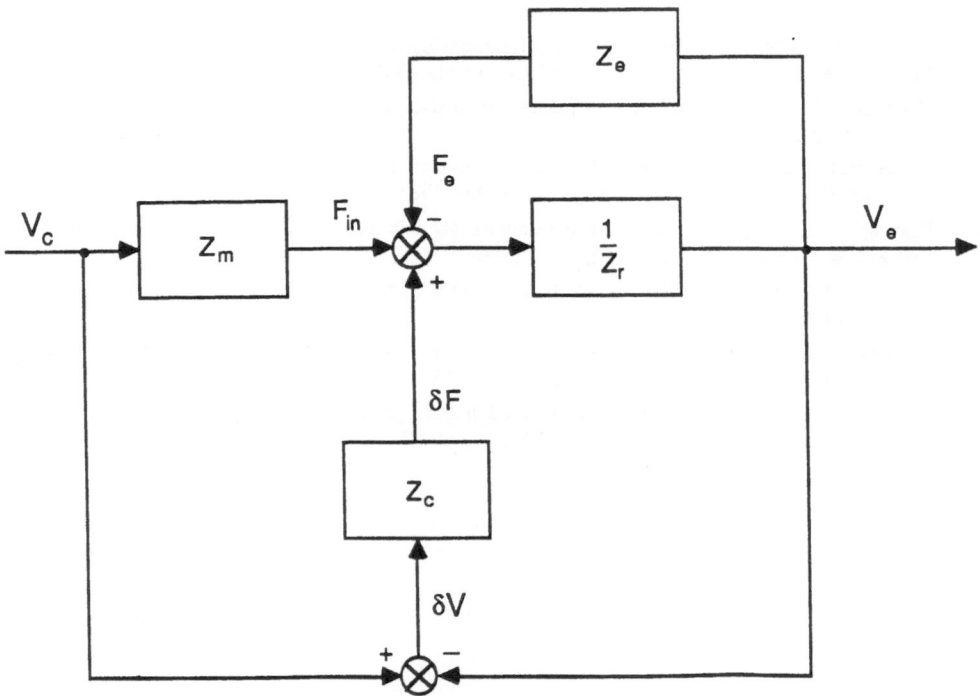

Fig. 3: Motion Control Mode During Contact

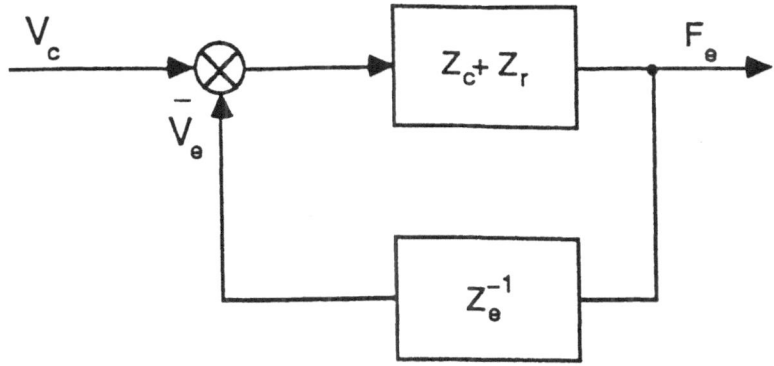

Fig. 4: Closed loop impedance control in flow loop

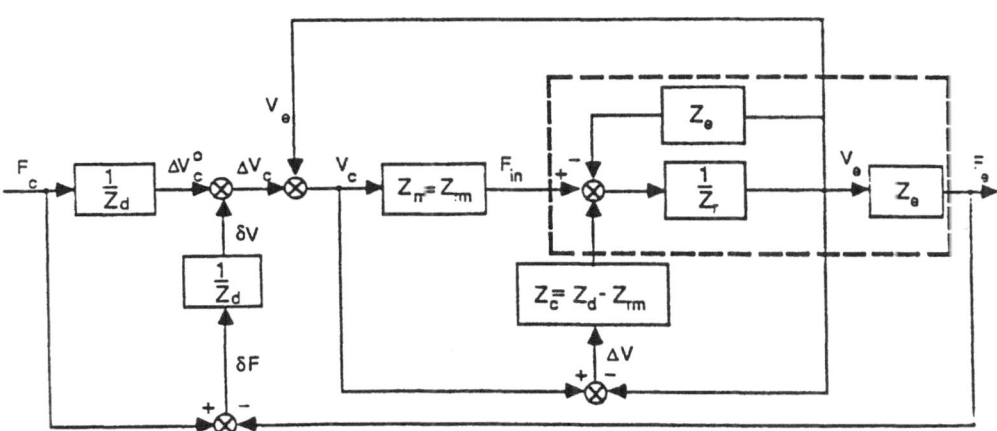

Fig. 5: Force and impedance control implementation

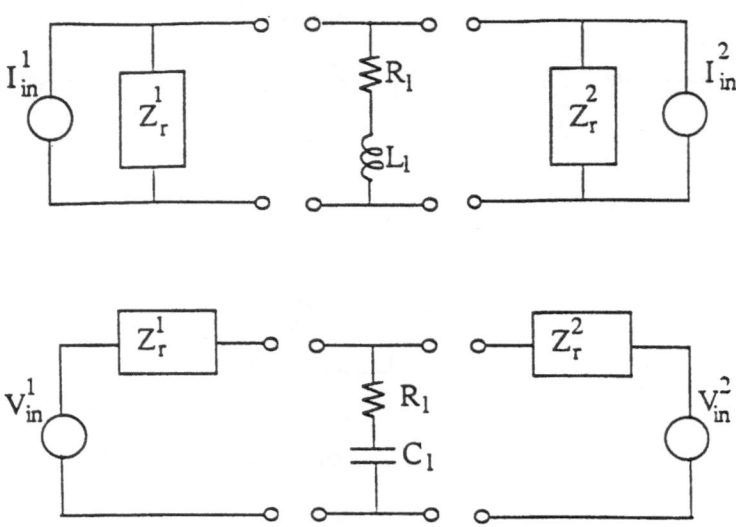

Fig. 6: Two fingers in contact with a compliant object

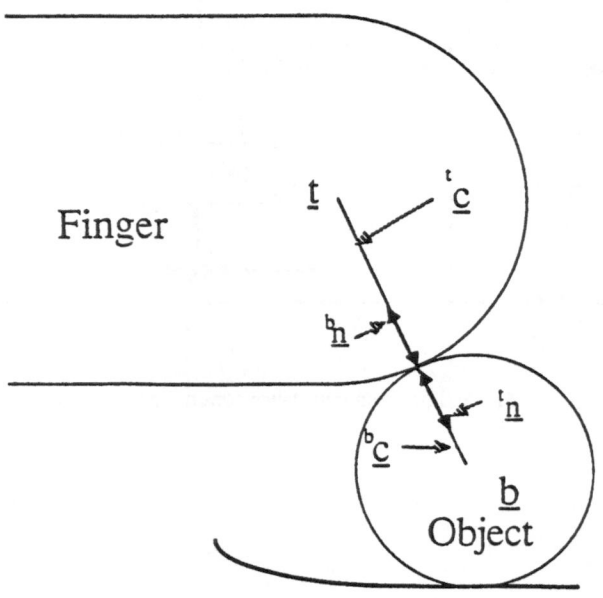

Fig. 7: One-fingered ball rolling on a fixed surface

PNEUMATIC SENSORS: THEIR USE AND PERFORMANCE IN FORCE, TACTILE AND POSITION SENSING AND IN SHAPE RECOGNITION

A. Romiti
Dip. di Meccanica, Politecnico di Torino
Corso Duca degli Abruzzi 24
Torino 10129   (Italy)

The application of pneumatic sensors on robots or manipulation systems is possibile and convenient in some cases when the use of compressed air is particularly suitable.

Pneumatic sensors may be air jet, contactless devices, or otherwise contact sensors.

Air jets may be arranged to work as proximity sensors, while pneumatic contact devices may be used as touch sensors or may be arranged to measure pressure or force.

Proximity sensors are used on the tip of grasping fingers for providing hand closure signals, or for measuring the distance between   fingers and workpiece.

The workpiece re-orientation devices, that often supply a robot, may also be provided with proximity sensors for defining workpiece position.

Proximity sensors may be requested to give out a threshold signal; for example, they must provide a digital signal when the measured distance is less then a given value; in other cases, they must give out a proportional signal.

The same sensor may often perform both functions; the threshold is established by pneumatically elaborating the output signal.

Some sensors are intrinsically digital, as the sensors based on jet obstruction by an interposed solid object.

A jet interruption sensor is made by an emitting nozzle and a receiving tube, where it is measured the recovery pressure, that is the stagnation pressure of the recovered part of the jet.

The distance between emitter and receiver is of the order of some centime-ters while the emitted jet diameter is of the order of one millimeter; the supply pressure value is generally greater than the minimum value that is necessary to create sonic conditions at the emitter outlet.

An interruption jet sensor may work also in a negative way; in this case, the flow between emitter and receiver is normally disrupted by a third perpendicular jet; the recovered pressure is then very low. If some object intercepts this jet, the flow to the receiver is re-established and the recovered pressure rises.

Other jet sensor types allow analogous measuring of distances in different lenght ranges.

Counter-pressure sensors act according to the flapper-nozzle principle. They are apt to precisely measuring very small, sub-millimeter distances, or to signal nozzle obstruction.

Induction sensors are made by two tiny coaxial tubes; the inner tube tip is made by a nozzle.

When the air jet emitted by the nozzle impinges on an object standing in front of it, the stream is diverted and a recovered pressure is sensed in the outer tube.

Different sensor types are actuated by tactile contact.

The counter-pressure sensors may act in this way when their nozzle is completely obstructed.

However, the name of contact sensors is generally applied to those that are actuated by the displacement of a moving part due to a contact force transmitted by an external body.

The motion of a mechanical element produces then a pressure change, either by letting some leakage to occur through an opening, or by letting the pressure to increase in a compressed bellow or balloon, or by the displacement of a valve spool.

The first two methods are essentially digital; the last one adapts itself both to digital and analog sensing. When the spool motion is elastically constrained, it is also fit for force measurement.

FLAPPER - NOZZLE JET SENSORS

A flapper - nozzle sensor is represented in Fig.1a.

The source pressure p_s supplies a flow in a duct, going through an orfice and, subsequently, through a nozzle, to the ambient athmosphere.

The orifice area, R_1, is lesser than the nozzle area, R_n.

The sensed pressure p^x is picked by a branch duct between orifice and nozzle.

The sensor may be used when the distances between the nozzle and the obstacle are in the range between 0 and 0,6 mm. A good sensitivity may be found if the distance is less than 0,2 mm; larger distances may be admitted by enlarging the sensor dimensions.

Diagrams of p_x pressure versus the normal distance h between nozzle and approaching object are represented in Fig.1b for a given nozzle area and different values of supply pressure and orifice area.

One can remark that the sensor provides linear response in a very narrow distance range.

The device is therefore inpractical for analog distance sensing, but for the case when a suitable feedback system is provided.

A common arrangement is described in Fig.2.

The displacement to be measured, y, is made equal to the displacement of the tip of a lever that acts as a flapper for a nozzle at a distance m from the lever tip.

The lever can pivot around a point at distance m+n from the tip. This point is fixed to the moving part of a bellow or membrane motor actuated by the pressure p_x.

Assuming that the flapper - nozzle distance h stays in the linear range of the sensed pressure p_x, one can write.

$$p_x = \alpha - \beta h \tag{1}$$

where α and β are constant.

If one calls c the bellow compliance, and ho the neutral value of h (for y=0, p_x=0), one can write.

$$h = ho - ym + c \, p_x \, n \tag{2}$$

Small displacements δy are related to the values δh by the expression:

$$\frac{\delta h}{\delta y} = - \frac{m}{c \beta n} \tag{3}$$

By letting $c \beta n \gg m$, one can reduce the values δh to a small portion of the values of the displacement y, in such a may that they can stay inside the proportional sensing range, A drawback of such arrangement is the time delay induced by the feedback circuit. In order to minimize such delay, the bellow volumetric capacity, its compliance and its moving mass must have low values. The orifice area must be as large as it is possible. Anyway, low system rise times are difficult to achieve.

Identification by counter - pressure sensors

Identification of workpieces may be achieved by letting the examined object rotate with respect to a plate supporting the nozzles of a group of counter - pressure sensors.

The workpiece may lay on the plate, and a mechanical fixture may be provided, that can rotate the piece while the plate stays still, or other-wise it can keep still the piece while the plate rotates. The fixture must always keep the piece inside an imaginary circular line, joint to the plate, corresponding to the circumscript circle of the workpiece.

In this case, the workpiece slides on the plate during the identification motion.

The sensors must be arranged so that the sequence of their signals is sufficient for identify the features (edges, holes, etc.) of the piece, and therefore for detecting the moment when a particular orientation is achieved.

The signals are due to obstruction of the counter - pressure sensors.

The identification regards a face of the object that lies on the plate; nevertheless the object does not need to be flat; if it has different equilibrium faces, it can be automatically turned if not identified in order to try a different face.

Obviously, each signal sequence is specific for a given workpiece design.

This technique may be modified by avoiding contact between workpiece and sensor plate; in this case, the sensors act like flapper-nozzle jet sensors, the flapper being made by the opposed part of the workpiece surface.

Induction sensors

These pneumatic sensors have the property that the recovered pressure is quite high even when the distances between nozzle and obstacle are markedly larger than the distances fit for counter - pressure sensors.

Fig.3 shows a diagram of the recovered pressure, p_x, versus the normal distance, h, for various supply pressure p_s. This diagram is relative to one particular induction sensor; by changing the sensor model there is some non essential variation of the characteristic curves.

Both analog and digital signals may be obtained.

Best digital induction sensors must have diagrams presenting some steep part, in order to obtain a precise and reliable switching of a pneumatic relay controlled by the pressure p_x.

Analog sensors should present a quite opposite behaviour, that is one must choose a distance range corresponding to smooth output pressure variations. Each sensor must be individually calibrated.

Induction pneumatic sensors may be used for tangential displacements sensing too. As an example, Fig.4a shows a particularly critical application; it was required to detect the position reached by a cylindrical body with blunt edges along an axis x, perpendicular to the sensor axis. Fig.4b shows the diagram of the output pressure p_x versus x, for various distances h between the nozzle tip and the cylindrical surface. At low speed, the detection accuracy was of the order of 20 µm, with statistical standard deviations of the order of 5 µm. The sensor actuated a Schmitt trigger acting on a bistable fluidic element. A 1 m lenght pipe connecting the sensor with the amplifier was provided for noise damping.

Pressure sensors

Pressure sensing on robot end-effectors, jaws or fingertips is essential for manipulation of fragile or delicate objects. It can be obtained by attaching to the jaws or fingertips some sort of bellow or balloon, filled by compressed air or by a liquid.

One may control the pressure in the elastic structure to a suitable level, by means of a pressure regulator.

If the robot hand has to pick delicate pieces having different dimensions, it is necessary to stop the jaws closure when the contact pressure reaches a given threshold. A sealed circuit can be made, mhere the fluid (generally a liquid) is kept.

The circuit is made by a poach and a tube having elasting compliance. The tube ends at the inlet of a fluid microvalve, that compares the pressure on a membrane versus a spring force or a regulated pressure force, and switches according to the force difference.

It will be shown here an application example where both air jet sensors and pressure sensors were used.

The problem regards a robotized system for assorted biscuit packaging. Various types of biscuits have to be orderly inserted each one in the proper plastic case inside a cardboard box. Each case must hold a given number of biscuit.

The gripper must be able to grip each biscuit according to its dimensions. The jaws must be very thin, so that they are made by sheet steel covered, on the contacting side, by a membrane forming a liquid fillet poach; when the pressure increases, a threshold valve is activated and the closure motion stops.

The gripper is illustrated in Fig.5. The jaws translate circularly by means of four link parallel linkages; they are actuated through slotted links by two pneumatic microactuators, that are kinematically in series.

The actuator CC controls the closure motion; the actuator CA, in series with CC, allows a limited jaw opening with respect to the closure position. The braking actuator CF is controlled by the sensor and acts on a stopping wedge.

Presence and roughness sensors

The same particular problem described above requires use of other sensors for obtaining biscuit conveyance according to a suitable orientation.

The biscuits are presented in homogoneous heaps where they are randomly oriented. They are transferred by moving ribbons to gravity channels and to the respective positioning stations.

At these stations, the biscuits lay on plates with holes through which compressed air is made to flow.

The pieces having only one stable plane bearing surface may be pneumostatically heaved and sent, by lateral air jets and mechanical deviators and channels, to the picking station with a correct orientation. If the pieces lay on curved surface on the supporting plate , they stay still because the lateral jets are not sufficient to overcome friction.

The air jet sensors E allow to determine if the biscuit is staying in the region "G" of Fig.6; in this case, a pneumatic actuator pushes the biscuit against a curved surface, overturning it with the help of suitably directed air jets. In this way the biscuit may lay on its plane surface and be conveyed by the air transportation system.

The air sensor is represented in Fig.7; a nozzle E_1 emits a jet that would impinge on the recaiver R but for a disrupting jet coming out from the nozzle E_2; when a piece intercepts this jet, the jet from the emitter E_1 may actually impinge on R, so that the recovery pressure in R increases.

It may occur that a piece has two dimensionally equal plane faces, on which it can steadily lean. The faces can have different superficial texture; in this case it may be required to identify the actual lay, because the object must be picked and placed keeping always the same face upturned.

Such case occurs, for example, for some wafer biscuits.

In this case, a surface texture identification system must be devised, that is based on a roughness sensor.

The roughness sensor that has been used was made by an induction air jet sensor, whose nozzle was flush with a plate on which the examined object was laying (see Figs. 8 a and b, where the sensor acts on different surfaces). The sensed quantity was the recovered pressure, varying according to the different leakege flow throughput corresponding to the different superficial textures. The transducing circuit is shown in Fig.9.

Contact and force sensors

Any displacement sensor can be turned into a force sensor by making it to measure a spring deflection. The same concept applies to pneumatic displacement sensors. Their sensitivity depends upon the spring stiffness and the displacement measurement sensitivity.

Force measurement may be carried out analogously, or otherwise a force threshold may be signalled, for example through end of run pneumatic pick-up.

Contact sensors must produce a signal at the instant when a robot hand comes into touch with a solid object. They may be used to control the stop or the slowing motion of the hand at short distance from the object; in this case they can be of the proximity type, if the object surface is sufficiently well defined with respect to the hand; otherwise, thay can be of the whisker type. These sensors are made by a long steel wire protruding from a small pneumatic valve, that normally is closed. When the whisker is touched, it actuates the opening of the valve; the upstream pressure decrease signals contact.

Both contact and force pneumatic sensors were mounted on a robot hand for picking randomly oriented workpieces in a bin, that will be described as an application exemple.

The method used was an orderly random search inside the bin, along a number of lines parallel to two sides of the container.

The bin holds mechanical workpieces that may be somewhat entangled. The workpiece upper lever in the bin changes with location and time.

The robot arm supporting the hand plunges vertically, and it must slow down approaching the workpiece upper level. A whisker controlled leakage valve, attached to the hand, provides a signal when a given distance from the workpiece surface is reached. The high whisker flexibility allows damage avoidance during the final approach motion.

After contact between hand and workpieces, some operations must be carried out, whose sequence depends on the signals coming out from some force sensors.

The hand is represented in Figs. 10a and b. It has two equal fore fingers and a shorter back finger, all actuated by pneumatic cylinders. The hand has a limited capatility of free swivel around the axis of the plunger; the hand palm may also limitedly rotate around an horizontal pivot.

The force sensors are located into the cases e and f, and they are made by pre-loaded springs. The rings b and b' allow to uncouple the upward and downward measuring springs d, d', held by collars e, e' around the rod a, joined to the plunger.

The ring displacements provide output signal through flush mounted transducer, so that up or down vertical forces and horizontal moments may be signalled.

Other "virtual" pneumatic sensors signal the actuation of the fingers by the measurement of the related cylinder pressures.

Sensors for multiple vacuum grippers

Vacuum gripping by sucking cups is easily signalled by pressure valves at the cup outlet. However, when multiple grasping is contemporarily actuated using a common sucking duct, one must provide a suitable set of valves for blocking the leaking cups and for signalling which cup has failed its grip.

An example is given by a vacuum gripper made for packaging, in one stroke, a complete set of chocolates in a box.

The gripper is shown in Fig.11. The cups are glued according to a suitable position matrix to a plate. Uncoupling of the cups is made by elements built in the plate over any cup location.

One of these elements is isolated in Fig.12. It is divided in two parts by the membrane 11, that is actually part of a rubber sheet sealing the top and bottom parts of the plate.

In Fig.9 it is represented the cup 4, joined by the duct 3 to the annular cavity 2. Over the membrane there is the chamber 1, joining, through the opening 7, the three-way valve 9 and the duct 5 to the sucking duct 3.

The chamber 2 is joined through the opening 6 to the ejector 8 and the valve 9.

When the gripper touches the pieces, the pneumatic valve 9 is switched by the rod 10; both chambers are connected to the ejector; the membrane heaves and let the duct 3 to be connected to the chamber 2, so that the cup is also connected with the sucking line.

When the gripper rises, the contrasting spring of valve 10 nullify the signal from the rod 10, so that the ducts 5 and 7 are joined.

A pressure signal is taken from duct 5. If the relative pressure is negative, the grasping has been successful.

If one or more cups and the contacting pieces are not able to show contact sealing, at the beginning of the hand raising the membrane closes the opening of duct 3 of such cups, while the well performing cups are still contacting their pieces.

The closure of duct 3 inlet is due to pressurization of chamber 1, while chamber 2 is still joined to the sucking line; therefore, the chamber 2 of any cup, after an unsuccessful gripping attempt, is isolated.

The signals from ducts 5 show the location on a position matrix of the unsuccessful graspings, therefore permitting automatic correction.

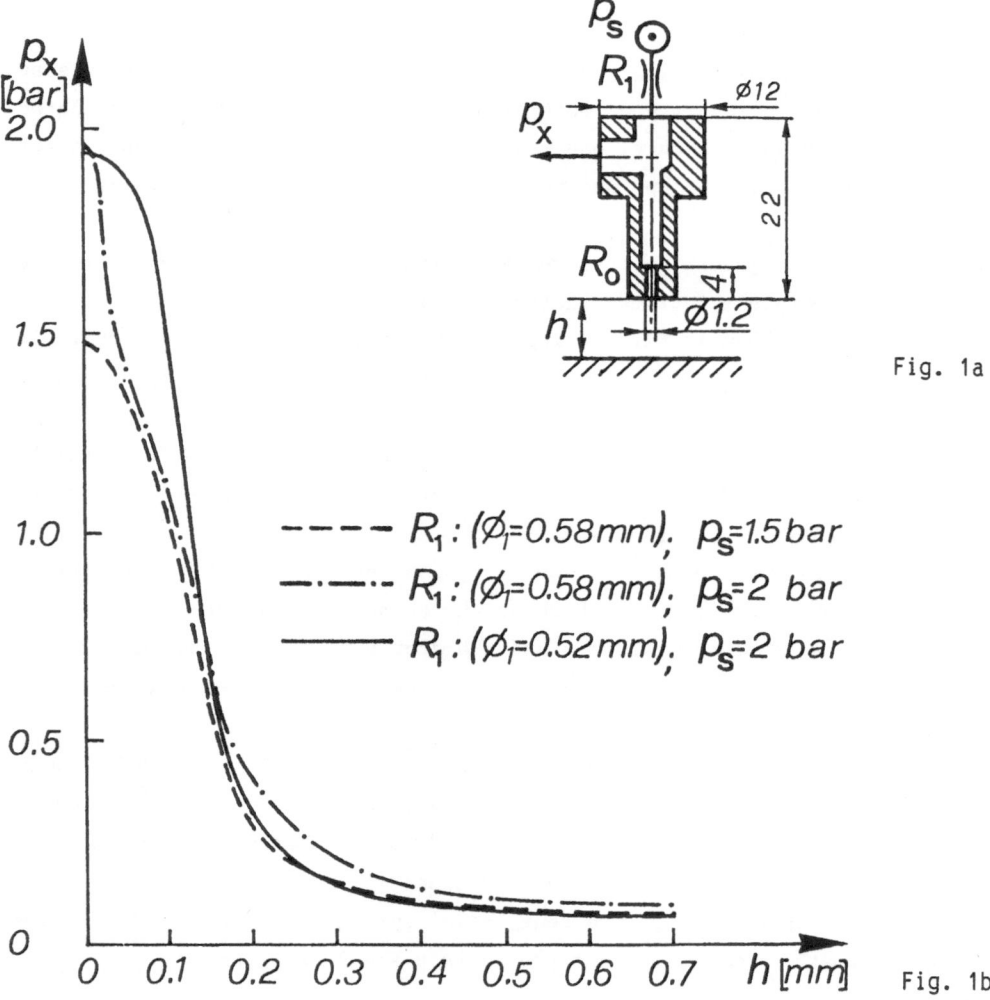

Fig. 1a

Fig. 1b

Fig. 1 Counter-pressure sensor characteristics for normal approach to an obstacle

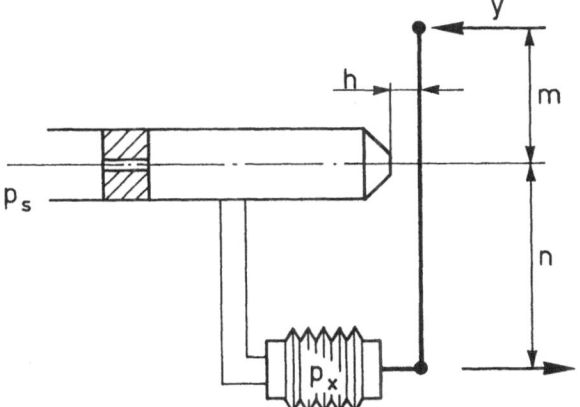

Fig. 2 Flapper-nozzle analog sensor

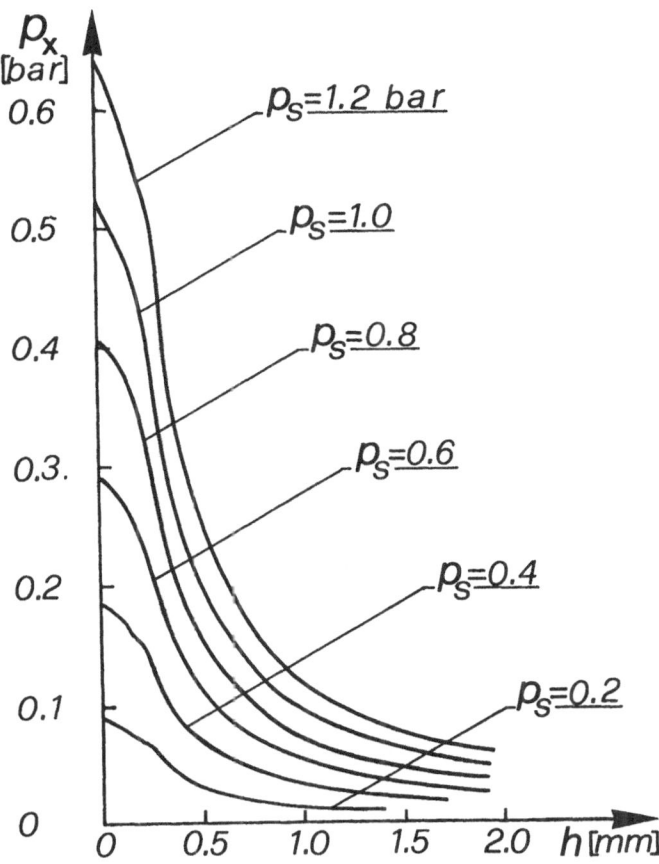

Fig. 3 Inductior sensor characteristics

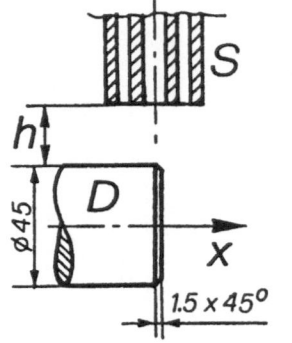

Fig. 4a

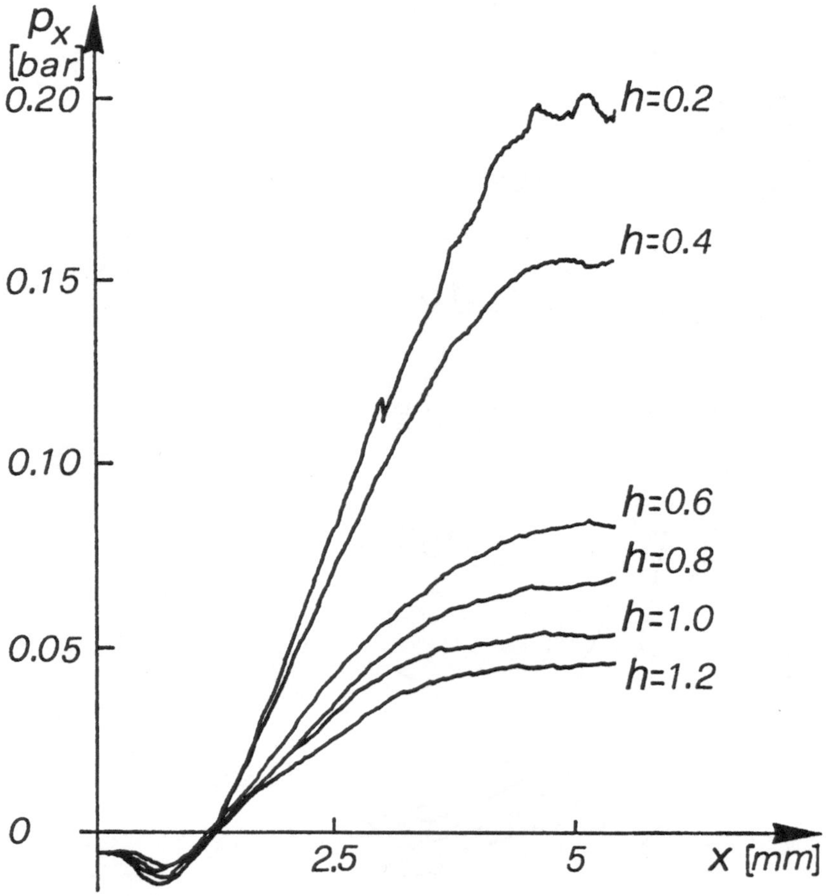

Fig. 4b

Fig. 4 Tangential displacement measurements with induction sensor

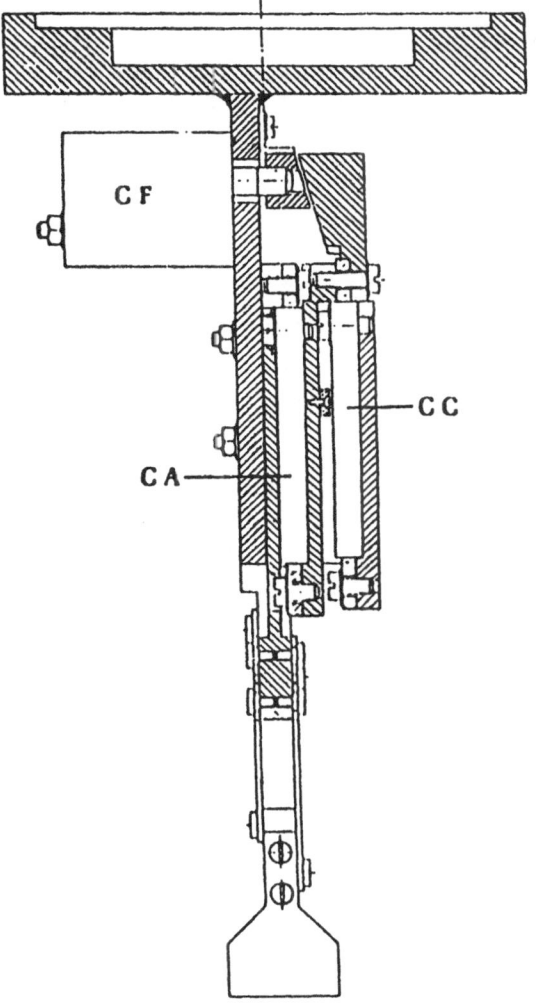

Fig. 5 Gripper with tactile sensing at fingertips

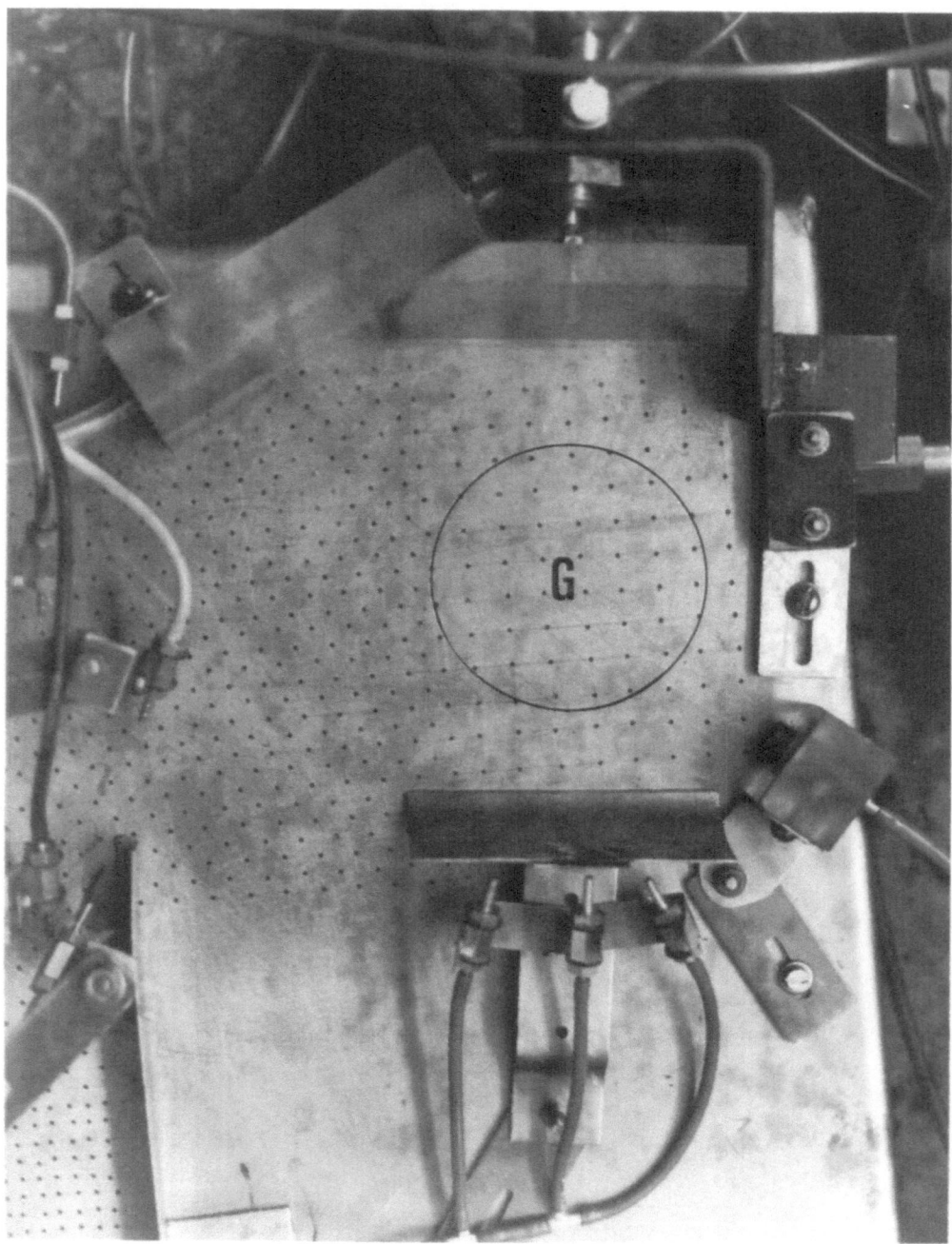

Fig. 6 Fluid bed and sensor setting

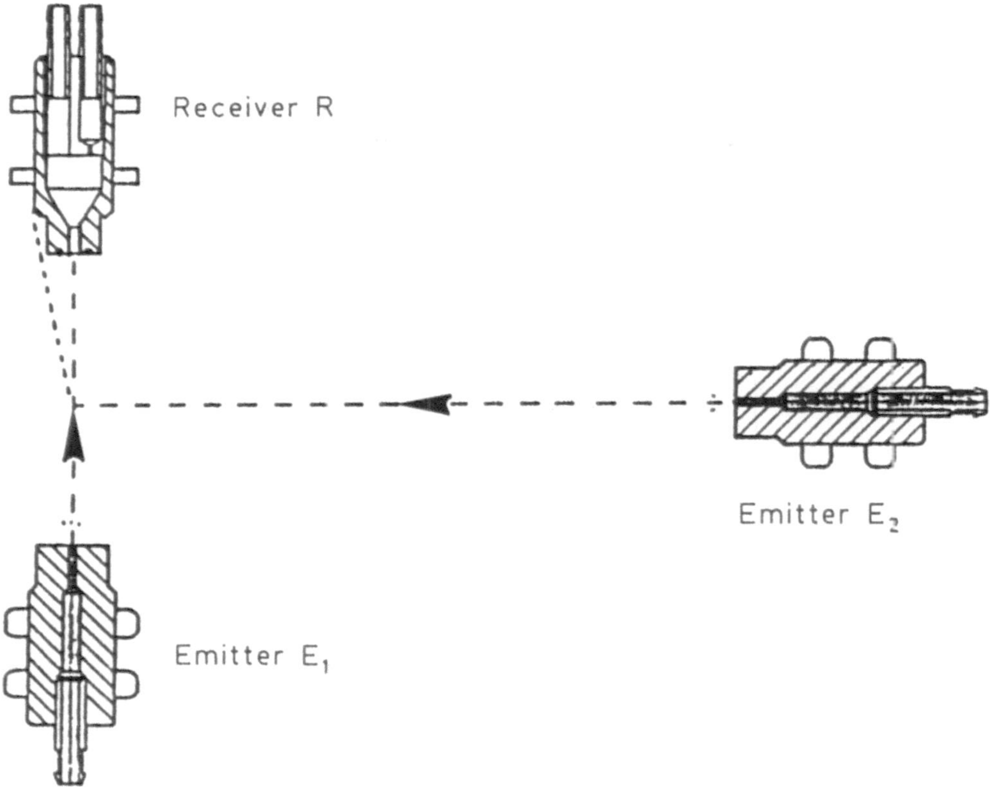

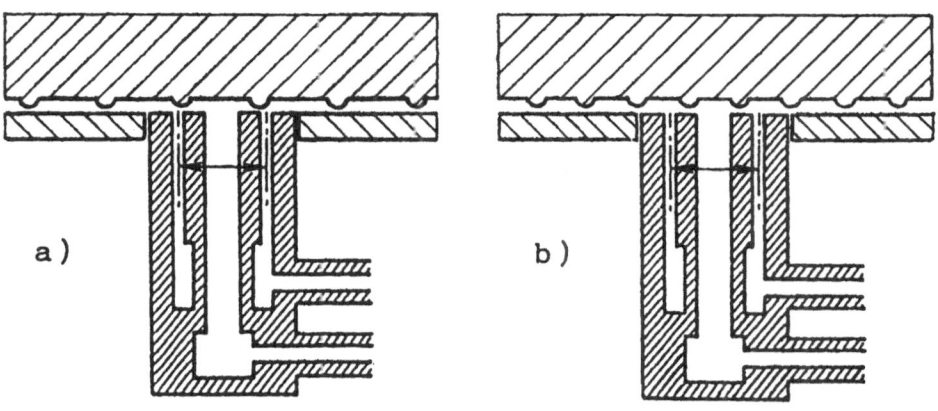

Fig. 7 Air jet presence sensor

Fig. 8 Roughness sensors

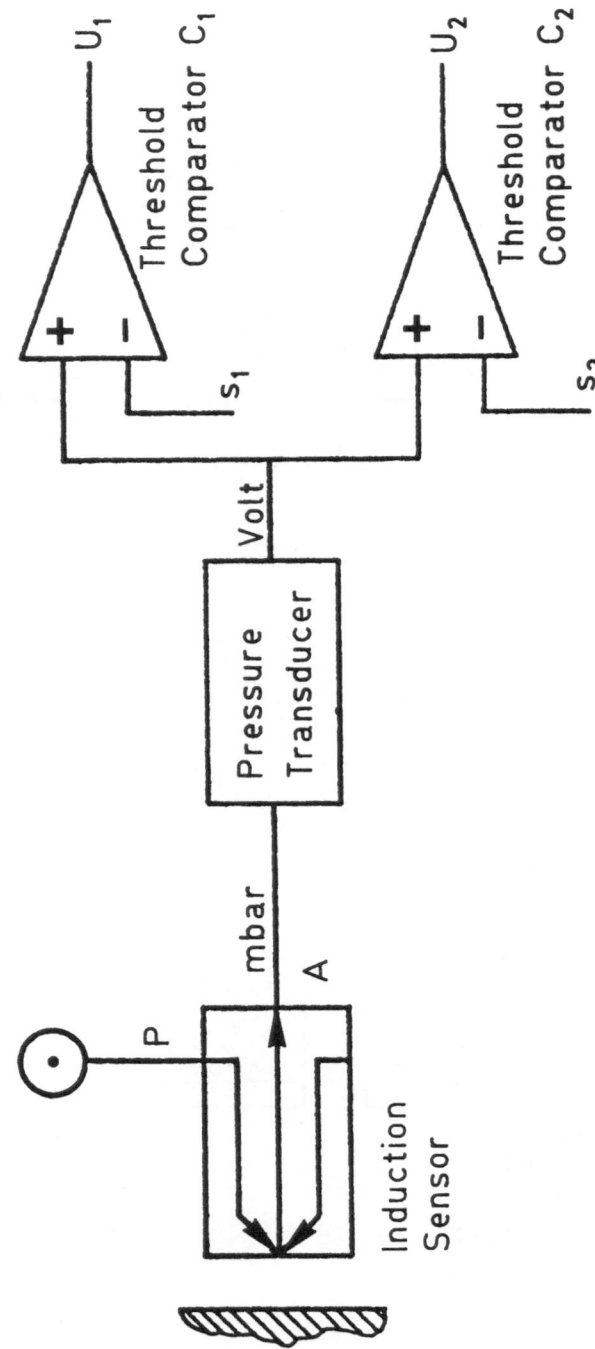

Fig. 9 Transducing circuit for roughness sensor

53

Fig. 10a

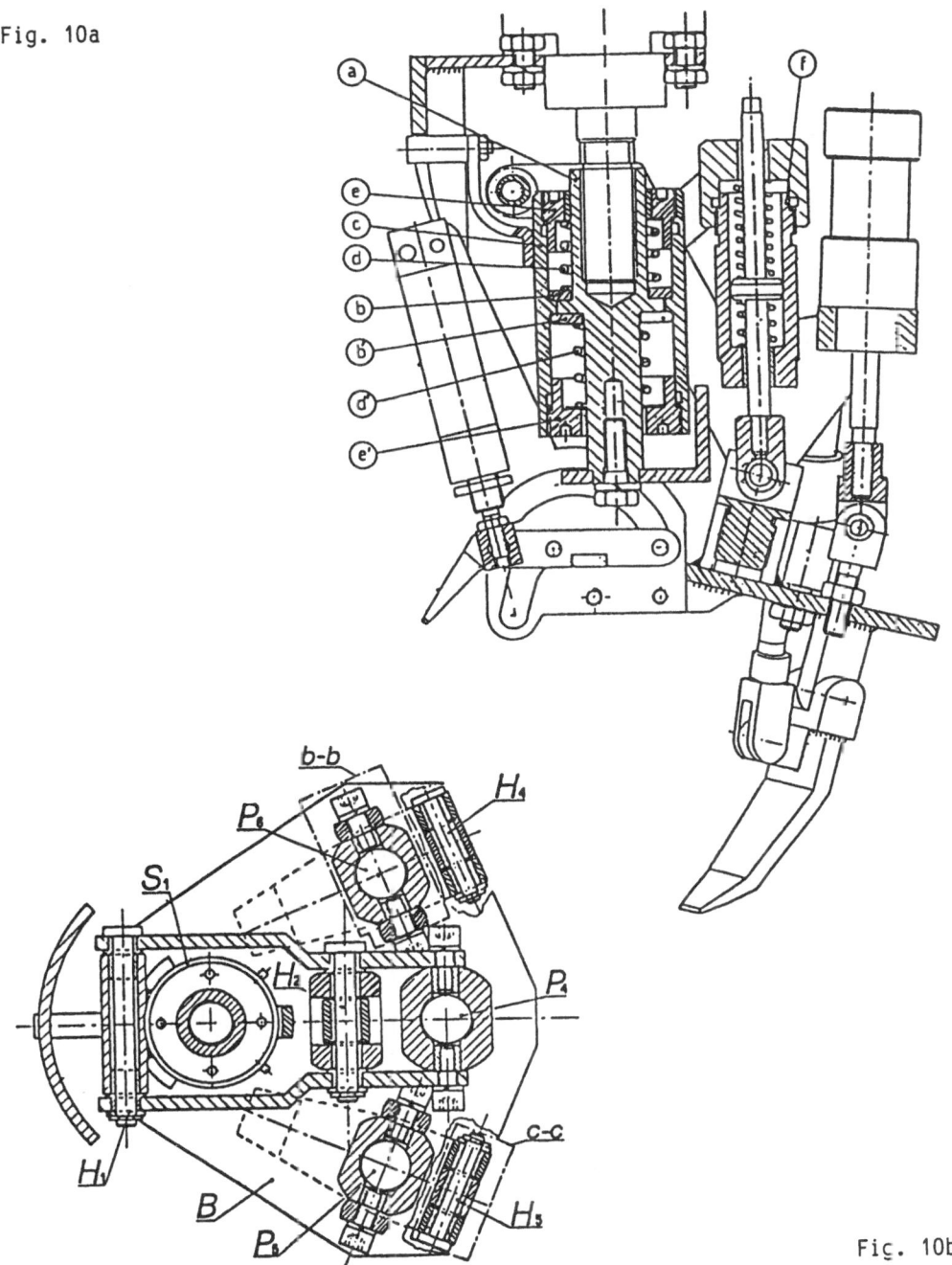

Fig. 10 Mechanical gripper and sensors

Fig. 10b

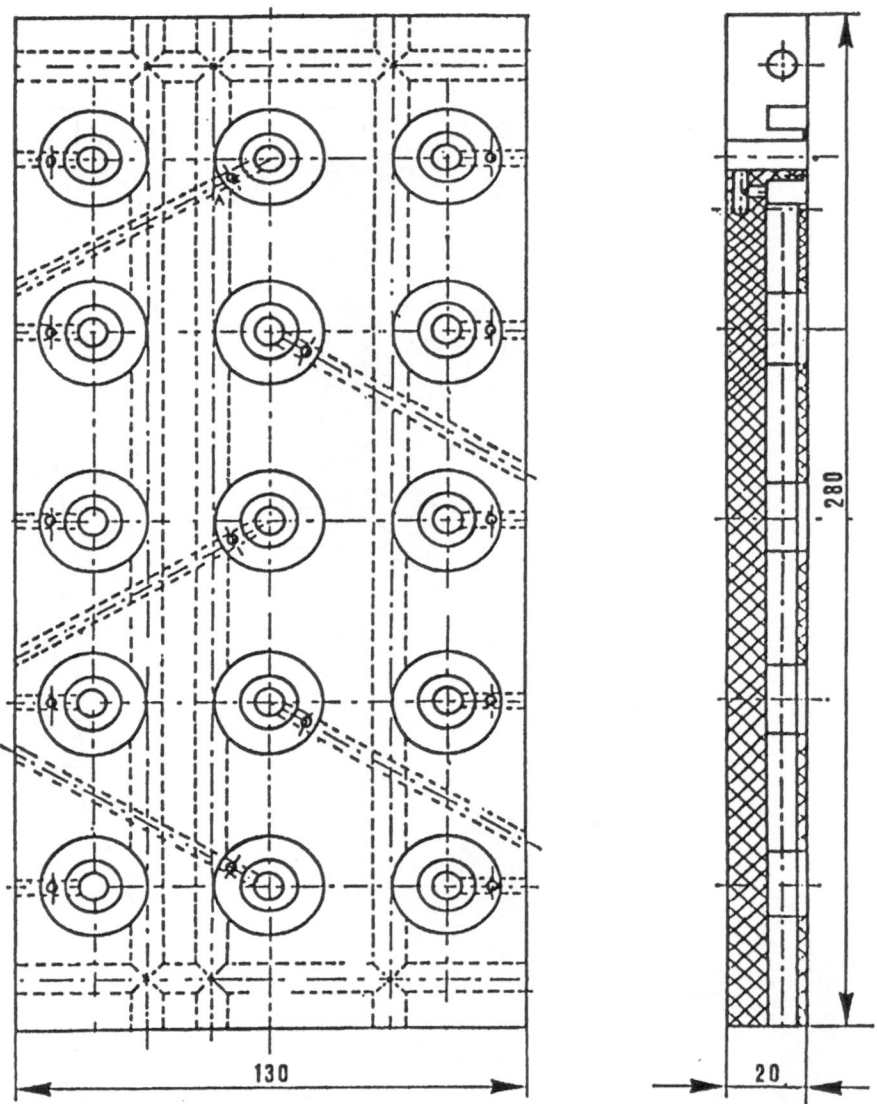

Fig. 11 Uncoupled multiple vacuum gripper

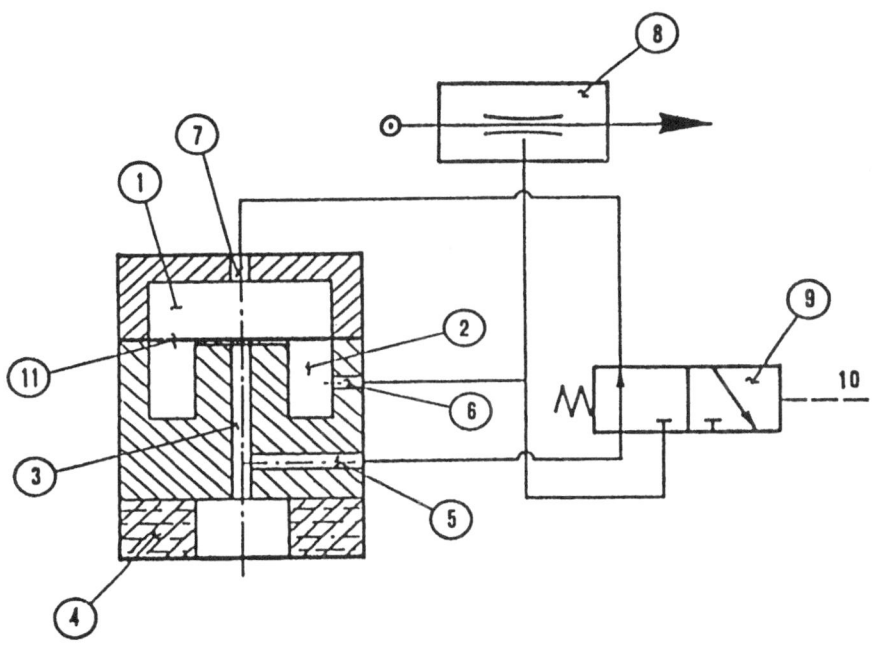

Fig. 12   Vacuum cup and grasping sensor

REFERENCES

1. Romiti, A.; Belforte, G.: Automazione a Fluido, Patron, Bologna, 1975
2. Belforte, G.; Romiti, A.; D'Alfio, N.; Quagliotti, F.: Some new Systems for Recognition and Positioning of Mechanical Parts, 10th ISIR/ 5th CIRT, Milano 1980
3. Belforte, G.; Romiti, A.; D'Alfio, N.; Ouagliotti, F.: Identification through Air Jet Sensing, 1st Int. Conf. ROVISEC, Stratford on Avon,1981
4. Romiti, A.; Belforte, G.; D'Alfio, N.; Quagliotti, F.: Picking from a Bin through Tactile Sensing, 11th Symp. on Industrial Robots, Tokyo 1981
5. Belforte, G.; Romiti, A.; Sorli, M.: A Sensorized Manipulation System for Food Industry Products, 9th BRA Symposium, Stratford on Avon, 1986
6. Belforte, G.; Bellosta, F.; D'Alfio, N.; Raparelli, T.: A sensorized System for Delicate Piece Selection, Handling and Packaging, 2th Int. Conf. on Robotics and Factories of the Future, San Diego, 1987.

II. TACTILE SENSORS

CARBON FIBRE SENSORS

J.B.C. Davies, B.Sc., M.Sc.
Department of Mechanical Engineering
Heriot-Watt University, Edinburgh   (U.K.)

Abstract

Carbon Fibre sensors possess the potential of providing robust, highly
sensitive transducers with poor respeatability and time dependent
characteristics. Dynamic response is limited to about 30 Hz.

1.   Introduction

The electrical properties of the element carbon have been known and
exploited for many years, notably in the early carbon microphones and in
present day motor brushes, commutator slip rings etc.   With the
development of carbon fibres by Phillips at the Royal Aircraft
Establishment, Farnborough, in the 1950's a new arena of application was
opened up before the carbon molecule.   The main applications of carbon
fibres as strength additives in various aeronautical applications, along
with its well publicised inclusion in golf clubs, tennis racquets, etc.
are well known.   In mid 1950's a Scientific Adviser in the navy,
recognising the limitations of the carbon brushes as used in electric
motors, Viz. a current transmission area of only approximately 30% of
the actual brush, suggested that a compressed group of carbon fibres,
with potentially individual filamental contact, may have superior
current transmission characteristics.   The brushes (a more accurate
description than hitherto) developed, did not, for various reasons,
exhibit the gains expected of them, but amongst the tests carried out,
it was shown that the electrical resistance between the brush and the
commutator was reduced as the axial load on the brush increased.   The
relationship is typified in Graph 1.

From this graph it can be seen that the compression/resistance
relationship is highly non-linear. This characteristic is thought to be
due to the individual switching action of the carbon fibre filaments.

In the unloaded state, the carbon filaments are electrically discrete, i.e. they conduct along their length but not transversly to their neighbours. There appears to be a transverse load threshold at which conduction occurs as a step function, the progressive characteristic of the whole bundle or tow being achieved by individual filaments sequentially reaching their conduction threshold.

Thus the graph can be split into 2 zones, 1 and 2. Zone 1 is the steep, sensitive portion ofthe curve when the vast majority of fibres are not conducting. Hence a small displacement or increase in load brings a dramatic change in the number of conducting filaments. In Zone 2 however, most filaments have crossed the conduction threshold and a large change in displacemnt or load only brings a small number of peripheral fibres into the transverse conducting state.

In an earlier paper, (1) the author described a robust, independent sensor making use of the variable resistance characteristics of carbon fibre tows or bundles. This device was sufficiently promising to encourage the author to investigate the basic properites of the carbon fibre tows.

2.   Areas of Investigation

2.1   Multi Point Sensor Operational Independence.

In order to establish the independence of the carbon fibre sensors, a simple arrangement was constructed, consisting of a single multi stand carbon fibre tow crossed intwo places by two other separate tows. This produced a sensor with two junctions approximately 20mm apart. The response of each junction was noted for various fixed applied loads at the other junction. At this loading density, no interference was experienced. Increasing sensor density may, however, introduce mechanical defection problems.

2.2   Dynamic Response

Manipulators making use of any external sensor must incorporate the signals provided by the sensor into its own control system. The signals supplied by the sensing system are the interpretation of the parameter or effect detected by the sensing medium and transmitted to the manipulator control system as a voltage, current, resistance, etc. variation. This variable may or may not be changing rapidly, but some knowledge of the response of the sensing medium under dynamic input conditions is essential.

Tests were carried out to determine the output variation and phase shift at various input frequencies. A small frame was constructed to support a LING DYNAMIC TYPE 201 oscillator and this was powered by a LING POWER OSCILLATOR TYPE 5VA. The frequency range of the type 201 oscillator is 5 - 10, 000 hz and that of Power Oscillator 5 - 50,000 hz. Output from the sensor was monitored by a Farnell 12 - 4 D2 Channel Cathode Ray Oscilloscope (C.R.O.) and phase shift and peak/peak amplitude were measured. To enable the C.R.O to monitor sensor output a different test circuit was designed as shown in Fig. (2) from the results obtained it seems unlikely that the carbon fibre sensors will find ready application in a tactile sensing system detecting high frequency inputs. In fact 30 hz probably represents the upper limit of the useable dynamic range.

These dynamic characteristics would not, however, effect the usefulness of the carbon fibre sensors in a quasi static tactile surface Fig (3).

2.3   Hysheresis losses

For any sensor that is to be used in a dynamic sensing system, good hysteresis characteristics are essential. If the sensor possess too high a hysteresis loss, its effectiveness in any system involving signal reversal will be very limited. Using a very simple experimental arrangement, the hysteresis loss was established as lying between 5 and 10% of full scale deflection.

2.4   Sensor Encapsulation

Carbon fibres, whilst possessing a high tensile strength are, by definition, extremely fibrous and prone to abrasive decay.   If, in addition, the encapsulating medium could provide not only protection and covering, but also act as the resilient member of the sensor, the load sensing range could be determined by the choice of encapsulating materials. With these criteria in mind, the following specification for the encapsulating material was drawn up.

Design Specification for Sensor Encapsulating Material

The material used must:-

1.   Adequately protect the Carbon fibres.
2.   Be an insulator.
3.   Provide suitable resilience within the sensor.
4.   Be stable and non-toxic
5.   Be compatible with the carbon fibre filaments.
6.   Be easily handled and moulded.
7.   Be readily available.

On the basis of satisfying the above requirements, and, in addition, being transparent and relatively quick setting, Silicon Rubber was chosen as the encapsulating material for the experimental work . This material is available in a high viscosity form, arriving complete with dispenser suitable for filling small moulds. In other applications, alternative moulding compounds could be chosen to enable various load/deflection Characteristics to be obtained.

3. Conclusions

Carbon fibre sensors, when moulded in an appropriate material, are reliable, robust sensors capable of providing an excellent non-linear Signal. No signal amplication is required and the characteristics of the output curve is such that the most sensitive portion of the range is the contact/no contact region. The sensor is not ideally suited to measuring small variations in load over extended periods as both time and temperature can cause significant, but as yet unquantified, variations in sensor output. The parameters affecting the working life are not yet fully understood, there being a large difference in operational stability between sensors manufactured by different processes. A manufacturing process that can provide a suitable compromise between sensor output characteristics, i.e. drift and repeatability, and an acceptable life time stability is required if the full potential of the sensor is to be realised.

References

1. DAVIES, J.B.C.   "Robust Carbon Fibre Sensor"
                    4th Int Conference on Vision and Sensoring
                    Controls London Oct 1984

2. LARCOMBE, M.H.E. Tactile Sensing Using Digital Logic
                    Report "Tactile Sensors for Robots"
                    University of Warwick, U.K.

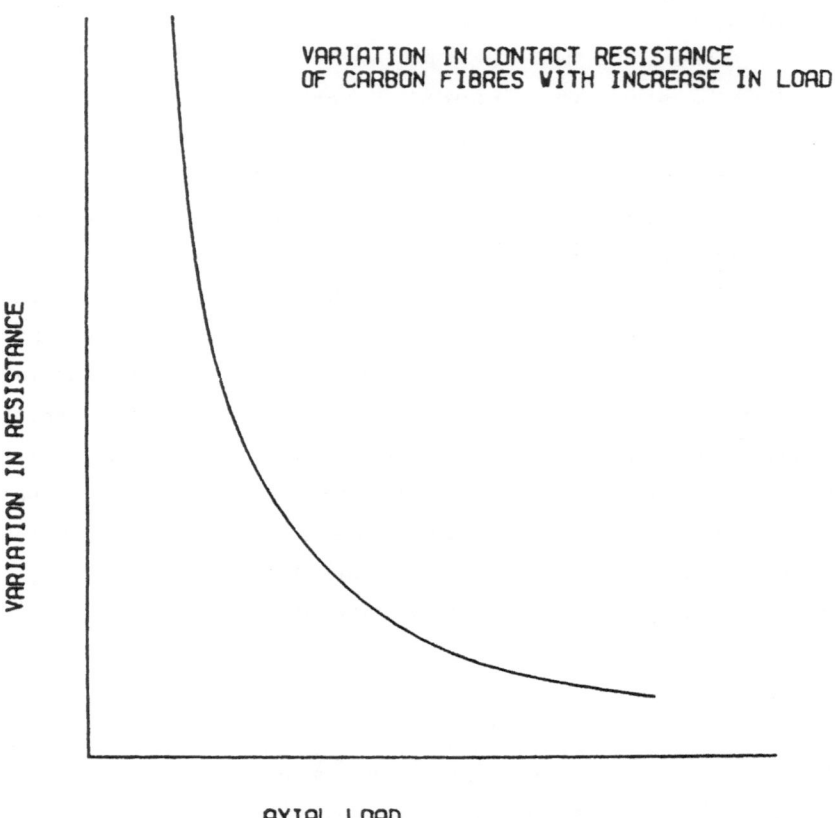

VARIATION IN CONTACT RESISTANCE
OF CARBON FIBRES WITH INCREASE IN LOAD

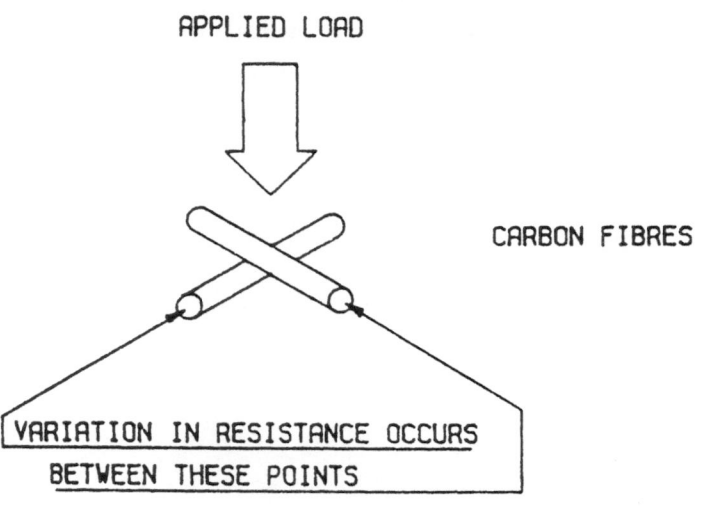

Fig. 1

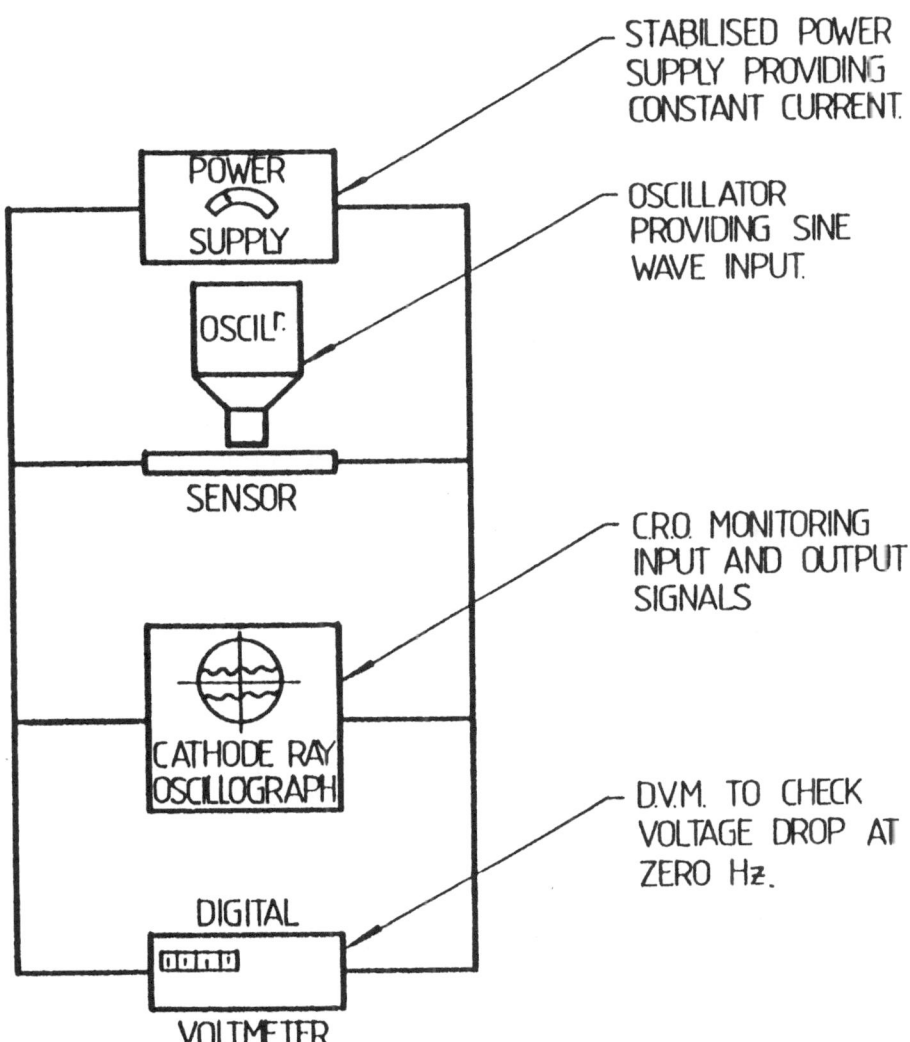

STABILISED POWER
SUPPLY PROVIDING
CONSTANT CURRENT.

OSCILLATOR
PROVIDING SINE
WAVE INPUT.

C.R.O. MONITORING
INPUT AND OUTPUT
SIGNALS

D.V.M. TO CHECK
VOLTAGE DROP AT
ZERO Hz.

Fig. 2

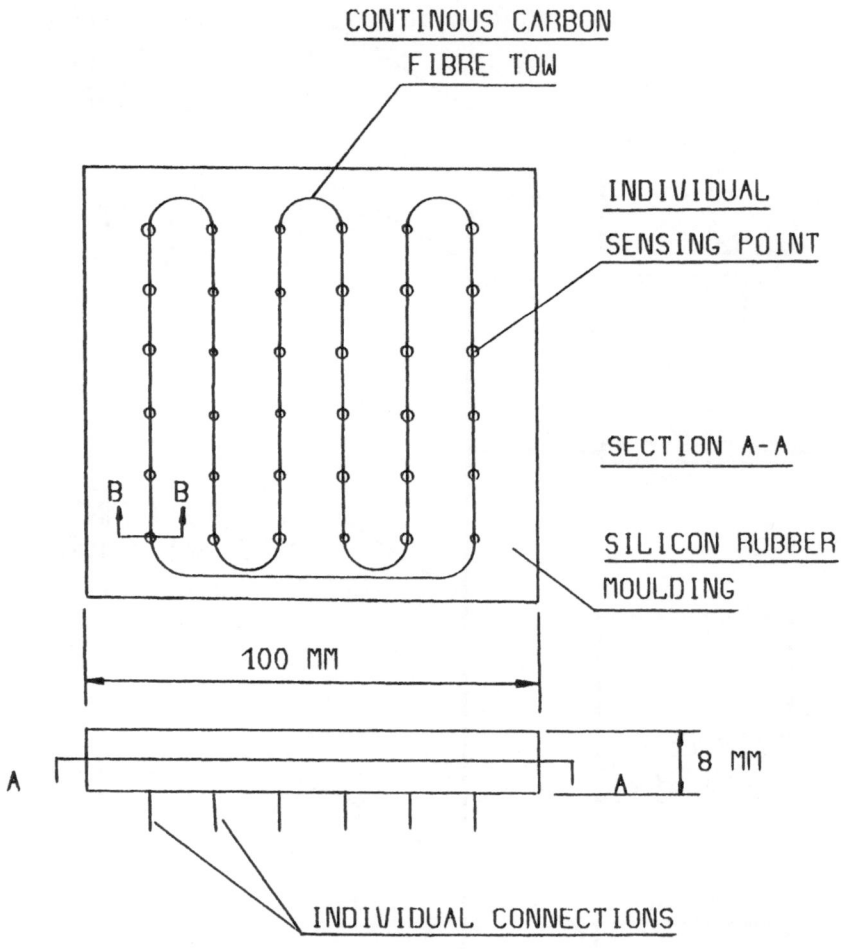

CONTINOUS CARBON
FIBRE TOW

INDIVIDUAL
SENSING POINT

SECTION A-A

SILICON RUBBER
MOULDING

100 MM

8 MM

INDIVIDUAL CONNECTIONS

Fig. 3

TACTILE GEOMETRY FOR IMAGES AND NORMALS

A. Cameron*, R. Daniel, H. Durrant-Whyte†
Department of Engineering Science
University of Oxford
Oxford OX1 3PJ   (U.K.)

Abstract

The shape of a tactile sensor imposes constraints upon its performance. This paper discusses the geometries of currently available tactile sensors for robots, and compares them with man's own tactile sensor, the finger. Possible alternative shapes are discussed along with the requirements for a versatile tactile sensor. Finally, a new tactile sensor is presented which appears to satisfy these requirements. Significantly, this sensor can be readily moulded to any desired geometry and degree of compliance.

1   Introduction

Robotics has been described as "the intelligent connection of perception to action" [2]. In order to be able to perform intelligent operations, robots will need to be able to perceive their operating environment, and react accordingly. This perception will rely on detailed information provided by a range of sensor systems.

Humans use several senses to collect information about their surroundings including sight, touch, hearing, smell and taste. Of these sight and touch are the most important for manipulation and object recognition operations, and these operations are the targets of robotics research.

Visual sensing for robots is a significantly more advanced field of research than tactile sensing, due to the availability of good transducers. In spite of this, tactile sensing will be central to an intelligent robot due to the limitations of vision in many environments. Particular need for tactile sensors lies in obtaining contact information to verify visual information, and in supplying information in environments in which visual data is unobtainable due to obscuration of the object, perhaps by a robot end-effector during a manipulation operation. Tactile sensors may also reveal information regarding object texture, hardness and elasticity which are not readily available from any other

---

*Supported by a Rhodes scholarship.
†Supported by a BP fellowship.

NATO ASI Series, Vol. F52
Sensor Devices and Systems for Robotics
Edited by A. Casals
© Springer-Verlag Berlin Heidelberg 1989

sensing technique. Tactile sensors will be vital in supplying feedback information for manipulation operations and in overcoming slip.

Although very few tactile sensors are, as yet, commercially available, surveys conducted in the United States have shown that a very high proportion of industrial manufacturers and users of robots believe that tactile sensing is needed in the next generation of robots [7]. This belief, coupled with the rapidly increasing level of research into this field, should ensure a significantly increased presence of tactile sensing technology in commercial robots in the near future.

This paper begins by looking at the human sense of touch. This approach is taken for two reasons. Firstly, this is the sense whose performance we are attempting to emulate in developing tactile sensors for robots, and the capabilities of the human sensor provide an intermediate goal for sensor research. Secondly, the human hand provides a working model of a versatile tactile sensor, and hence may be expected to provide pointers to possible design considerations for a robotic sensor.

Section 3 outlines some industrial applications for automation, to which it is hoped tactile sensing may make an important contribution. The nature of the sensing required for these situations is investigated, and methods of extracting the desired information are considered. The shapes of tactile sensors that are currently being developed are discussed along with their advantages and disadvantages for different tasks. Envisaged trade-offs between different geometries are outlined.

Section 4 draws on the findings of the previous two sections to propose some design criteria for a tactile sensor and suggest a new geometry.

The final section introduces a new tactile sensor currently being developed which will potentially meet the design criteria. This sensor offers the important advantage of being able to be cast into different geometries for different tasks, and hence can utilise any geometry determined to be optimal.

2 Human Touch

In attempting to develop tactile sensors for automation, it is instructive to look at the human sense of touch, as this is essentially the attribute that we are attempting to model. Unfortunately, the manner in which man uses his hands to explore and manipulate objects using touch is not particularly well understood by physiologists. It is clear, however, that human touch is an inherently active sense, and that we obtain most of our information by rubbing our fingers across the surface of an object [17].

The tasks of manipulation and object recognition appear to be the most important uses to which man puts his hands. For both these tasks, however, other senses are also used, including proprioceptive sensing, thermal sensing and vision. Indeed, tactile sensing is rarely used on its own, but nearly always in conjunction with other sensors.

In almost all observations of human tactile sensing, the fingers are used to actively explore the object, and it has been shown that the resolution of the tactile sensing elements is significantly finer when used in this dynamic mode than when simply pressed against an object to obtain a passive image [12].

The search patterns used to gather information in object recognition appear initially random, but become more structured as an impression of the object is built up. Readily identifiable features such as edges, corners and holes, appear to play a significant role in both manipulation and recognition tasks. Different recognition behaviour is followed if the object under examination is significantly larger in size than the finger or hand, as opposed to the situation in which the object is relatively small. Tiny objects are usually rolled between the fingertips while gathering tactile data, whilst larger objects are analysed by grasping with the whole hand.

Some of the physical characteristics of the human sensor which deserve examination are its shape, elasticity and compliance, density of sensing elements, and the nature of these elements.

The shape of a finger is a "slightly-squashed" cylinder, with diameter approximately 1.5cm, and with a hemispherical end. This shape is particularly useful for the hand's function as an end-effector for manipulation tasks. It also enables the tactile sensors located on the fingertips to reach into occluded regions to collect tactile data.

The cylindrical shape is not however ideal for collecting large quantities of tactile data as a very small portion of the sensor lies in contact with the object at any particular time, and hence only sparse data can be collected. This disadvantage is partially offset by the compliance of the finger, which enables it to mould itself to the surface of the object under examination, and as a result some dense tactile data can be obtained. Compliance in the surface of the probe also proves to be advantageous in manipulation operations and the mechanics of this field has been examined by Fearing [3].

The density of the sensing elements is greatest in the fingertips, corresponding with the flatter portion of the finger [17]. This is the region that is rather compliant and can be pressed against an object to obtain a tactile "image". In other regions of the finger, the sensor density is reduced, as is the compliance, and these areas are used only for collecting sparse tactile data.

A relation emerges between shape, compliance and sensor density in the human finger, corresponding with the type of tactile information that can be obtained. For dense data, a flat (or curved but compliant) shape is utilised with a large number of sensing elements. Sparse data is readily obtained with curved shapes, and neither compliance nor a high sensor density is required.

3   Machine Handling

Tactile sensing offers to extend the scope for automation in many varied tasks. These range from manipulation tasks for end-effectors including parts assembly, insertion, and compliant grasp, to artificial intelligence tasks such as object recognition and orientation. These different tasks will require different tactile information and a sensor that provides excellent data for recognising objects from a class may not be very useful for helping overcome slip in manipulation.

Typical information that will shortly be readily available from tactile sensors may include normal and shear force data measured at the surface of the sensor, and a tactile "image" of the points of contact between sensor and object. Some prototype sensors have been reported which offer some portion of these capabilities [6,10,13,15,16]. The resolution initially available will be limited by materials considerations and by the bandwidth constraints of the medium used to transfer the obtained data to a processing unit.

With increasing miniaturisation of components and "in situ" parallel processing (as has already been investigated by Raibert & Tanner [15]), these problems will become less significant.

In order to maximise the quality of the data available from a tactile sensor for a particular task, one parameter that can be optimised (but has received negligible analysis to date) is the sensor geometry. Currently available tactile sensors are shaped either as planar surfaces (arrays) or curved in cylindrical form (probes). Both these shapes have intrinsic advantages and are useful for different tasks, but have pronounced disadvantages.

Tactile arrays are useful for obtaining passive images of the object under consideration, especially if the object has a predominantly planar surface. If the face of the sensor can be pressed against that of the object, a tactile impression of the surface can be obtained. The impression will be similar in many aspects to the visual image that is returned by a camera, and it is thus hoped that visual image processing techniques may be applicable to the tactile array.

Tactile arrays are less useful for determining the shape of a complex object. Generally they are only capable of measuring forces applied normal to the face of the array (although Hackwood & Beni have designed an array that measures torque and shear forces [6]).

The use of tactile sensor probes (or fingers) implies dynamic sensing, as very little of the sensor will be in contact with the object at any instant in time, thus yielding very little data in the passive case. This requires a high level of feedback control from the sensor to the positioning device to enable useful data to be collected. Once the data has been accumulated new image processing techniques have to be developed to extract the desired information as the similarity with visual images no longer exists. (Such algorithms have been developed by Gaston & Lozano-Pérez [4] and Grimson & Lozano-Pérez [5]).

Probes are very useful for determining the shape of an object by curve tracing, as they can readily detect the normals to the object surface from the point of contact on the probe. They are also generally more manœuvrable than arrays by virtue of their shape, particularly for exploring concavities of a complex object.

Information, other than shape, which may be discerned from tactile sensor data includes texture, elasticity and stiffness of an object. It is not clear which sensor will prove superior in collecting this data. In her work with the "French finger" (a probe), Bajcsy was able to determine these parameters [1]. Her techniques for elasticity and stiffness may be just as readily applied to an array sensor, whilst texture information may be able to be derived from a passive array image.

The major difficulties for developing probe sensors probably revolve around the control, data processing and image recognition tasks. Because it may suffice for the array sensor to be less active, and because of the potential overlap with visual image processing, the major difficulties at present for their development probably lies in the area of transduction, to enable higher resolution impressions to be obtained and to also extract vector information.

Given that both shapes have merits in determining different quantities the optimum shape for a sensor will be task-dependent. A worthwhile exercise will be to study several robotic operations which it is hoped will be aided by tactile sensing and to study the nature of the sensing required. (Such a study for several industrial tasks has been undertaken by Harmon [8].) The sensing required for slip detection and manipulation may differ markedly from that required for object identification.

Another aspect that will be important in the choice of sensor shape is the algorithm that is to be implemented, as the data required by the algorithm should also give leads as to the optimum sensing strategy. (For instance, the algorithms of Gaston & Lozano-Pérez and Grimson & Lozano-Pérez use sparse data which is well provided by probe sensors, however feature extraction algorithms, as used in visual processing and the algorithm of Hillis [10], require denser data.)

An associated design consideration with the shape of the sensor is its compliance. Generally "softer" materials have been included with array sensors enabling them to mould themselves to the sensed object surface, and hence extract maximal information from heavily textured surfaces. (The compliance of the sensors located on an end-effector can also serve to aid in manipulation tasks [3].) As the emphasis with probe sensors has been on acquiring sparse data, stiffer surfaces have been employed, providing a greater degree of certainty in sensor location, and hence more accurate measurements of position.

Before any shape can be judged as being optimum however it is necessary to clearly define the nature of the sensing required. This is an important aspect that has been neglected by many researchers in this field, who appear to have developed sensors without defining the task for which these sensors will be used or will prove optimum.

4 What can machine touch learn from human touch

The study of the human finger, undertaken in Section 2, should provide leads in resolving design considerations for a robotic tactile sensor. The uses envisaged for tactile sensors are as aids in the same manipulation and object recognition operations for which man uses his hands.

As discussed in the previous section, the specifications for such a sensor should depend on the specific task required, however a study of typical operations should lead to some general guidelines. Several papers have sought to define general specifications for a desirable tactile sensor [9,7]. Harmon's specifications are listed below:

1. A touch-sensing device is assumed to be an array of 10×10 force-sensing elements on a 1-sq-in flexible surface, much like a human fingertip. Finer resolution may be desirable but is not essential for many tasks.

2. Each element should have a response time of 1-10ms, preferably 1ms.

3. Threshold sensitivity for the element ought to be 1g, the upper limit of the force range being 1000g.

4. The elements need not be linear, but they must have low hysteresis.

5. This skinlike sensing material has to be robust, standing up well to harsh environmental conditions.

In light of the discussion in the previous section regarding the optimal shape for a tactile sensor, another specification that may be added to this list is the requirement for the sensor to be readily available in any shape or size required for the task.

In seeking a shape for a general purpose tactile sensor, it is appealing to attempt to combine the intrinsic advantages of both array and probe sensors. This is indeed the technique adopted by nature which employs both a probe (the fingertip) and an array (the front of the finger) on the human tactile sensor. This shape deserves examination as a tactile sensor on a general purpose robotic end-effector.

Other areas in which the human hand might provide leads for sensor research are in the associated areas of compliance and sensor density. As noted in Section 2, the combination of a flat (or curved but compliant) sensor surface, with a high sensing element density, is used by nature to examine small or intricate objects and obtain tactile "images". Sparse data, obtained from less compliant regions with lower sensing element densities, are quite sufficient for object recognition of larger objects. Not only are different sensors used in these situations, but also different sensing strategies, and robotics research will probably also obtain better results by adopting different strategies for both situations.

It was noted in Section 2 that the human hand is more sensitive to transient than steady state signals. This enables man to detect small perturbations in a tactile environment in the presence of larger signals. To enable robotic tactile sensors to respond to high frequency signals in a similar fashion a short response time is required, and this is reflected in Harmon's suggestion of a response time of 1ms.

It is also clear that man rarely uses his hands alone, but nearly always in conjunction with other senses, in particular vision. In attempting to extract the maximum advantage from robotic tactile sensors, a high-level of integration with vision and other sensors should be actively pursued.

5 Description of Sensor

A new tactile sensor is being developed which shows promise in meeting the specifications listed in the previous section. This sensor (shown diagramatically in Figure 1) involves several stages of transduction.

The sensor utilises a layer of photoelastic material as the primary transduction element. The force pattern at the surface of the photoelastic layer results in a characteristic stress pattern within the material. This stress pattern will reveal not only normal forces, but will also enable shears and torques to be identified. Standard techniques of photoelastic analysis can then be used to extract the stress pattern from the material as a visual image. The technique involves shining circularly-polarised, monochromatic light through the photoelastic material and onto a reflective surface (see Figure 1). The reflective surface is a compliant mylar material and provides the interface between the sensed object and the sensor. The reflected light passes back through the photoelastic material and the circular polariser to a camera where the image can be observed. The camera thus constitutes the final transduction stage where the visual image is converted into an electrical signal for processing.

If the photoelastic material is in an unstressed state, it will have no effect on the

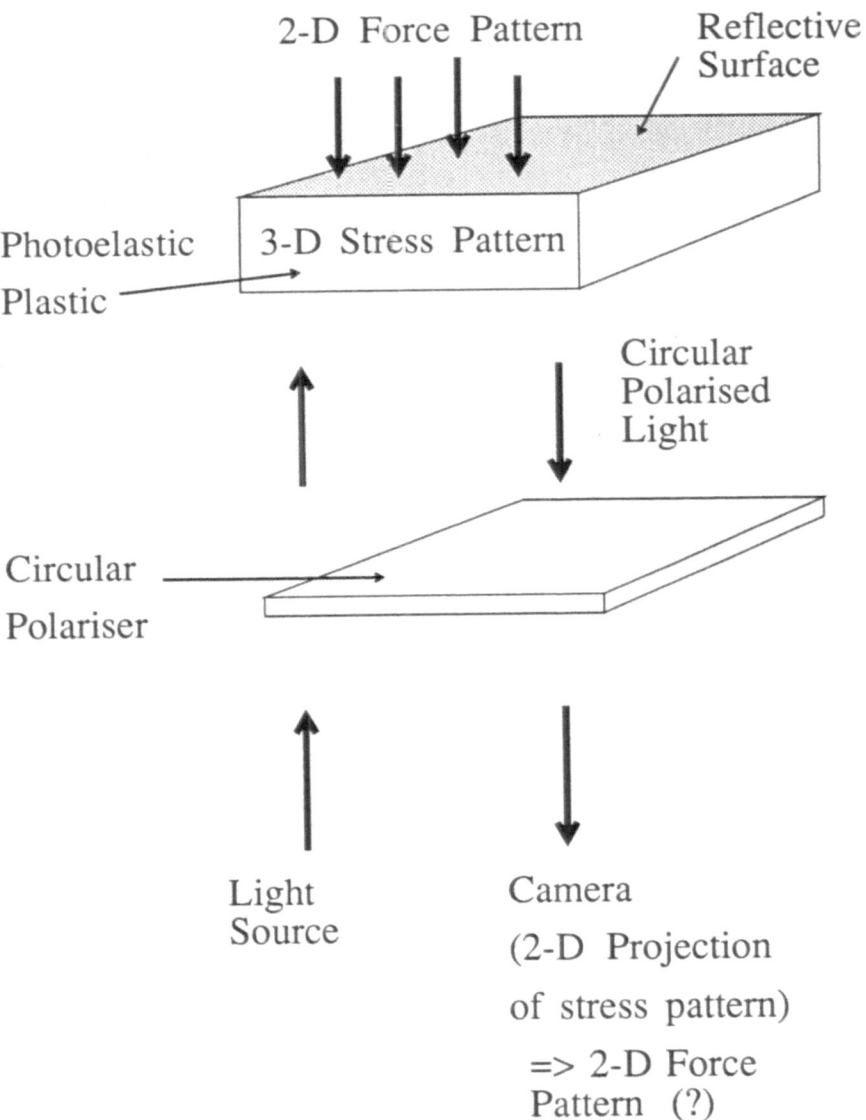

Figure 1: Schematic diagram of the tactile sensor.

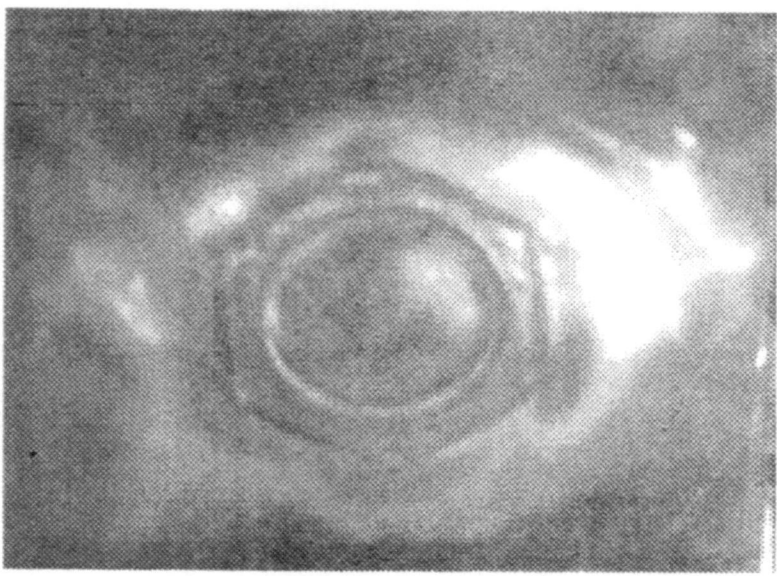

Figure 2: Tactile image showing outline and stress pattern from hexagonal nut.

optical properties of the transmitted light. The resulting image observed by the camera will be uniformly dark as none of the light will pass back through the circular polariser. If, however, the material is stressed, due to the presence of normal or shear forces at its surface, a characteristic pattern will be observed due to the effect of the optical properties of the stressed material on the polarised light. A brief discussion of the optical principles involved with this technique of photoelastic stress analysis is given in Appendix A.

A typical image obtained using a prototype sensor is shown in Figure 2. The fine resolution evident represents one particular advantage offered by this sensor. The resolution of the sensor is only limited by the resolution of the camera used, and is hence markedly finer than most other sensors currently being developed. The potential application of visual processing techniques to the obtained tactile image is another enticing aspect of this sensor. Other sensors have attempted to take advantage of the more advanced field of visual sensing by converting the tactile image into a visual pattern with encouraging results [13].

Most significantly, the material can also be obtained in liquid form and cast into a more suitable shape than the array that has so far been investigated. Although the prototype sensor used a mylar reflective surface as the interface between object an sensor, a more suitable non-reflective material may be chosen for its robustness and compliant properties, as reflective glues are available for bonding the surface to the photoelastic material. The choice of a suitable "skin" for the sensor has not been investigated as yet.

Several design considerations have yet to be fully investigated, including the optimum thickness for the photoelastic layer. A thicker layer leads to a more compliant interface with the object with the associated advantages and disadvantages discussed in the previous section. The compliance can also be varied by choosing a material with a different modulus of elasticity, as the photoelastic material is available in a range of thicknesses and moduli.

In measuring the performance of this sensor against the specifications listed in Section 4, results obtained so far indicate that specifications 1 and 2 (with regard to resolution and response time) can be easily met. Further, the sensor does not suffer from hysteresis effects, and hence satisfies specification 4. The range of sensitivities of the sensor have not yet been measured, but these can be selected by a suitable choice of the stress-optical coefficient of the photoelastic material. Finally, the durability of the sensor can be improved by an appropriate choice of skin material.

The ability to mould the sensor to any desired shape, and to select its compliance, plus the potential applications of visual processing techniques, represent significant advantages of this technique over other transduction methods currently being developed.

Other researchers have investigated the use of photoelastic materials in tactile sensors [11,14]. The most similar work to that reported in this paper was the preliminary research conducted into tactile sensors for the Utah/MIT dextrous hand project. Their prototype sensor also used a photoelastic layer as the transduction element which was sensed using polarised light. In their sensor, however, the polarised light was passed through the layer parallel to the sensing surface, rather than perpendicular to it. This restricts the amount of tactile data that can be extracted from the sensor, and, as a result, an array of such sensors would be necessary to build up a two dimensional tactile image.

6 Conclusions

Considerations of the functions of the human hand and the requirements for robotic tactile sensors have indicated that shape should become an important consideration in tactile sensor design. The need for a sensor which can be moulded to the optimal geometry for a particular task may be met by a photoelastic tactile sensor currently being developed.

Preliminary research into this new tactile sensor have been encouraging, although many further aspects require investigation. These include research into associated design criteria such as the optimal compliance and elasticity for the photoelastic material and a protective skin. The required resolution of the tactile "image" for a particular task, and the most suitable information extraction algorithms will also be studied.

A Photoelasticity

Photoelastic materials have long been used as an aid in stress analysis. Such materials become birefringent when stressed, and the changed optical properties can be observed with polarised light. (Birefringence is the phenomenon whereby the material has a distinct optical axis and light travels through the medium at a different speed when polarised in the direction of the axis than when polarised orthogonal to it.) Thus by interpreting the light pattern obtained after it has passed through the photoelastic material, the stress pattern can be deduced.

The method used in the prototype sensor described in this paper used monochromatic light passing through a circular polariser and a layer of photoelastic material before striking a reflective surface and returning along the same path.

Assuming the photoelastic material is unstressed it will have no effect on the properties of the light. The light will still be circularly-polarised when it strikes the reflective surface, where its "handedness" will be reversed (ie right-circularly polarised light will become left-circularly polarised, and vice-versa). The light will then be unable to pass back through the circular polariser, being of the wrong "handedness" and no light will return to the camera.

If, however, the photoelastic material is stressed, the components of light resolved in the direction of the stress axis will travel at a different speed, and the $\pi/2$ phase difference between orthogonal components required for circular polarisation will not be preserved. The emergent light will be elliptically-polarised, and this effect will be doubled when the light passes through the material a second time following reflection. The degree to which the light will become elliptically- (or even linearly-) polarised will depend on the thickness of the material, the magnitude of the internal stresses, and the stress-optical coefficient of the material. The maximum amount of light will be passed back to the camera when the effect of the two traversals through the photoelastic material is to completely reverse the "handedness" of the circularly-polarised light (ie a relative phase change of π between the two orthogonal components). In this case there will be no attenuation and all the transmitted light will arrive at the camera. Intermediate cases, between this and the case of no phase change due to an unstressed medium, will result in a corresponding loss in light intensity.

Thus the light pattern observed by the camera corresponds exactly to the stress pattern within the material, and can be used to extract the force pattern at the surface.

References

[1] Ruzena Bajcsy. Shape from touch. In *Advances in Automation and Robotics, Vol. 1*, pages 209–258, JAI Press, 1985.

[2] Michael Brady. *Artificial Intelligence and Robotics*. AI Memo No. 756, MIT, February 1984.

[3] Ronald Steven Fearing. *Touch processing for determining a stable grasp*. M.S. Thesis, MIT, September 1983.

[4] P.C. Gaston and T. Lozano-Perez. Tactile recognition and localization using object models: the case of a polyhedra on a plane. *IEEE Trans. Pattern Anal. Mach. Intell.*, 6(3):257–266, May 1984.

[5] W.E.L. Grimson and T. Lozano-Perez. Local constraints in tactile recognition. *Int. Jour. Robotics Research*, 3(3):3–35, Fall 1984.

[6] S. Hackwood, G. Beni, and T.J. Nelson. Torque-sensitive tactile array for robots. In A. Pugh, editor, *Robot Sensors Vol. 2, Tactile and Non-Vision*, pages 123–131, Springer-Verlag, 1986.

[7] Leon D. Harmon. Automated tactile sensing. *Int. Jour. Robotics Research*, 1(2):3–32, Summer 1982.

[8] Leon D. Harmon. Robotic taction for industrial assembly. *Int. Jour. Robotics Research*, 3(1):72–76, Spring 1984.

[9] Leon D. Harmon. *Touch-Sensing Technology : A Review*. SME Technical Report, 1980.

[10] W. Daniel Hillis. A high-resolution imaging touch sensor. *Int. Jour. Robotics Research*, 1(2):33–44, Summer 1982.

[11] S.C. Jacobsen, J.E. Wood, D.F. Knutti, and K.B. Biggers. The utah/mit dextrous hand: work in progress. In Michael Brady and Richard Paul, editors, *Robotics Research: The First International Symposium*, pages 601–653, The MIT Press, 1984.

[12] R.S. Johansson and A.B. Vallbo. Tactile sensory coding in the glabrous skin of the human hand. *Trends in Neurosciences*, 6:27–32, 1983.

[13] D.H. Mott, M.H. Lee, and H.R. Nicholls. An experimental very-high-resolution tactile sensor array. In A. Pugh, editor, *Robot Sensors Vol. 2, Tactile and Non-Vision*, pages 179–188, Springer-Verlag, 1986.

[14] Brain W. Prahl and Peter M. Tracy. Pressure/tactile sensing with intrinsic fiber-optic sensors. *Sensors*, 3(8):48–52, August 1986.

[15] Marc H. Raibert and John E. Tanner. Design and implementation of a vlsi tactile sensing computer. *Int. Jour. Robotics Research*, 1(3):3–18, Fall 1982.

[16] J. Rebman and K.A. Morris. A tactile sensor with electrooptical transduction. In A. Pugh, editor, *Robot Sensors Vol. 2, Tactile and Non-Vision*, pages 145–155, Springer-Verlag, 1986.

[17] A.B. Vallbo and R.S. Johansson. The tactile sensory innervation of the glabrous skin of the human hand. In G. Gordon, editor, *Active Touch*, pages 29–54, Pergammon Press, 1978.

A VIDEO SPEED TACTILE CAMERA

P.W. Verbeek, P.T.A. Klaase*, A. Theil [+]
Technical University Delft, Department of Applied Physics
Lorentzweg 1, 2628 CJ Delft (Holland)

Abstract

A tactile image sensor is described that applies piezoelectric films of
PVDF (polyvinylidene fluoride) to map a 2dimensional pressure distribution.
The sensor consists of two layers of PVDF. One layer is used to generate,
the other layer to sense acoustical vibration. Driven at fixed frequency
the configuration can locally switch between resonance and non-resonance at
a change of the acoustical impedance of the surrounding media. An object is
detected as a blob of locally deviating acoustical impedance. The spatial
distribution of acoustical impedance is resolved by linewise driving and
columnwise sensing. Application of pressure impedance converters makes the
sensor pressure sensitive. Measurements on an experimental tactile image
sensor with nine elements are discussed. These experiments have shown that
for a tactile image sensor with a binary response a spatial resolution of
at least 25 elements per cm^2 can be realized. The acoustical frequency is
sufficiently high to allow video-speed electronic read-out of the detected
amplitudes.

* National Organisation of Applied Physics (TNO)
Plastics and Rubber Institute
Schoemakerstraat 97, 2628 VK Delft, The Netherlands

[+] National Organisation of Applied Physics (TNO)
Physics and Electronic Laboratory
Oude Waalsdorperweg 63, 2597 AK The Hague, The Netherlands

NATO ASI Series, Vol. F52
Sensor Devices and Systems for Robotics
Edited by A. Casals
© Springer-Verlag Berlin Heidelberg 1989

1. Introduction

Tactile sensing may record many different quantities such as pressure, shear, slip and even temperature and humidity. Our sensor essentially measures acoustical impedance. Usually a tactile sensor determines a single local pressure and its development in time. A tactile image sensor measures a spatial distribution, e.g. a spatial pressure distribution. Such a sensor consists of an array of pressure sensitive elements. Patterns of pressure can be determined. Dynamic patterns need image sensors of corresponding speed. Compared to visual image sensors existing tactile image sensors are relatively slow. The tactile camera we propose matches the speed of video camera's.

A tactile image sensor has interesting applications in the field of automation and rehabilitation. The sensor can be used to supply a robot with tactile sense which, for instance, can be utilized in assembling or sorting. Sensing the object in its gripper the robot can perform tasks like classifying the object, positioning, texture and surface inspection, slip detection and force measurement. For many of these applications a sensor with only an on/off response (a binary image sensor) will satisfy.
Today, few tactile image sensors are commercially available and there are several research activities in this field. An overview is given by Harmon [1], [2].

This paper is organized as follows: In section 2 the requirements are discussed for a tactile camera to be used in an assembly robot or in a prosthesis. In section 3 the choice of PVDF as working material is argued. In section 4 the working principle of the tactile image sensor is presented. Section 5 is devoted to the multiplexing problem. In section 6 measurements on an experimental sensor are reported. Conclusions are given in section 7.

2. Requirements

The design and the requirements of a tactile image sensor depend strongly on its application. However, when the sensor is applied in a robot gripper or in a prosthesis, humanlike tasks are involved and more specific demands can be formulated. Some of these requirements are:

- Tactile
  - Fast hysteresis-free response matching video frame rate.
  - Fingertip-like high spatial resolution (2 mm).
  - Dynamic range 1 to 1000 g/cm^2.
- Mechanical
  - Flexibility.
  - Robustness.
  - Light weight.
- Electronic
  - In- and output through few leads (multiplexing).

3. PVDF as working material

The sensor described here uses piezoelectric films of polyvinylidene fluoride (PVDF) as working material. The polymer PVDF is not brittle like most commonly used piezoelectrics such as the ceramics PZT and Barium Titanate or the crystalline quartz. It is cheap, easy to produce and it can be made in very thin (some μm's), flexible films which gives the opportunity to provide a complex form with a pressure sensitive layer. Therefore PVDF is well suited for use in a tactile image sensor. For details of the chemical and physical properties of PVDF we refer to Kepler and Anderson [3].

4. The working principle

The tactile image sensor of our design has two layers of piezoelectric material (see figure 1). One layer, with electrodes connected to an alternating voltage source, (the driving layer) is used to generate longitudinal acoustical thickness vibrations in the layer package. Due to this motion an alternating potential difference results across the second layer (the sensing layer) which can be measured with an AC-voltmeter. The grounded common electrode between the two layers shields the driving

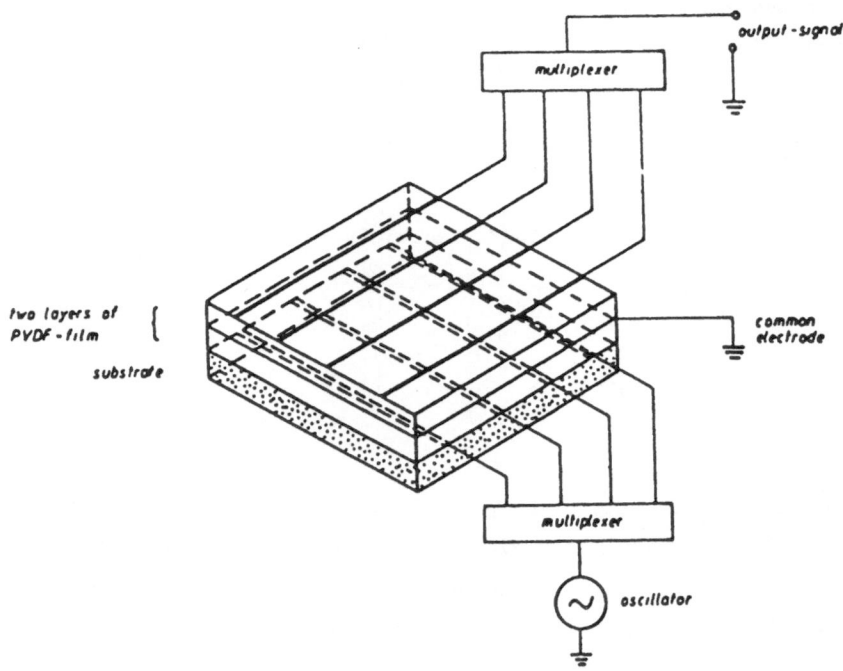

Figure 1. A 4 x 4 element dual layer PVDF tactile image sensor and its scanning circuit.

electrode from the sensing electrode. The working principle is based upon the fact that the acoustical impedances of the surrounding media determine the resonance conditions, which on their turn determine the acoustical motion, its amplitude and the voltage sensed.

Let, for instance, a package of two equal layers of total thickness d be surrounded by a heavy metal backing at one side and by air at the other. The package will then have a simple $\frac{1}{4}\lambda$ resonance mode ($d=\frac{1}{4}\lambda$) which can be excited by a driving voltage of frequency $f=c/\lambda$. (In PVDF the speed of sound c is 2.4 km/s.) An appreciable voltage will be sensed across the second layer. If a heavy (e.g. metal) object is pressed against the sensor and replaces the air, the resonance frequency doubles while the driving frequency remains the same and can no longer sustain resonance. The voltage amplitude across the second layer drops: the presence of the object is detected. For a sensor divided into a matrix of separately driven and sensed elements the map of voltage drops constitutes an image of object presence.

Heavy metal backing or objects have a high acoustical impedance, air has a low acoustical impedance, all compared to PVDF. The sensor detects the spatial distribution of acoustical impedance. In order to get a tactile image sensor that measures the spatial distribution of pressure, pressure-to-impedance converters must be inserted to make the sensor pressure sensitive. As a P-to-I converter one can use a layer of a material of which the acoustical impedance varies with applied pressure. Alternatively a layer with pressure dependent coupling to a heavy layer or to the object can be used. The pressure dependence can be established by a density variation (some foams have this property) or by a contact area variation (as in dented layers).

Though for PVDF the mechanical damping is larger than for most other piezoelectrics, the dual layer structure shows several distinct longitudinal resonance modes. When mounted on metal and facing air, i.e. between an acoustically high impedance medium and a low impedance one, resonance occurs for driving frequencies that satisfy the condition that an odd multiple of a quarter of the acoustical wavelength λ equals the total thickness of the two layers. When operated in the key-tone, one has the $\frac{1}{4}\lambda$ resonator discussed above. When mounted on a sufficiently low impedance medium there will be resonance if package thickness equals an even multiple of a quarter of the acoustical wavelength. Now the lowest frequency mode is the $\frac{1}{2}\lambda$ mode.

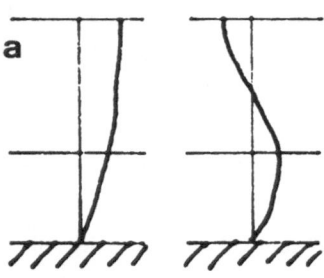 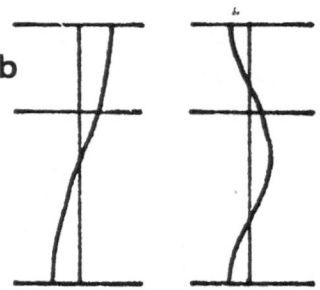

Figure 2. The acoustical amplitude in the cross-section of the piezo-
         electric films.   2a: $\frac{1}{2}\lambda$ and $\frac{3}{2}\lambda$ resonances when mounted on a
         high impedance medium. 2b: $\frac{1}{2}\lambda$ and λ resonances in free air.

The acoustical impedance sensitivity is maximum in a resonance mode.
Therefore, the sensor is designed to operate at a resonance frequency. The
various resonance modes are illustrated in fig. 2. An extensive simulation
program was performed to evaluate several impedance and layer thickness
configurations. We have implemented the $\frac{1}{2}\lambda$ configuration. For a given
material thickness (strength) it doubles resonance frequency and hence
camera speed compared to the $\frac{1}{4}\lambda$ mode. Moreover, heavy backings are avoided.

5. Multiplexing

For two-lead elements row-column multiplexing is a standard solution to
reduce the n x n lead pairs of an n x n array to n + n leads. All
elements in a row share one lead, all elements in a column the other.
Signals from elements on the corners of the same rectangle are difficult to
disentangle. DC signals can be kept apart by diodes but PVDF does not
supply reliable and lasting DC signals. This is the reason why three-lead
rather than two-lead elements and hence a double rather than a single layer
has been chosen: all elements in a row are driven simultaneously; all
elements in a column add their sensed signals but only the one on the
driven row really contributes to the column output. Figure 1 shows how an

array of tactile elements is realized. The elements simply consist of those areas of the layer package where a row and a column electrode overlap. The electrodes can be deposited on the PVDF layers in any arbitrary shape.

A multiplexer allows each driven row of elements a fixed number of oscillations. During this time interval the column response of the sensing layer elements must be detected which can be done either in parallel or serially. In this way the entire array can be scanned. The frequencies attained by PVDF allow video-compatible image generation. This is possible by employing a total scanning time that equals the video refresh rate.

For the dual layer PVDF tactile image sensor and its video-compatible read-out patent has been applied for.

Figure 3. The experimental tactile image sensor.

6. Experimental results

Figure 3 shows our experimental 3x3 tactile image sensor with three driving and three sensing electrodes. Both the width of these electrodes and the intermediate distance are 1 mm, resulting in a spatial resolution of 25 elements per cm^2. The electrodes have been deposited by metalizing through a mask. The sensor is mounted on a PVC substrate which has low acoustical impedance. The PVDF (lower) layer that is fixed on the substrate is 50 μm thick, the second (upper) layer is 49 μm thick.

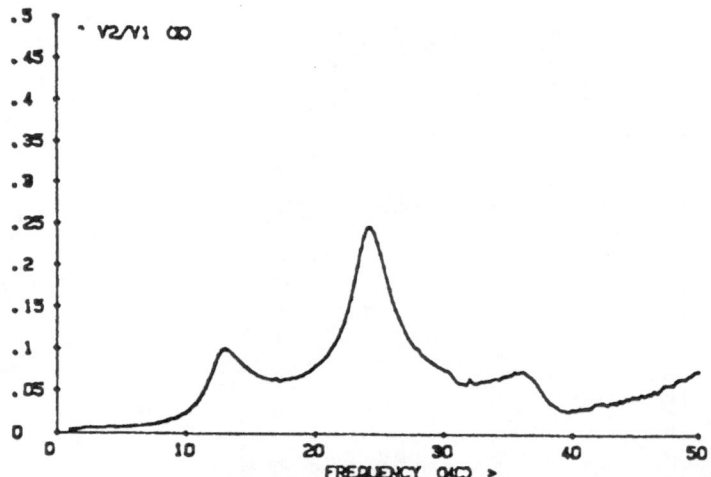

Figure 4. Frequency characteristic of one of the nine elements of the experimental tactile image sensor. ½λ and λ resonances occur at 12.5 and 25 MHz.

Figure 4 shows the frequency characteristic of the sensed voltage to driving voltage ratio of one of the nine elements when the lower layer is driving and without an object on the sensing layer. The peaks at 12.5 and 25 MHz correspond to the ½λ and λ resonance modes (d=99μm, c=2.4 mm/μs, ½λ=d resp. λ=d). The frequency characteristic shows that the sensed to driving voltage ratio is approximately −60 dB at the $\frac{1}{2}\lambda$ resonance frequency (12.5 MHz). This can be improved to become −36 dB through electric tuning i.e. compensating the capacitive parts of the (electric) impedances of the driving and sensing layers. The effect of electric tuning and the acoustical impedance sensitivity is illustrated in figure 5.

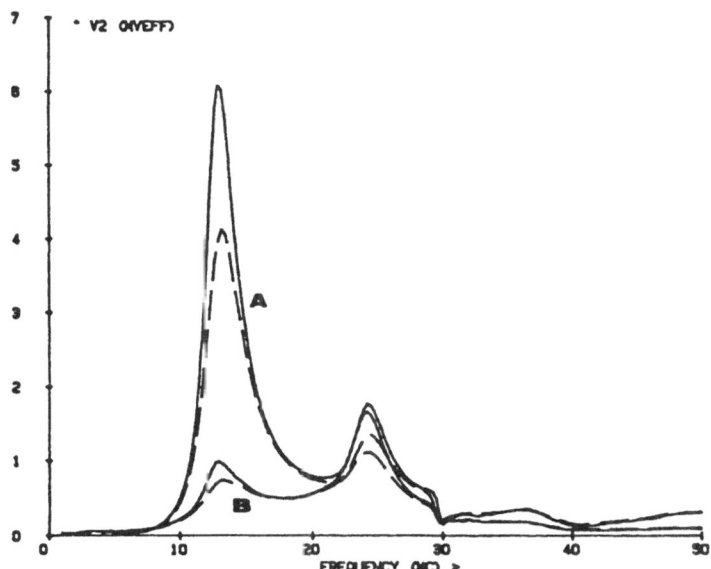

Figure 5: The effect of electric tuning and of coupling an object.
A: tuning, B: no tuning; _____ no object, --- with object.

Pressing a 1x1 mm2 massive silicone rubber object directly against the sensing layer causes a 30% decrease of the sensed voltage. A pressure of 2 atmospheres (.2 N/mm^2) was sufficient to ensure the necessary acoustical coupling. Coupling through a P-to-I converter layer will considerably reduce the gripping pressure needed.

An experiment in which a 1x1 mm^2 point object was moved over the sensor face showed that the sensitive areas of single 1x1 mm^2 elements at 2 mm centre distances do not overlap.

In another experiment silicon foams with an open cell structure were used as pressure to impedance converter layers. The principle of P-to-I conversion was confirmed. However, these P-to-I converters showed considerable hysteresis effects and their pressure response did not reproduce very well. With these P-to-I converter layers the sensitive areas of single elements overlapped. Consequently the image was blurred.

7. Conclusions

The experiments have shown that until suitable pressure to impedance convertors become available the tactile image sensor described in this paper is restricted to binary sensing. A spatial resolution of over 25 tactile elements per cm^2 can be realized. The working frequency of 12.5 MHz allows reliable detected signals with 2 MHz bandwidth, i.e. 128 elements per video line. The technique presented appears readily extendable to a resolution of .5 mm . Thus, a 300 lines video frame could accommodate two 128x128 element handsized sensors.

8. References

[1] Harmon, L.D., "Touch Sensing Technology: A Review", Robots V Conference, October 1980.

[2] Harmon, L.D., "Automated tactile sensing", The International Journal of Robotics Research, vol.1 nr.2, 1982.

[3] Kepler, R.G., Anderson, R.A., Piezoelectricity in Polymers, CRC Critical Reviews in Solid States and Materials Sciences, November 1980.

PRESENT AND FUTURE OF TACTILE SENSORS

Panel Chaired by P.W. Verbeek

In this discussion it was pointed out that there is not a clear definition of "tactile", but this word seems to encompass some qualities such as : pressure, temperature, shear, heat conductivity and humidity. An intuitive antropomorphic interpretation of tactile sensing is feeling or touching.
Non-human tactile sensing may be more useful in some applications but less in some others from human pressure sensing.
Single elements tactile sensing is well-known therefore the discussion may be limited to tactile imaging, pressure imaging and possibly transient.
A clear definition of "image" is again hard to give.Certainly essential is a more dimensional neighbour relation between data.
Rather than abstract definitions an operational definition can be given. Consider the task of a human operator feeling the inside of a car fire with his finger tips in order to find reinforcement cords sticking out of the surface. Also consider the task of gripping an object and perhaps recognizing it from the pressure distribution in the gripper. Also moving the object in a hand, manipulation is a possible task. These tasks are at different hierarchic levels, from high to low level the tasks are : recognition, manipulation,gripping control and inspection.
In all these applications a two dimensional pressure image sensor would be adequate. At the moment hardly any of these are available but they will be within the next five years. It is therefore time to consider the processing of pressure images. Some in the audience believe that standard image processing techniques will be sufficient, others that the pressure images being typically sparse (few contact areas) need specialized processing. However, image processing also tends to a handling of regions of interest only, which may lead to a convergence of processing needs for tactile and visual images. Other quantities than pressure may be added as tactile color.
The role of tactile imaging in manipulation and gripping control-like limiting maximum local gripper pressure seems to have no direct visual counterpart. Hence developers in these fields should not wait for image processing specialisor but try to use tactile image information rightaway.

NATO ASI Series, Vol. F52
Sensor Devices and Systems for Robotics
Edited by A. Casals
© Springer-Verlag Berlin Heidelberg 1989

III. ACOUSTIC SENSORS

ACOUSTIC RANGE SENSING FOR ROBOTIC CONTROL

J.S. Schoenwald
Science Center
Rockwell International Corporation   P.O. Box 1085
Thousand Oaks, CA 91360   (USA)

1. INTRODUCTION

Manufacturing systems often require frequent changes in product shape or assembly positioning. Effective use of robots in production lines require sensors to adapt the path of the end effector to the task. In addition variations in part details in many batch process or low volume production runs may arise which are not incorporated in the off-line programming assembly procedures. It is therefore desirable to incorporate the adaptive ability to follow a surface from a precisely fixed offset distance of a few inches/centimeters in real time to track true surface contours, while still satisfying global task goals.

This motivation has led us to research in sensors, signal processing systems and control algorithms that enable us to develop and test strategies for responsive behavior compatible with task objectives. Real time behavior allows for path modification without violating the goals of the robot's programming. Indeed, one of the principal requirements may be accuracy in tool placement relative to an object whose location and orientation is not rigorously predetermined or specified.

2. PVF$_2$ ACOUSTIC RANGING SYSTEM

The approach taken in this early research had been to develop an acoustic sensor system that is mountable on the robot gripper. A key requirement is that the sensor package itself not add significantly to the mass of the gripper, since this will reduce in direct proportion the load capacity of the robot for its performance specifications. Out of consideration for this limitation, we focussed our efforts on polyvinylidene fluoride (PVF$_2$), a piezoelectric polymer available in sheet form as the transducer element in an acoustic ranging device. A system was demonstrated in which the pulse-echo delay time was measured to compute the distance from robot gripper to target. The distance data was transmitted to a host computer which issued motion commands to the robot arm to adjust gripper position.

The electrical/mechanical properties of PVF_2 are listed with quartz and PZT for comparison in Table 1. Commercially prestretched, metallized and poled films with a thickness of 110 microns were used.[1] PVF^2 has excellent linearity and low hysteresis, making it superior for transducer applications to presently available conductive elastomers used in several touch sensors for robots. It requires no bias voltage either, as is required for electrostatic transducers. Further, because the effect is piezoelectric, its stability is superior to electret tranducers, which rely on surface charge effects.

TABLE 1

COMPARISON OF PVF_2, PZT AND QUARTZ

PROPERTY	UNITS	PVF_2	PZT(Z)	QUARTZ
DENSITY	kg/m^3	1.78	7.5	2.65
STIFFNESS	N/m^2	3	83.3	77.2
REL. DIELEC. CONST.		12	1200	4.5

PIEZOELECTRIC CONSTANTS

d	10^{-12} m/V	23	110	2
e	10^{-2} coul/m^2	6.9	920	15
g	10^{-3} Vm/N	200	10	50
h	10^7 V/m	65	90	380
k	%	12	30	10
Acoustic Impedance	10^6 kg/sec-m^2	2.5	30	14.3

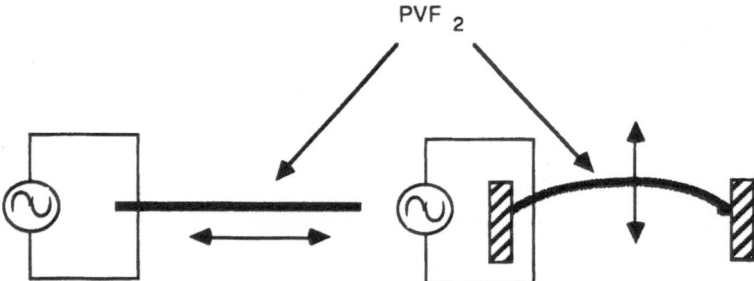

PVF$_2$

Figure 1. Electromechanical displacement of PVF2 (left) unclamped, and (right) clamped at two ends.

In the case of piezoelectric crystals and ceramics, bimorphic type structures have been used to decrease the high mechanical impedance. However, advantage may be taken of the thin film format of PVF_2 to devise a simple and efficient structure[2]. Referring to the left side of Figure 1, when a sinusoidal field is applied to a sheet of PVF_2 along the normal axis (z), the sheet vibrates in a transverse direction (x-axis). By curving the sheet as shown on the right, the transverse vibration is converted into a pulsating longitudinal motion. When the sheet is firmly clamped at the two ends, compressional waves can be excited or detected.

The electrical impedance of transducer elements on the order of 1 cm x 2 cm is high, about 100 kohms. However, a simple, inexpensive coil transformer matched the impedance to 93 ohms, so that a very respectable piercing whistle is produced by a 5 volt signal at 10 kHz from a 50 ohm signal generator. Pulsed transmission between two such transducers results in a broad bandwidth waveform, approximately 30%.

Figure 2 is a plot of the radial power distribution vs angle from the transmitter. Figure 3 is a plot of the power transmitted vs angle between both transducers when one element orbits the other, but is held in fixed orientation. The plot in Figure 2 gives power amplitude information pertinent to a single transducer response, while Figure 3 illustrates the angular position dependence of the power response of the pair.

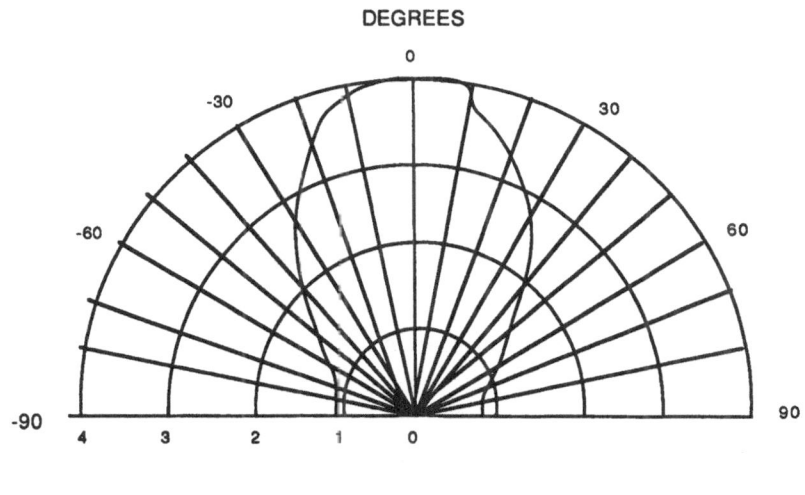

Figure 2. Radial power distribution vs angle from the transmitter.

An acoustic ranging system has been devised based on PVF_2 acoustic transducers Two such elements have been mounted on the gripper of a PUMA Unimate 560 robot. The transmitter has a coil transformer for impedance matching, as described above. The receiver feeds its signal to a FET preamplifier. The signal is then further amplified and a detector circuit produces a positive envelope pulse. By measuring the time delay T between the transmitted pulse and the received echo, we may

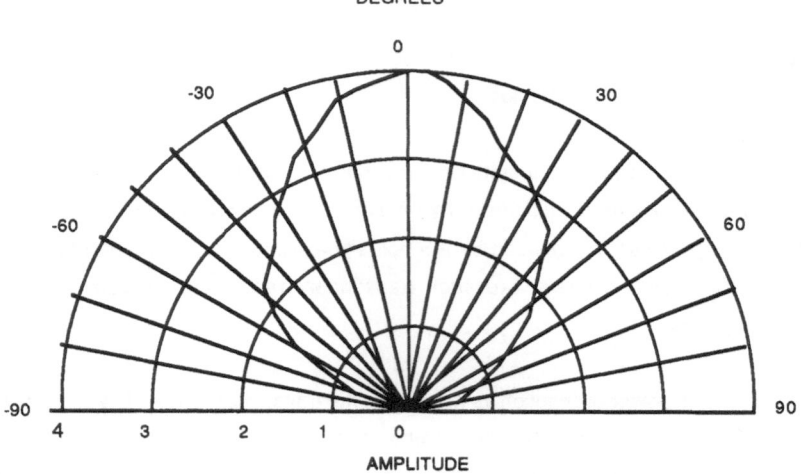

Figure 3. Angular position dependence of the power response of the pair.

compute the distance from the transducer pair to the target. A block diagram of the entire robotic system is shown in Figure 4. The echo at the receiving PVF_2 sensor is amplified 20 dB by a high impedance FET preamplifier. It is further amplified another 40 dB. A detector outputs the envelope of the waveform to a one-shot Schmidt trigger, which provides a 1.5 V 1 microsecond pulse to trigger the counter, oscilloscope, and retrigger the pulse generator. The counter is equipped with an IEEE-488 interface. On command from a microcomputer, the counter outputs the measured time interval through the data bus. The microcomputer (HP 216S) performs the distance computation and continuously outputs the computed value to the local host computer (VAX 11/780 or Omnibyte) via RS-232C serial data link. The host serves as the local workstation supervisor, and contains the command set to control and alter the motion of the robot.

The system can be made more sophisticated by using the received echo to retrigger the pulse generator. This "sing-around" technique, familiar to radar systems, is interesting to observe. The pulse signal produces an audible click from the transmitter. At a distance of about one foot, the delay period is 2 milliseconds, which results in a 500 Hz buzz. As the range to the target closes, the pitch of the buzz increases in inverse proportion to the distance. In the absence of any target, there would be no echo pulse to retrigger the transmitter. To overcome this problem, the pulse generator is simply operated in a gated mode, i.e., it retrigger internally at some preset low rate unless externally triggered sooner. The internal rate is set low enough to correspond to a distance greater than any realistically anticipated for the workspace.

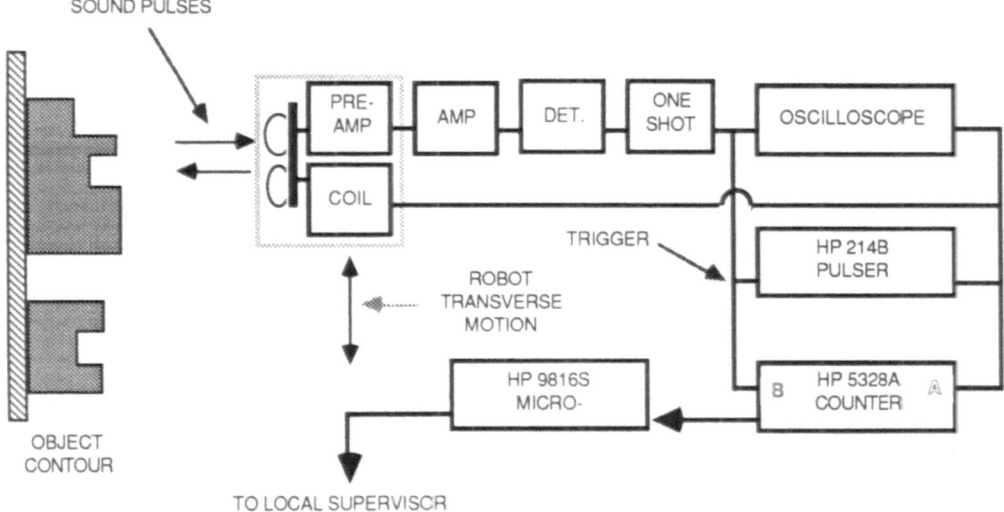

Figure 4. FVF$_2$ range sensing system.

The presence of noise will cause the system to compute erroneous time intervals incorrect distance values. The detection and measurement system is partially protected from such behavior n its current configuration. By choosing the counter's averaging period to be greater than some multiple of the correctly measured delay time, then the measured time intervals are averaged and the effect of a noise triggered event is diluted. Then no sudden and disastrous motion is likely. Even then, such noise would result in the robot arm backing off from its objective or slowing in its approach.

Our demonstration of acoustic ranging for control of robotic motion took the form of a terrain following experiment. The objective was to maintain a fixed distance above a collection of objects over which the robot gripper performed a horizontal excursion. If the distance data differed from a desired value by more than a preset limit, a vertical motion command was issued to the PUMA controller to correct the separation. This technique is also applied in the subsequent ultrasonic systems described below.

Accuracy is determined by several factors: environmental noise, air thermal gradients and time measurement fluctuations, triggering stability, acoustic wavelength, etc. At 10 kHz, the acoustic wavelength is approximately 3 cm, implying a range accuracy of perhaps one fourth of this distance. In fact, because it is possible to retrigger on the maximum of the detected envelope of the echo waveform, some further improvement is possible. Changes in target position as small as 50 mils are detectable. At ultrasonic frequencies, say 200 kHz, two advantages accrue: the background noise spectrum is lower, reducing any false distance information, and resolution improves since the wavelength is then approximately 1.5 mm.

3. RANGING AND CONTROL WITH NARROW BANDWIDTH TECHNIQUES

A technique is now described by which the range from a robot gripper to any target of interest is determined by means of continuous wave (CW) acoustic signal processing. The acoustic wave consists of two pure tones of equal amplitude. The resulting transmitted signal is a carrier at the average of the two tones, and the amplitude undergoes 100% modulation at the difference frequency. Two geometries were considered: bistatic ("pitch-catch") and monostatic ("pulse-echo"). The received signal is delayed by the time of flight. At the carrier frequency a 2π ambiguity exists at multiples of the acoustic wavelength; at 40 kHz, this ambiguity occurs about every quarter-inch.

By using a balanced two-tone CW signal, it is easy to observe and measure the phase delay in the envelope of the received wave, which appears to slide relative to the reference transmitted signal. Furthermore, the beat frequency is the inverse of the null-to-null time interval, and this is the amount of time-of-flight delay that must occur before the first 2π ambiguity occurs in the observed display. Clearly, since the beat frequency should be much lower than the carrier, the nonambiguous range of this system corresponds to the acoustic wavelength at the beat, not the carrier.   A 1000 Hertz beat allows absolute range determination to about one foot. The detected power in each signal is a sine wave at the beat frequency. The phase between these two processed signals is used to compute distance. By use of phase detection methods and a microcomputer, it is possible to keep track of 2π repetitions by indexing, and thus extend the absolute range measurement.

A computer automated system performed this range measurement and communicated the result to the process supervisor which, in turn, generated real time path modification commands to the robot controller. In addition, provisions were made to test and adjust the beat frequency for stability on a periodic recalibration schedule. Setup procedures include a determination of sound velocity, which may vary with environmental temperature, and which can be periodically remeasured from phase measurements made between two positions of precisely known displacement.

Consider a single-tone cw acoustic wave transmitted between two transducers over a distance l.   If λ is the acoustic wavelength and ω is the angular frequency ($2\pi f$), then the wave number $k = 2\pi/\lambda = \omega/c$. The transmitted signal (of amplitude V_0) will be detected at the receiver as

$$V = \alpha V_0 \exp[-i(wt-kl)], \tag{1}$$

where α is a term accounting for propagation attenuation. If we were considering a monostatic case, the distance l would be replaced with the quantity 2l. The power in the received signal is the RMS value of V,

$$|V|^2 = (\alpha V_0)^2 \qquad (2)$$

While Eq. (1) permits very accurate measurement of <u>relative</u> displacement, because of the phase ambiguity occurring every 2π interval of the argument kl, there is no absolute determination of distance.

An alternative approach is described in which a transducer broadcasts a CW signal consisting of two pure tones of equal amplitude and differing in frequency by a relatively small fraction of the average of the two frequencies. Two signals of equal amplitude at frequencies f_1 and f_2 superpose:

$$V(t) = V_0[\exp(-2\pi i f_1 t) + \exp(-2\pi i f_2 t)] = 2V_0 \exp(-i2\pi\{f\}t)\cos([2\pi\delta ft/2]), \qquad (3)$$

where $\{f\} = (f_1 + f_2)/2$ is the carrier frequency with a value equal to the average of the two original ones, and $\delta f = f_1 - f_2$ is the beat frequency. The carrier frequency is then modulated by the cosine factor in equation (3) with a period equal to half the difference between the two frequencies f_1 and f_2. The reflected wave will have the form

$$V = \alpha V(t')\exp[i\{k\}l] \qquad (4)$$

where $V(t')$ is given by Eq. (3), $2\pi\{f\} = c\{k\}$, and $t' = t + \delta t$, and δt is the transmission time delay. The power received varies as

$$|V|^2 = (2\alpha V_0)^2 \cos^2([2\pi\delta ft'/2]) = (2\alpha V_0)^2(1+\sin[2\pi\delta ft'])/2 \qquad (5)$$

The detected power therefore behaves like a sinusoid with frequency $\delta f = (f_1 - f_2)$ with a DC bias, delayed by Ît from the source signal.

The delay time δt is related to the propagation distance by $l = c\delta t$ through the velocity of sound c, so that Eq. (5) becomes

$$|V| = 2(\alpha V0)^2(1+\sin[2\pi\delta f(t+\{l/c\})]). \qquad (6)$$

Thus, by obtaining the power density waveforms for both signals and comparing the relative phase shift, it is possible to obtain the actual distance without ambiguity up to a maximum distance L_{max} given by $L_{max} = c/\delta f$. Now the 2π ambiguity depends on the beat frequency, not the carrier, and is therefore much longer. It may be argued that resolution is sacrificed by operating with the beat frequency, but as will be shown in the reduction to practice described in the following section, resolution can be quite excellent, and the reduction in speed requirements to track the phase at the lower beat frequency reduce the need for large bandwidth phase detectors, with consequent economy of design.

The transducers used are an inexpensive transmit/receive pair operating most efficiently at about 40 kHz.[3] Both the transmitted and received signals have a characteristic "string of pearls" appearance. The separation between nodes in either trace is $T = 1/\delta f$, and the period of the cosine envelope function is 2T. The displacement in time between a node in the received wave and the corresponding node in the reference waveform is a direct measure of the acoustic time of flight.

While the acoustic signals occur in the 40 kHz regime, all signal processing and phase detection is performed on the waveforms produced by square law detection of both the reference and received signals. Eq. (6) is then an accurate representation for both channels, with $I = 0$ and $\alpha = 1$ for the reference channel.

Figure 5 is a block diagram of the functioning ranging system. Two signals , at 39.5 and 40.5 kHz, are combined and directed to the sensor head for transmission, and a square law detector (SQLD). The detector has a 2 kHz bandwidth to produce a smooth sine wave at the beat, which must remain below this cutoff or suffer reduction in sensitivity. The received acoustic signal is conditioned by preamplification, attenuation, and subsequent detection before presentation to the test channel of the gain phase meter. A digital multimeter functions as an ADC and can be read by the IEEE-488 interface. The micro computer manipulates the phase data to compute the range, which it then formats and transmits to the next higher level of control in the robotic workcell.

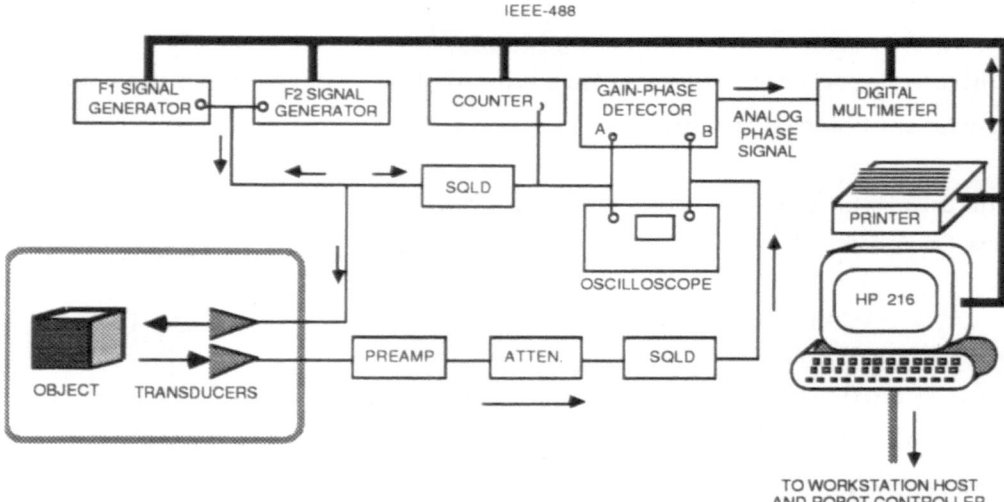

Figure 5. Block diagram of the two-tone narrow band ranging system.

The range L_{max} can be made large if δf is made small. The velocity of sound in air is approximately 331 m/s. With a beat frequency of 500 Hz the unambiguous range is 66.2 cm (~26 inches). Ambiguity can be avoided altogether by ramping $\hat{f}$ upward from zero until a 2π radian phase shift is detected, which establishes the maximum range. Thus, chirp techniques common to radar such as CTFM may also be implemented in this system.

Since the phase is continuously monitored by the computer, it is a simple matter in software to keep track of successive cycling through 2π, and extend the range indefinitely without loss of accuracy, apart from limitation due to dynamic range. This is precisely the case, and ranging out beyond one yard without change of reference signal amplitude or received signal level (through the attenuator) is easily accomplished and very robust.

Both signal generators are programmable in amplitude as well as frequency, so that output levels can be continuously adjusted to compensate for larger attenuation at larger distances, avoiding degradation of the phase stability of the received signal.

The counter periodically monitors the beat frequency (which serves as a standard) for relative drift in the two signal generators. Any measured drift in the beat standard generates a correction order to one of the generators, which is a high stability synthesizer. In this fashion startup drifts are compensated for.

An accurate knowledge of the sound velocity is required for maintaining range accuracy over the distances of interest and over a reasonable time interval. This is accomplished by a computer controlled procedure, whereby the robot moves the gripper-mounted sensor through a known precise displacement, and the velocity is computed from the measured phase shift. Thus no model of sound velocity as a function of temperature, pressure or humidity is needed.

Phase resolution is limited to 0.1 degree by the gain-phase meter. The spatial resolution is then set at one part in 3600 of the value of L_{max}, which is set, in turn, by the beat frequency $\hat{f}$. Thus, at 1 kHz beat the resolution is 3.8 mils (where we have assumed a sound velocity in air of 13600 in/s). A two kHz beat would improve the resolution to less than 2 mils.

Short term (~1 s) deviations about an average reading are typically +/- 4 mils at two inches separation, and +/- 12 mils at 20 inches. These numbers are quoted for a 1 kHz beat in bistatic mode. At two inches, the scatter is equal to the resolution limit of the phase-gain detector. At longer distances, however, fluctuations due to air currents are to be expected. Such short-term fluctuations are of little concern at larger distances; knowledge of position becomes critical only at close distances for most applications requiring manipulation between and assembly of parts.

Long-term drift is due to variation in sound velocity. In a 24 hour test, bistatic perception of an 8 inch distance experienced a total swing variation of 200 mils. This would correspond to a 11º centigrade (~ 20º Fahrenheit) excursion, which seems large, but not grievously so. Note that barometric pressure and humidity have not been taken into account. Such long-term drifts can be eliminated altogether by periodically recomputing velocity with a recalibration procedure as described above.

When care is taken to perform the setup and calibration procedures properly, all range measurements at distances from 2 inches (minimum set by current fixture design) to 38 inches (the maximum tested) were found to be well within the scatter limits observed due to the resolution of the phase-gain meter or fluctuations observed at larger distances due to air currents, which are cumulative in effect.

A striking feature of the design of the system is the immunity to noise, which is typically transient or incoherent in nature. The measured phase was found to be virtually immune to severe impact made on the test fixture, when the transducers were held at fixed separation. While this immunity degraded somewhat at longer range, it could be easily compensated by decreasing the attenuation in the receiving channel or directly increasing the amplitude of the two ultrasound tones.

4. IMPROVING KINEMATIC / DYNAMIC MOTION CONTROL WITH ULTRASONICS

We now describe a sensor system based on acoustic ranging at 215 kHz which has been used to control the offset distance between a robot's end effector and an object surface with considerable accuracy. The range sensor and end effector were mounted on a PUMA 560 robot programmed in VAL II and equipped with a real time I/O port (ALTER). Data from both the acoustic range system and an eddy current proximity sensor were first collected under static conditions to establish noise levels. Both sensors were then used to monitor open loop linear robot motion between two teach points over a precision ground test plate. The deviation from a true straight line path was observed to be sometimes greater than 10 mils.

Closed-loop servo control of the normal distance from the test plate was then implemented using the acoustic sensor for control and the proximity sensor as a performance monitor. Long term average path deviation was reduced to less than 2.0 mils, and fluctuations around the desired path were characterized by a standard deviation of +/- 3 mils when the control algorithm acted on single acoustic range measurements. This is consistent with the fluctuations observed in static (motionless) measurements, and is attributed mostly to fluctuations in the acoustic impedance ($Z = \rho v$) of the air path. ρ is the density of air and v is the velocity of sound.

A statistical analysis of both the static and dynamic range data indicate that averaging as few as ten measurements can reduce the fluctuation in readings to +/- 1.2 mils or better.

The noise limit of the system is attributed primarily to thermally induced fluctuations in the air velocity and the inherent stability of the sensor electronic design. This limits distance measurement accuracy to +/- 3 mils in data taken at 10 Hz rates. Signal averaging ten readings improves the precision to +/- 1.2 mils. This determines the limit of accuracy of position control using the acoustically measured range for position control of the robot arm. Open loop (non-servoed) position accuracy was normally +/- 6 mils for trajectories over one foot. Thus kinematic accuracy was improved by a factor of 2 to 5 times, down to the noise limit of the sensor. Dynamically changing force loading on the robot arm causes deflection of the links and error in the true position. A 1.5 kg mass produces an 8 mil sag vertically of the robot arm end point. Using the acoustic sensor for servo control, vertical position error was reduced to +/-3 mils. The performance of the acoustic sensor was verified on a test surface accurate to +/- 0.1 mil over two feet with an eddy current sensor accurate to +/- 0.1 mil.

Figure 6 is a block diagram of the acoustic and eddy current sensor systems. The primary element of the acoustic system is an ultrasonic ranging module with separate transmitting and receiving transducers and interface electronics module.[4] Detection range is presently 3 to 24 inches in this demonstration system. A 215 kHz narrow beam acoustic pulse is transmitted toward the cbject surface. Upon triggering the output pulse, a latching voltage is generated and stays on until the echo is received. A universal counter measures the pulse length, which is the two-way transit time. The counter interfaces to a laboratory microcomputer (HP 216) via IEEE-488 communications, where the distance is computed. Trigger control of the ultrasonic module is accomplished by a function generator trigger output, which may be programmed from the HP 216 over the same IEEE-488 interface bus. This allows for variable triggering as distances change.

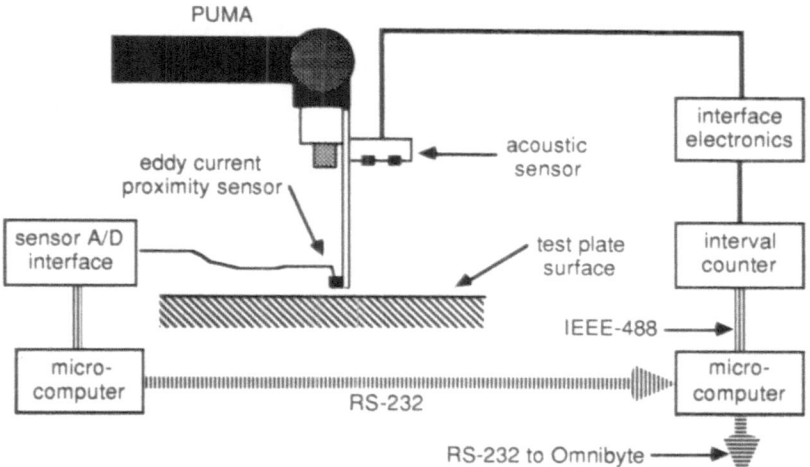

Figure 6. Acoustic and eddy current sensor and signal processing systems.

When the acoustic ranging system is engaged in real time adaptive control, the computed distance is converted to binary data for output to a real time process control microcomputer (Omnibyte) via an RS-232 serial link. The Omnibyte hosted the control algorithm to generate and transmit move commands to the real time I/O port (ALTER) of the PUMA controller via RS-232 (Figure 6).

The eddy current proximity sensor[5] was attached to the end effector with a specialized fixture allowing simultaneous position monitoring by both sensors. In control mode, the acoustic sensor data was not archived for off-line analysis. Subsequent sections will show that the two sensors tracked manipulator motion identically so that eddy current sensor data alone was more than sufficient. Absolute accuracy from this sensor is reliable to 0.1 mils.[6]

In the first experiment the PUMA arm was positioned toolplate face down normal to a test plate precision ground to +/- 0.1 mil. The acoustic sensor was approximately 7 inches from the test plate surface, while the eddy current sensor was nominally 30 mils from the test plate. Range data were collected as paired readings from both sensor systems, and absolute offsets were subtracted out. The resulting matched arrays of distance variation were stored in mass memory for off-line analysis. Measurements were obtained by simultaneously triggering both systems. Eddy current data variation is about an order of magnitude smaller than the acoustic sensor, making it a reliable reference against which to measure the acoustic sensor performance. The absolute range of the eddy current sensor is about 100 mils, so that the dynamic range of that device was about 1000:1. A rough comparison with the acoustic sensor shows that the noise variation in raw acoustic data is about 2.5 mils over 7 inches, with a dynamic signal to noise range of approximately 3000:1.

When groups of ten successive readings were averaged, the averaged "points" in each data set reflect a smoothed reduction in variation, which is not surprising. We attribute the major portion of the acoustic signal noise to fluctuations in the acoustic impedance of air, which results in velocity fluctuations. The result is that, with a minimum of signal averaging, it is possible to resolve position to within +/- 2.0 mils using the acoustic sensor.

The PUMA was taught a linear path between two endpoints separated by 10 inches with the eddy current sensor at a nominally fixed distance of 30 mils above the surface of the precision test plate and the acoustic sensor at 7 inches, as before. Both the acoustic and eddy current sensors monitored the vertical axis motion as the robot made the 10-inch traverse over the plate at the rate of 6 inches per minute. A systematic drift exceeding 8 mils was observed. Larger errors are known to occur when the trajectory is more complicated. Both sensors show identical results, enhancing confidence in the reliability of the acoustic range method. Averaging the data over groups of 10 readings, indicates that submil accuracy is possible in tracking robot motion using the acoustic sensor.

With the actual robot path being monitored accurately, then it remains to show whether a control loop can act on the data to correct, in real time, the trajectory derived from off-line programming.

In the next experiment the taught path was repeated. The range data acquired by the acoustic sensor system, however, was passed to the Omnibyte computer. A comparison of the desired and measured vertical distance generated a difference error signal. A correction command proportional to the difference error was transmitted to the robot controller through its ALTER I/O port every 28 milliseconds. If the difference error was below a threshold value (+/- 3 mils), no correction was sent. Since the actual single measurements were being used, it made no sense to lower the threshold below the typical variation in sensor readings. Typical accuracy in acoustic range readings were on the order of +/- 2.5 mils. Reduction of the threshold much below this window resulted in an observable oscillation in the vertical direction, caused by the noise in the acoustic measurements. The resulting improvement in path linearity measured by the eddy current sensor is evidenced by the fact that the path deviation measured by the eddy current sensor is approximately half of the previous result, with a variation comparable to the acoustic sensor system noise level as verified by a statistical analysis of the data.

The next step in assessing the effectiveness of acoustics as a means of adaptive control deals with the positional error that may arise when mass is added to or detracted from the end effector of the robot arm, such as when an object is grasped or released. To simplify the procedure, and avoid system perturbations that might arise from the mechanical gripping activity of the end effector, a basket was attached to the manipulator, in which successive weights were dropped. Range information was gathered from both the acoustic sensor and the eddy current proximity sensor. The acoustic servo control loop was not engaged and the robot arm was programmed to maintain a rigid configuration.

TABLE 2

	Arm static		Arm under acoustic control	
	SD	SD of averages of 10 readings	SD	SD averages of 10 readings
Proximity sensor	0.18	0.05	2.60	0.77
Acoustic sensor	2.73	0.34	*	*

Statistical data on the measurements made by the acoustic range sensor system and the eddy current proximity sensor. Units quoted are in mils. SD = standard deviation. *When the robot arm is under acoustic range servo control, only proximity sensor data is available.

In a repetition of earlier measurements, the PUMA arm was kept static, with the manipulator pointing down vertically toward the precision ground surface, without any acoustic servo control. Acoustic data showed a standard deviation in measured distance of 2.73 mils (Table 2). This is consistent with sound velocity fluctuations due to ambient thermal conditions in the laboratory. Proximity sensor data, more accurate by one order of magnitude, had a standard deviation of 0.18 mils, indicating that the arm is more stable than indicated by the acoustic sensor. These two figures give a measure of the noise level and accuracy to be expected from the acoustic sensor.

With the arm again in the same position, but with vertical position under servo control from the acoustic sensor, the arm is now free to move as directed by the control algorithm. Data were not accumulated from the acoustic sensor, but were ported directly to the workcell controller which generated position command for the PUMA controller. Data were also simultaneously recorded by the proximity sensor system. The proximity data looks like the original acoustic data because the arm receives commands to move based on range data obtained acoustically, which has an order of magnitude more noise in it than the proximity sensor. Indeed, the proximity sensor now shows a standard deviation in position of 2.60 mils, nearly identical to that found for acoustic data with the arm in static condition. Thus, the arm is responding to noise in the acoustic system. If time averaging of the acoustic data were performed before issuing a motion command, the standard deviation of the arm position would be only 0.77 mils.

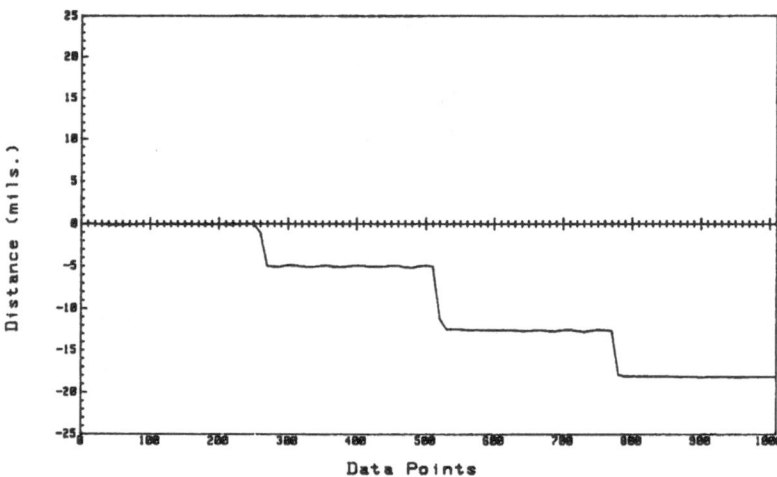

Figure 7. Displacement of the manipulator arm (at the end effector) as three masses of 500 grams each are added. The arm is locked in a static position and is not under acoustic range servo control.

The next step consisted of sequentially adding three 500 kg weights to a basket attached to the manipulator, with the arm in static (nonservo controlled) position. The arm experiences a net sagging of 18 mils without recovering, as measured by the proximity sensor. This exceeds the specified accuracy of the robot, disqualifying it from some applications.

When the acoustic ranging servo control loop is activated, corrective repositioning is achieved. In Figure 7 each instance of mass addition is noticeable by a significant shift in arm position. Figure 8 shows the arm behavior under servo control from the acoustic sensor. The end effector is returned to its original height, apart from a net shift that is now only 3 mils after the full 1.5 kg load is added. The standard deviation in position as read by the proximity sensor is 2.34 mils, in close agreement with previous readings taken under acoustic servo control. This places the arm's positioning accuracy within the +/- 4 mil position *repeatability* specification claimed by the manufacturer.

The short-term (or high frequency) noise in the detection systems limits the accuracy to which the robot arm can be monitored or controlled. To the extent that averaging can be tolerated without compromising the tracking speed, tracking accuracy can then be improved. We recommend that the entire problem of sampling rates and control would benefit from a detailed analysis of the noise spectral density of such systems and characterization by parameters such as an Allan variance.[7] In much the same way that long- and medium-term drift and short-term variation in frequency sources may be

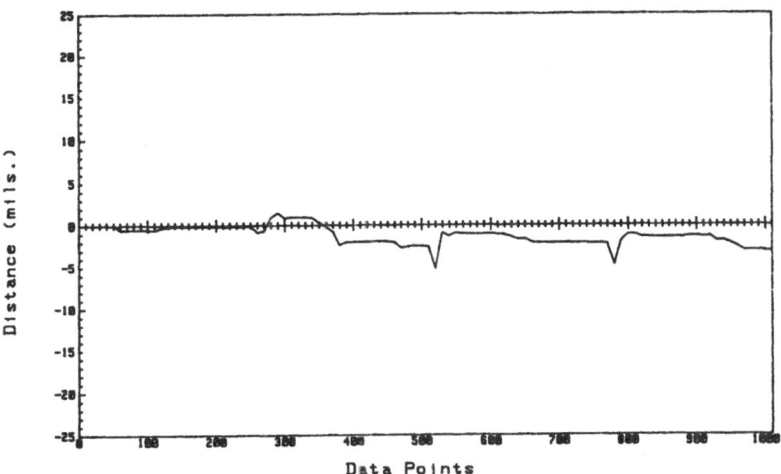

Figure 8. Position as measured by proximity sensor with PUMA under acoustic range servo control. Three masses of 500 grams each are added to the end effector. The net sagging is limited to 3 mils.

characterized by single side-band phase noise and Allan variance, it may be possible to derive generic parameters for defining the accuracy/speed trade-offs and figure-of-merit for robotic sensors and control systems.

It is clear from the last experiments that a standard industrial robot is limited in its ability to maintain its position or configuration accurately when a changing mass load dynamically alters the moment arm of the structure. The acoustic servo control loop provides the means to effect precise relative positioning between a manipulator endpoint and a certain limited class of objects, without placing strict requirements on the location of the objects. The control process may be extended to include control of the rate of closing between a manipulator and a moving object. This might be the case for a robot arm on a fixed base tracking objects moving on a conveyor belt, or for a mobile arm which translates and rotates as the arm is moved from station to station. When payloads or angular momentum change dynamically acoustic sensor based control can improve tracking performance.

The set of experiments described above were performed with laboratory instrumentation to demonstrate concept feasibility. The system bandwidth (or servo control loop cycle time) is therefore limited by the speed governing the parallel data communications protocols between the test instrumentation and the sensor data processor (IEEE-488 standard), and serial data communications between both the sensor data processor (HP 216) and the Omnibyte workcell controller and between the workcell controller and the PUMA controller, both RS-232C. In particular, the interval counter is limited to data transmission every 100 milliseconds. This places an upper limit on the speed of data acquisition and control of 10 Hertz, which is less than the bandwidth of the PUMA I/O port (Alter) of 35 Hertz. A further reduction to practice would eliminate the communications bottleneck inherent in test instrumentation (which is generally not intended for high speed real time process control) and raise the control loop bandwidth considerably. The next step would be to link the workcell controller directly with the joint controllers, eliminating communications through the PUMA controller.

5. CONCLUSIONS

We have described a series of acoustic sensor-based control systems interfaced to a PUMA 560 robot equipped with a port permitting real time path modification. The acoustic techniques consisted both of pulse-echo and continuous wave narrow band methods, borrowed shamelessly from radar. We have demonstrated the effectiveness of such methods to maintain positional accuracy and surface tracking despite the effect of dynamic mass loading or changing moment arm of the robot. The positioning accuracy is based on ranging between the end effector and the object of interest, and is independent of the robot's posture or source of structural deformation, e.g., joint free play and structural deformation induced under load.

ACKNOWLEDGEMENTS

This work was supported by the Rockwell International Independent Research and Development Program.

REFERENCES

1. KYNAR, Pennwalt Corp., King of Prussia, PA
2. M. Tamura, "Piezoelectric Polymers, Properties and Potential Applications," proceedings, 1977 International Ultrasonics Symposium, IEEE Cat. # 77CH1164-150, pp. 344-346
3. Projects Unlimited, Dayton, OH; part #SQ-40T/R-16
4. Massa Products Corp., Hingham, MA. 02043
5. Kaman Instrumentation, Colorado Springs, CO 80933
6. D. E. Whitney, et al, "Short- and Long-Term Robot Feedback: First Report," NO. NSF/MEA-84010, 1984; available from NTIS
7. D.W. Allan, "Statistics of Atomic Frequency Standards," Proc. IEEE, Vol. 54, No. 2, pp. 221-230, February 1966

SONAIR ULTRASONIC RANGE FINDERS

S. Monchaud
INSA-LATEA
35043 RENNES Cédex (France)

KEYWORDS/ABSTRACTS :

ROBOTIC/RANGE FINDERS/ ULTRASOUND/ CALIBRATION/ MULTISENSORY/
3D PICTURE ANALYSIS/ 3D PICTURES SYNTHESIS

Rangefinders, particularly SONAIRS, constitute an important part of new robot sensory systems. We review the different measuring techniques. The attained performances depend mostly on the :

- physical properties of the air and ultrasonic transducers

- electronic and its coupled microprocessing

- movment of SONAIR in the discovery of its surrounding.

Some of them are contradictory. We must make a good compromise. We present our choices in our field of application : the mobile robot sensory equipment and those of the 3D synthetic vision system. In order to improve their performances, we conclude our paper by considerations on the ideal transducers that would be produced industrially and we present the concept of multisensory range finding systems using the present possibilities of optics and acoustics.

Telephon : 99.28.64.99   Telecopy : 99.63.67.05 Usager Telex Public : 730 800 R

NATO ASI Series, Vol. F52
Sensor Devices and Systems for Robotics
Edited by A. Casals
© Springer-Verlag Berlin Heidelberg 1989

DEFINITION

Range finding is a distance measurement obtained by optical, radioelectric or acoustic means. The word "range finding" must not be used in the sense of "remote measurement" which relates to the creation and transmission of a signal representing the result of a measurement. The official word for a range finding system is a "range finder" (French Official Journal of June 24 1982).

1- CALLING BACK (1)

Ultrasonics is widely used for remote control in TV tuning and garage door activation but they can also measure speed and distance. Ultrasonic air systems SONAIR operate with principles similar to those used in radio, radar and sonar. In this last case, the system is used in the water. The major difference is that ultrasonic systems transmit high-frequency sound ($\sim$40 Khz) through the air to the receiver which converts acoustic energy into electrical signal. We have many means to solve this problem.

- Echo : continuous-wave (CW) ultrasonic systems are used as intrusion sensors and for general object detection. The microphone continually receives echos from surrounding objects. An amplitude sensitive detector responds to change in the echo-signal magnitude caused by a moving target. This system used two transducers. However one transducer can be used by **duplexing**. It is not a real range-finder (see definition).

- Pulse-echo : in that case the measure is made by pulsing the transmitter and **measuring the time interval** between the transmitted pulse and the received echo. For example the velocity of sound is approximately 345 m/sec. The echo from a target 5 m away will arrive 10/345 = 29 ms after emission of the transmitter pulse. To avoid error in associating an echo with its correct parent pulse, usually only one pulse is in the air at any given instant. Also in that case we can use one transducer. Automatically switching a common transducer between transmitter and receiver eliminates the cost of using two separate transducers. There are several options of using this technique, which is called duplexing. In addition to requiring only one transducer, duplexing saves space and ensures that the receiver and transmitter are pointed in the same direction, a feature especially important with beams diverging less than 10°. The pulse-echo technique is a real teachnique used for sonair.

- Sophisticated ultrasonic system use frequency-modulated, continuous-wave (FW/CW) beams to measure both distance and relative relocity. The transmitter emits a continuous signal that cyclically increases and decreases in frequency. The echo signal is shifted in frequency with respect to that being transmitted by an amount proportional to range. Since phase shift can be converted into time units T, then

T = 2 R/V where R = target range and V = Velocity of sound. The Doppler effect of a target moving toward the transmitter shifts upward the range of frequency in the echo ; a receding target produces the opposite effect. The receiver converts this frequency-range shift to relative velocity. This technique is very attractive but, as I know, not really used in robotics applications. Why ? perhaps the problem of these ultrasound transducers is not solved in a compact form.

In conclusion of this technique survey we know that we have means to measure distances in the air but many problems are not yet compleatly solved.

2- ULTRASOUND RANGE-FINDING IN ROBOTICS (2)

The indirect measurement of a length i.e. its determination without the use of a standart, has been employed for centuries and in various domains. For example in astronomy where the evaluation of distance could not be made by direct measurement.

There are many important industrial fields using range finding techniques:
- in civil engineering : stabimetric process (or tachymetric), the theodolite, electronic length-measuring instruments etc...
- in photography : the autofocus device for example, in balistics etc...

From this rapid presentation, it can be seen that civil industries first developped several range finders but with specifications (for example large-distance measurement) different form those needed in Robotics. This fact led us to study the problems set by robotics in the field of range finding, to list the different techniques and systems used and to define their salient features. A man and a robot with vision organ (with 3 D possibilities) do not see a real scene in the same manner. In artificial vision, the scene is seen as a group of plane and rounded surfaces, with many ambiguities due to shadows and covered parts of objects. As these effects vary with the vieving angle and lighting conditions, there is a great risk of interpretation errors which complicate the problems of artificial vision by computer. In an industrial envirornement the 3 D picture is imperative when the robot must, it self, find the position of an object or a part of an object (protuberance, hole...etc).

On account of the orientation of our research, we have worked on the problem of the third dimension (or depth) with the following constraints :
- the distance pictures memorized in the computer must contain all the points which determine the intersection of horizontal planes situated at different arbitrarily chosen heights and all the objects contained in the scene.
- the explored distances ranged from to 10 meters (middle range).
Many methods exist for solving this sort of problem. They may, in general, be classified in two bigfamilies, the active methods and the passive ones. We work only in the active optical and acoustic methods. In the present paper, we describe only the acoustic ones.

Ultrasonic waves have wave-lengths ranging from a few centimeters to a few microns. Their amplitude decreases exponentialy with distance propagation speed is greatly influenced by the surroundings but is practically independent of frequency for a given type of surroundings. The energy efficiency is very much lower in air than in water. The radiation diagram shows that the intensity is maximum on the source axis. The frequency of the emission signal modifies the aperture angle of the beam. The emission and reception angles of the sensors are fairly large (between 10° and 120°) but not enough to permit a panoramic vision (named 360° vision) which can be obtained with several "emission-reception" systems or by putting the sensor on a adjustable rotary support. We choice this second solution. Range varies with the **nature of the transducer** (piezoelectric céramic, electrostatic transducers...etc) and with the frequency (20 to 200 Khz), and can be more than 10 meters. The **poor directional properties** of the ultrasound beam practically impose range finding techniques based on measurement of the time of flight. It may be noted that, on account of the reversible properties of the acoustic waves, just one transducer is necessary, used alternatively as emitter or as receiver. But this result in a **blind zone** (emitter-object distance corresponding to the time during which the sensor sends but cannot receive) of about 50 cm.

Ultrasonic range finders have several advantages :

- they are, in general, of small size, making them easier to use especially in robotic applications and give, in real time, a measure.

They have some disavantages :

- atmopheric movments disturb the measurement and can reduce the range. Working can be disturbed by noise (industrial for example) which induces detection errors.

- the collection of measurement may be affected by the sensor (**movment** of a manipulator arm or a mobile robot). For precision measurements or for high speed robot displacement, it is necessary to take the displacement into accound.

- emission power increases almost linearly with the input voltage. Each manufacturer specifies a maximum voltage for his transducer. Above this value, the power of the acoustic signal remains constant.

- the largest source of range measurement error is the effect of temperature [1] upon the velocity of propagation V. This effect is expressed by :

$$V = Vo \ \sqrt{1 + t/273} = 331.5 + 0.607\ t$$

where Vo = velocity of sound at 0°C and t = temperature (°C)

Humidity has a smaller effect on velocity, but the calculation is more compli-cated. This effect is defined by :

$$Vh = \frac{Vd}{1 - \frac{e}{P}(\frac{w}{a} - 0.625)}$$

Where Vh = velocity of sound in humid air (m/s) ; Vd = velocity of sound in dry air (m/s),
P = barometric pressure : e = vapor pressure and w et a are the ratios cf specific
heat at constant pressure to specific heat at constant volume for water vapor and
air, respectively. At 20°C the velocity of sound increases 0.3 % when humidity increases
from O to 100 %. This factor is negligible compared to the effects of temperature
and wind. Where as a crosswind can blow an ultrasonic beam off target, wind blowing
parallèl to the beam shifts the apparent sound velocity and frequency. In a one-way
system, the frequency arriving at the receiver is shifted by the ratio of wind velocity
to the speed of sound in still air. This frequency shift is, of course, negative if the
wind is moving opposite to the direction of propagation and is positive for same direction
wind. Either way, as wind velocity increases, the effect is equivalent to detunning
the receiver. Frequency errors introduced by parallel wind in echo systems are
essentially cancelled out. Parralel wind in either direction slightly increases apparent
range because the wind slightly increases round-trip time of the pulse. However, range
error for normal wind velocities is insignificant for most range-measuring applications.

In conclusion, to this first part, il can be stated that :

- range finding techniques are now necessary for a large number incustrial
applications particularly in robotics.

- with acoustic methods, the interpretation of the measure is acute. Fortunatly
the present state of technology, mainly integrated circuits, permits the productive
of the **necessary microprocessor systems** which pilot the SONAIR.

- the existing commercial systems cannot satisfy our goal of a 360° panoramic
vision system in a common natural surrounding and in the same time a good target
resolution and optimum use of available transmitter power. The first hope asks a wide
ultrasonics beams, the second require narrow beams. In this last case we have sharply
defined patterns. All this justifies the construction of laboratory prototypes.

3- CONCEPTION AND CONSTRUCTION OF OUR SONAIRS

The several prototypes, that we have built, are divided in two groups :
monotransducer (duplexing technique) and bitransducer (pulse-echo technique). Each
one has a different field of application and possesses qualities and disavantages that
will be reviewed.

- In the case of monotransducers, the detection is punctual and discriminating
(detection of contours and small targets). The coincidence condition between emitter
and receiver is thus automatically verified which makes subsequent interpretation
of the measures easier. There was also a tendency to reduce the emission one by the
addition of an external cylinder or parabola. A reduction factor of 10 can be obtained
for an initial emission one of apprcximatly 30° solid angle.

- On the contrary, the bitransducer system allows the detection of non-planar targets, less reflective, not perpendicular to the propagation axis and cuts out **blind zone** (situated at a short distance and in which no-object can be detected).

3.1- Hardware

In the following description, it should be noted that the first prototype was named US RTC, the second US VERNITRON and the US POLAROID after the manufacturers of the transducers used.

1- Separate emission-reception transducers (US VERNITRON)

For emission, the oscillation of the transducer is obtained by a periodic square wave signal, of 24 Vdc (obtained from batteries) with a frequency suitable for the transducer. This frequency and the signal emission time are variable. These two parameters (frequency and time of emission) are **programmed by microprocessor.** The excitation frequency can be modified, according to the selected transducer, between 614 Khz and 16 Khz and the time of the emission between 0.4 and 1.5 ms approximatly. The repetition rate of the pulse trains and the emission time vary proportionately to the distance which separates the sensor from target. This **adaptation of the emission circuits** is an original feature of our system. It entirely eliminates the blind zone which handicaps the other systems presented, while maintaining good range. The hardware interface is very simplified. The advantages for robotic applications (3) :

. the output of the ultrasonic distance sensor is very simple

. the network is quite general and adjustment time very short.

For the reception part two operational amplifies A2 and A3 are used which amplify the signal. But the output signal has a positive and a negative part with a slightly negative DC component. The amplifier A3 also compensates for the DC component. The gain is controlled by A2 and the DC component by A3. Finally a precision voltage comparator was added and the decision level controlled. Adjustement are made experimentally. First, a target is placed at the extreme measurable range and the gain of A2 adjust (offset = 0). Next the target is removed. If there is still detection, the noise level is too high and we must adjust the offset of A2. This calibration is typical of this type of sensor (4).

2- Monotransducer emitter-receiver

The hardware difficulty occurs at the level of the analog-interface between the microprocesseur card and the ultrasound transducer. The manufacturers of transducers propose special interface cards. The interface and the double wire connexion must be bidirectionnal.

. In the case of the US RTC (5), the interface is a specialised MSI chip, proposed by the firm National Semiconductors some passive components (resistors, capacitors, inductors...) and some active components (transistors...) must be added and the chip is thus adapted to the transducer choosen closely and in a rigid manner. Adjustement of this interface is very critical [4]. The end product is very compact and not expensive.

. In the case of the US POLAROID, the manufacturer proposes an interface card with very attractive properties [6]. The emission signal lasts for a little more than 1 ms. It is composed of 8 periods of a 60 Khz signal followed by 8 periods of a 56 Khz signal, 16 periods at 52,5 Khz and 32 periods at 49,41 Khz.

The amplifier circuit made by POLAROID is a programmable integrated circuit. The aim of this programmation is to compensate for the attenuation of the signal with distance. The variation of gain can reach 60 dB.

This multifrequency emission signal and it reception by an amplifying circuit with a variable gain and bandwidth are the two original features of this hardware interface. It must be noted that there is compatibility of the level of the connection between the hardware and the microprocessor during it (TTL level) in the case of the US RTC. On the contrary, with the US Polaroid, a matching circuit must be added to the TTL level.

3- Data presentation with the microprocessor system

Two types of microprocessors have been used successively. INTEL 8085 is connected two peripheral chips INTEL 8755 (Reprom memory and parallel I/O and Timer). The serial INPUT-OUTPUT which perfects the necessary functions for this type of range finders is made directly at the chip level. The presence of Timers (clock generators with programmed frequency) is fundamental to generate the different types of useful signals. In addition the development of the different ultrasound range finders has been made easier by the use of keyboards and displays at the level of the evaluation kits.

Available with these two microprocessors, the use of 2 Kbytes of Reprom memory and 512 byts of RAM memory is suitable for the dimensions of the software that may be met. At the same time the 3 interrupt-levels available are well adapted to this type of application. The US RTC range finder was driven by a microprocessor INTEL 8085. The US VERNITRON and US POLAROID range finders were driven first by an INTER 8085 system then by a monochip 8751. From this last experiment, we deduced that it is preferable to use the 8051, because the compact multifunction form of this chip allows the hardware to be reduced to only one chip (integrating the 8 bits central unit, the Ram and Reprom memory and the timers) with and that is a new and very useful feature a programmable up/down counter function. With its three interrupt levels and serie input, this chip is the most suitable one for this type of application.

3.2- Software

The general flowchart of this software is based on the frequency of the operations made by the central unit for a complete measurement cycle. First it generates an emission signal at an ultrasound frequency for a sufficient time (determined by a clock function in the central unit), then it starts an internal chronometer which counts the time separating this emission from a possible echo reception. Two situations can occur :

. there is detection of an echo. The central unit must stop counting, memorize this value and after calibrating the system (2), translate this count into a distance value that is also memorised.

. there is nodetection. After an arbitrary time, the system must initiate a new cycle of measurements. This software organization shows that the tasks are very numerous and varied, they are purely serial, except perhaps at the level of the watchdog function, which acts as an interrupt level. The software is closely dependent on the type of hardware interface used. In fact, our experience shows that the simplest and most general driving software is that of the US VERNITRON (separate emissionreception transducers). It has been seen that, in this case, the hardware interface is also very simple and very general. On the other hand, in the other two cases : US RTC and US POLAROID (monotransducer emitter-receiver), the command software is also very simple but it is not general. For emission, in the case of the US RTC, it is sufficient to apply to the interface circuit a square signal of level 5V and duration equal to the emission time. In the case of the US POLAROID, the interface card is powered during the required emission time. For reception, in the case of the US RTC and US VERNITRON, it is only necessary to detect the echo which has the form of the interrupt level for the central unit of the microprocessor. On the contrary, the dialogue is more complex, in the case of the US POLAROID. It is necessary, in fact, to detect signals which indicate the arrival of the echo.

In every case, especially that of the monotransducer, the different operations must be well synchronbed so that the emission and the reception times do not overlap.

The sequencing of these operations needs precision which takes into account the large constraints due to real time measurements.

4 - RESULTS

We have two types of results :

4.1- Fixed Sonair

The importance of three fundamental parameters for the quality of the target detection has been shown : the range, the emission cone and the floor effect (or many horizontal flat surfaces). The nature of the material and the value $\underline{S}$ are also important. For example, the sonair will detect a piece of wood having a surface area (5) of 50 cm^2 at a distance (d) of 4 cm. But, if d = 300 cm, the piece of wood must have a surface area of approximately $0.5 \ m^2$ to be detected (case of US VERNITRON). All our measures show that the sonair has many dead angles. The problem has been simplified with is a real case, the portion of space which is not seen, is larger. It must be remembered also that the target's material modifies the measurements other papers [7] noted that the pitch of the return echo is proportional to the range of a target. The form (timbre) of the echo was reflected, hence indicating the nature of the target.

Because soft and hard targets reflect differently attempts have been made to use this property to obtain textural information. But unfortunately, the angle from which the ultrasonic beam is reflected from an object affects echo strength and thus appears to change textural information e.g. a hard object at an angle could look like a soft object viewed directly. Other information is needed to resolve such ambiguities.

The existence, in all space, of the principal and secondary lobes of the ultrasonic transducer's emission cone, particularly in the horizontal plane and in the vertical plane (with respect to the sensor plane) induce, in some cases, parasitic detect ons.

The results of our measurements show that to avoid these parasitic detections, a monotransducer range finder must be at a minimal range of 165 mm from the flat surface. For a bitransducer range finder the minimal range is 240 mm with a sensor plane parallel to the flat surface and 275 mn when this plane is perpendicular. Placed at this minimum height, the range finder can detect a wodden cube with 42 mm sides (test target) at a minimum distance varying between 300 mm and 465 mm according to the type of range finder used and its position with respect to this flat surface. Of course, these minimal distances of detection and the dimensions of the detected target may be reduced even more by decreasing the height of the range finder or inclining towards the flat surface. The range of the Sonair is then reduced and little by little, the ultrasound range finder is transformed into a proximeter which detects small targets placed on the flat surface.

These effects appear when the flat surface is the floor or in the case of a rotary sensor-holder arm. The range finder must thus be situated at a sufficient height above this plate.

4.2- Mobile Sonair

In our case, the distance measurement is made between the ultrasonic sensor with a rotation around the vertical axis and one or more target(s). In the case of only one object, various types of class of objects must be distinguished : plane surfaces, angular surfaces, rounded surfaces (e.g. cylinders).

Each one of these classes of objects has been examined by the monotransducer ultrasonic range finders (US RTC and US POLAROID) or bitransducers (US VERNITRON).

The results can be seen in the litterature already published [8]. Rotational scanning of cornered obstacle, in the case of concave corners will now be presented. (Figure 1).

It can be seen that if the ultrasonic range finder is used alone, positions of planes or cylindrical obstacles can be obtained, but the precise location of the corners or the widths of the targets cannot. The panoramic view obtained by the sonair rapidly locates the space zones encumbered with targets, but the zone is deformed. Particularly, the smaller the projection of the object surface on a plane perpendicular to the axis of the ultrasound beam, with respect to the intersection of this surface with the ultrasound emission cone, ·the worse the observation of the object profiles. These deformations prevent object recognition and can event suppress the free zones between obstacle groups.

5- OUR APPLICATIONS

Our sonair is an element of a multisensory system. The multisensory system is composed of a mixed optical and ultrasound range finder, mounted at the mjdpoint of the rotating arm supporting the range finder ladar (active triangulation laser). The main goal of our turntable is to obtain panoramic views of the surrounding world for further application, for example the navigation of a mobile robot in a unknown world strewn with obstacles or the animation of synthetic pictures in audiovisual applications. Examination of the fields of current applications where distance pictures are necessary, shows that a panoramic of 360° is rarely needed. In general, a zone of interest can be anticipated.

5.1- Selective mapping

When the turntable searches closely and very quickly detects the presence or the absence of objects in a zone. Each time an object (or group of objects) is detected by the Sonair, the laser reconstructs the shape of the objects whose mean position has been very quickly determined. It is possible that the choice of the detection limits is such that no object is found. Then the panoramic exploration will very rapidly indicate "zone free of target". It is of interest to note that if the Sonair and the Ladar has not been coupled, the use of the Ladar alone would need the acquisition and treatment of thousands of bytes of measurements before concluding that objects were absent. The time gain is spectacular. With our coupling method, the increase in the number of measurements, and consequently in acquisition and computing time, is directly proportionnal to the number of objects in the explored zone. It is possible, automatically ir manually, to explore a specific zone, or to display or erase certain scene planes rather than others, so long as they do not overlap in the scanned zone. At present, the explored surface may be more than 100 m^2.

5.2- Incrustation of objects or actors in a 3 D Synthetic scene

The range between the object (or the actor) and the Sonair is measured permanently. At the same time a TV camera films this moving object (or human) with no background (or with so called "blue ground"). In the picture synthesing machine, the decoration is memorized in three dimensions. Each pixel of this real picture has its geometrical coordinates and to each pixel of the synthetic picture, the operator also gives geometrical coordinates. These two pictures are superposed, compared and mixed in real time. These modern techniques of 3 D audiovisual creation also raise the problem of the location in space of objects and human beings on the one hand and, of the real camera (or imaginary camera in the case of the synthetic decoration) on the other hand. Solving this problem assumes that range finding techniques are employed to locate the position and the orientation of the objects with respect to each other.

6- CONCLUSION

The SONAIR concepts are not really new ideas. Ten years ago appeared on the market a monochip SONAR system. Proposed by National Semiconductor this chip the LM 1812, offers an attractive solution for the detection of targets but the resolution is bad.

Good target resolution and optimum use of available transmitter power require narrow beams that have sharply defined patterns. Presently most ultrasonic transducers for transmitting airborn beams use a piezoelectric crystal mechanically coupled to a diaphragm. These transducers have the same coherence limitations suffered by all diaphragm transducers that are mechanically coupled to a driving element. This problem can be avoided by using an electrostatic transducer in which all portions of the diaphragm are driven simultaneously by a common electrical signal which has negligible spreading time. The solid angle of a sonic beam generated by the end of a cylindrical transducer is approximately :

$$\Omega = \frac{0.37 \quad \pi \cdot \lambda^2}{r^2}$$

where Ω = angle of divergence λ = wavelength of sound in air (m) and r = dix radius (m).

Transducers also can be oval and rectangular. In that case a transmitted beam is widest in the direction of the transducer's smallest dimension and vice versa. We insist on the problem of transducers because we think that it is the key of future advances in the field of Sonair. Recently a german firm announced ultrasonic condencer transducers for the frequency range from 30 to 50 Khz with a wide bandwidth, highly efficient transmitter **but with two design options – wide – angled or narrow beam directivity.**

That is the second part of the question. In robotic applications it is offen necessary to have, at the same time, a good pattern and a wide-angled view.

We think that is impossible to solve for many physical reasons. In France, many groups of researchers proposed solutions for this problem. In Toulouse, the robotic group of LAAS working around the mobile robot HILARE, proposed to multiplicate the number of transducers. Each one has a good directivity and the wide-angle is reached by the proposed to take a plane area of many ultrasound transducers dedicated electronic drives each emitter transducer with a phase shift. Theoriticaly, this solution is attractive, pratically it is very hard to realise and reach the goal.

We propose another solution, the concept of the multisensory turntable. All these studies take into account the limitations and deficiencies that showned up in our research work on the range finding techniques.

As a result of the experiments made on the multisensor turntable, we estimate that our two primary goals should be :

. to increase the volume of space explored by these 3 D range finding devises. We tried to reach a volume of 100 m^3.

. to reduce the acquisition time and the processing time by a better specialisation of the type of sensory data acquired.

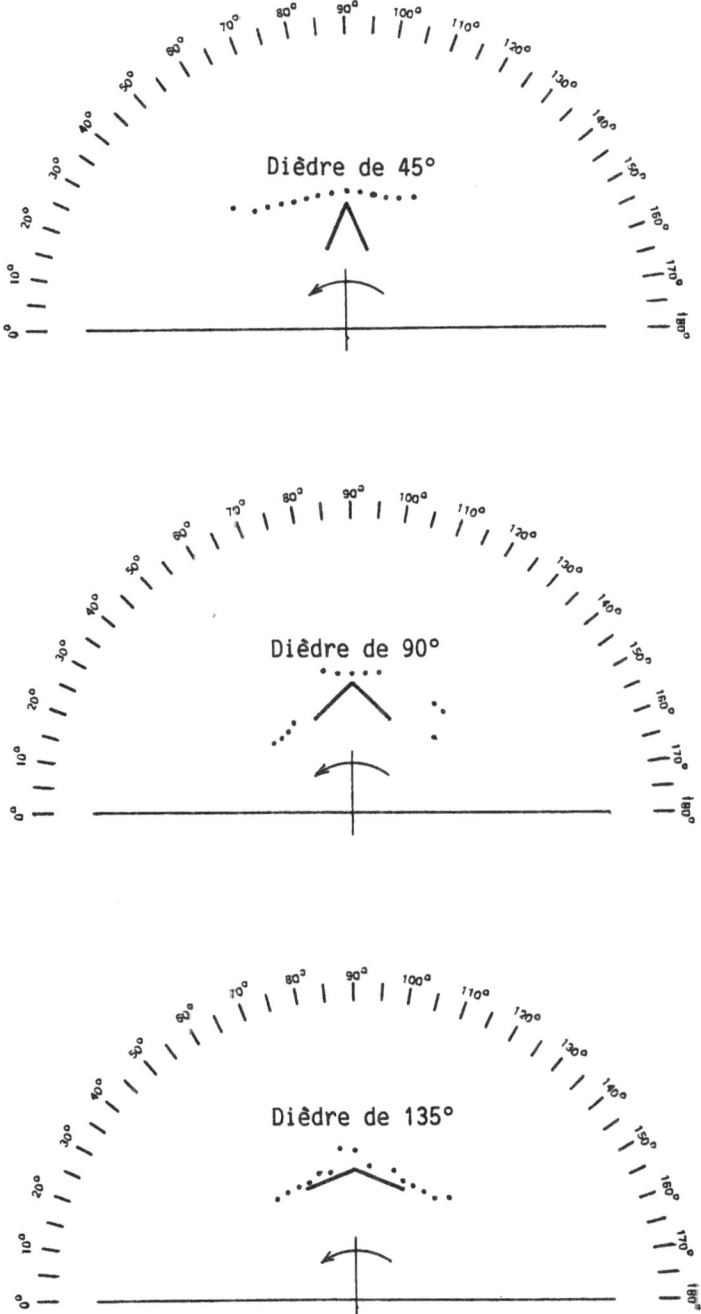

Figure 1 : Rotational Scanning of a concave corner

(US Polarcïd monotransducer emitter-receiver)

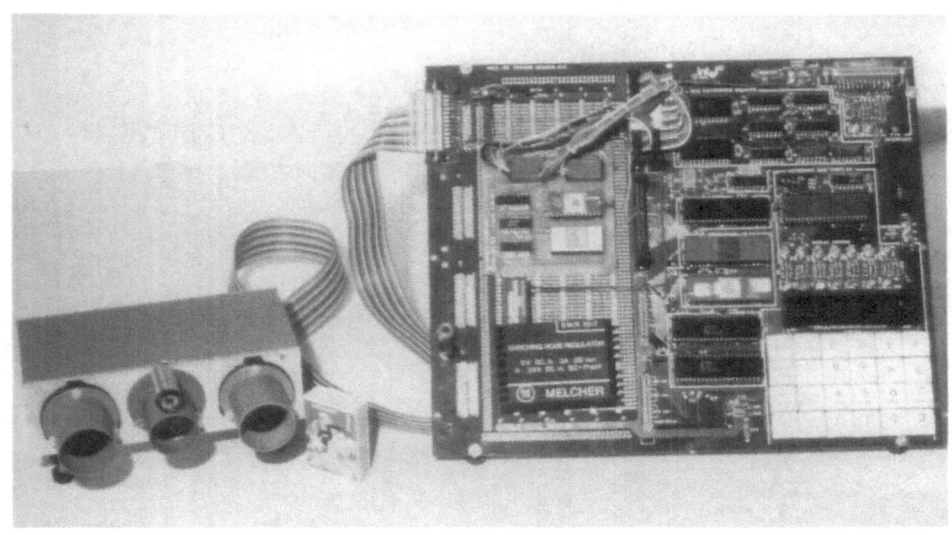

Our first SONAIR (US Vernitron)

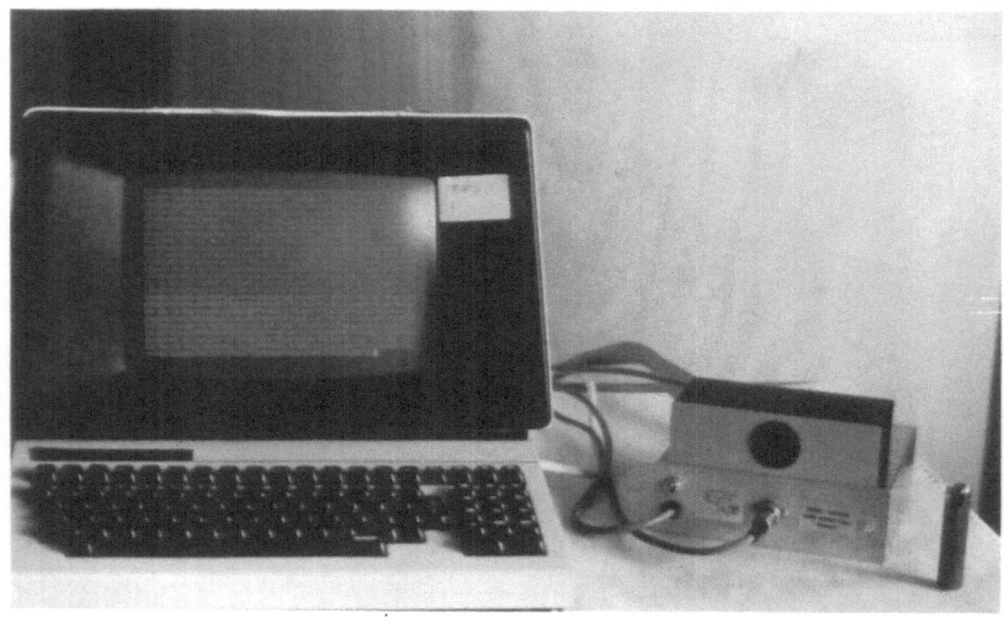

Our SONAIR (Us POLAROID) used in audiovisual applications

Our SONAIR mounted in the middle of a multisensory turntable

LITERATURE REFERENCES :

1- CROSS TAO : Controlling with ultrasonics Machine
   Design March 9. 1978. PP. 90-96

2- MONCHAUD Serge : Contribution à la télémétrie pour robots de troisième
   génération. Doctorat d'état es-Sciences Institut National des Sciences Ap-
   pliquées de Rennes et Université de Rennes I
   IDA 86-4 juillet 1986 (France)

3- MONCHAUD S., MERGUEN H. and LEMAIRE B. A self adapting low cost
   Sonair for use on mobile robots   Octobre 1981
   Sensor Review. pp. 180-183.

4- LOOFBORROW T. How to built a computer controlled robot.
   Hayde Book Compagny Inc.

5- MONCHAUD S., PRAT R. Détection d'obstacles par capteur simple de mesure
   de distance à ultrasons.
   Le Nouvel Automatisme. Janvier-février 1981. PP 61-66.

6-   Ultrasonic Ranging System
   Polaroïd Application note (1983).

7- SMITH M.and COLES. Design of a low cost general purpose robot.
   Proc. of 3rd IJCAI pp. 324-335

8- MONCHAUD S. and PRAT R. Mobile robot turns to colour
   Sensing Sensor Review. July 1982. pp. 134-137.

ULTRASONIC IMAGING FOR INDUSTRIAL SCENE ANALYSIS

H. Urban
Krupp Atlas Elektronik GmbH
Sebaldsbrücker Heerstr. 235
D 2800 Bremen 44   (F.R. Germany)

ABSTRACT

An overview on the complex field of acoustical imaging is presented. In the field of industrial scene analysis mostly optical systems are employed. Acoustical methods are inferior to optical methods with respect to resolution and to data acquisition time. However acoustical systems can measure the distance of objects easily thus giving a three dimensional representation of the field of view. This is difficult to obtain by visual methods and for a limited class of applications acoustical imaging may be complementary or even competitive to optical imaging.

1. INTRODUCTION

Acoustical imaging appears to be a practical means for obtaining visual images of objects located in media with poor or no visibility. The application of acoustical imaging has seemed to be significant mostly in the biomedical field, the non destructive testing (NDT) of mechanical structures or in underwater imaging in the ocean. By means of acoustic waves objects can be examined for internal structure without inflicting the radiation damage which is often a hazard if x-rays are employed. On the other hard acoustic waves often propagate in media which are impervious to other forms of radiation. So far few publications have been known which discuss the application of acoustic imaging in air. However, with the increasing interest in robotics and in the related field of computer vision the three dimensional imaging capability of acoustic systems gains increasing interest. Vision systems designed for robots or for automatic manufacturing have certain limitations in distance measurement and speed of processing

NATO ASI Series, Vol. F52
Sensor Devices and Systems for Robotics
Edited by A. Casals
© Springer-Verlag Berlin Heidelberg 1989

of image data. The use of acoustical image systems has the potential to overcome some of these limitations. Therefore acoustical imaging could either operate with vision methods in a complementary role or for some applications be a competitive sensory system.

In almost all cases the imagery obtained by acoustic radiation is inferior to the imagery we are all familiar with via optical radiation. The reason for this is the longer wavelength of acoustic radiation and related to this the smaller aperture compared to optical systems. As a consequence radial and lateral resolution of acoustical systems is inferior compared to optical systems.

2. ACOUSTICAL IMAGING SYSTEM CONCEPTS

Image Formation

The propagation of acoustic radiation between objects to be visualized and the detector or array of detectors produce a transformation of the spatial information. If an object is illuminated by coherent radiation an echo wave field is generated which consists of a superposition of waves reradiated from the various reflection points of the object. The echo wave field forms a two-dimensional distribution in the system input plane. It can be shown that in the Fresnel region ($D < R < D^2/\lambda$) and in the far field the back transformation of the wave distribution on the system input plane essentially is a Fourier transform, with an additional quadratic phase term for the Fresnel region (see fig. 1).

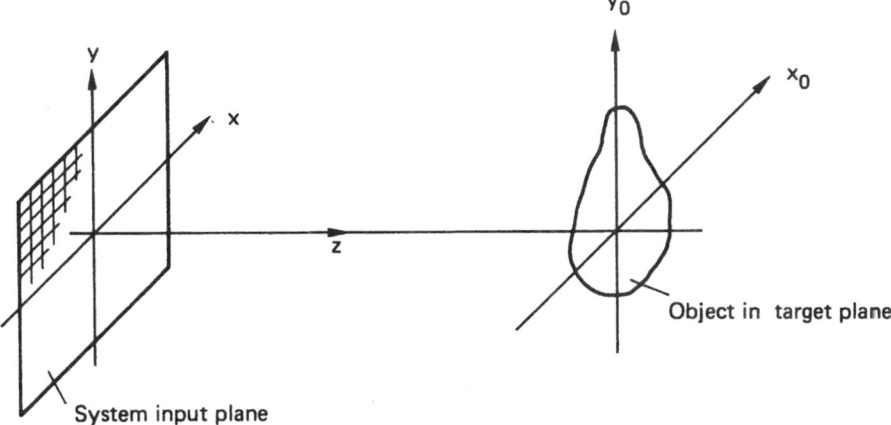

Fig. 1: The object is illumineted by a plane wave and the echo wave field forms
a two-dimensional distribution in the system input plane.

The Fourier transformation of a monochromatic wave field yields the image
distribution:

$$I(x_o, y_o) = A_o \exp\left(j \frac{k}{2z}(x_c^2 + y_o^2)\right) \int\int_{-\infty}^{+\infty} H(x,y) \exp\left(j \frac{k}{2z}(x^2+y^2)\right) \cdot \exp\left(-j\frac{k}{z}(xx_o+yy_o)\right) dxdy \quad (1)$$

H(x,y) is the field distribution in the system input plane

$I(x_o,y_o)$ is the distribution in the image plane, i.e. the object function
in the target plane

$k = \frac{2\pi}{\lambda}$ the wave vector

z distance between object and input plane.

The term A_o is made up of a constant and a term which accounts for spherical spreading loss. A simpler notation of the above integral gives

$$I(x_o, y_o) = C \mathcal{F}\left\{\exp\left(jk\left(\frac{x^2+y^2}{2z}\right)\right) H(x,y)\right\} \quad (2)$$

This is the basic mechanism on which many monochromatic imaging systems are designed. Acoustic imaging consists of four basic processes which are applied to the acoustical wave field. These are:

(1) conversion of acoustical energy to electrical energy
(2) spatial processing to obtain an image from the wave field distribution
(3) detection of high frequency signals and conversion to a down modulated display signal
(4) display of signals

The different approaches for technical systems differ mainly in the order in which the first three basic processes are applied.

Beamformed Acoustic Imaging

A beamformer is a device which does the spatial processing and transforms the transducer signals into one or more beams. Each single beam can be considered as a spatial filter which probes the field of view in one single direction. Beamformers can be built for transmission or reception. In reception systems one single beam or a whole fan of beams can be formed.

In single beam systems the image function is accomplished by scanning the field of view and display of beam energy in a proportionally scaled form. The field of view can be scanned in one, two or three dimensions. The image function is called then A-scan, B-scan or C-scan respectively. Fig. 2 shows a B-scan system and the related image display.

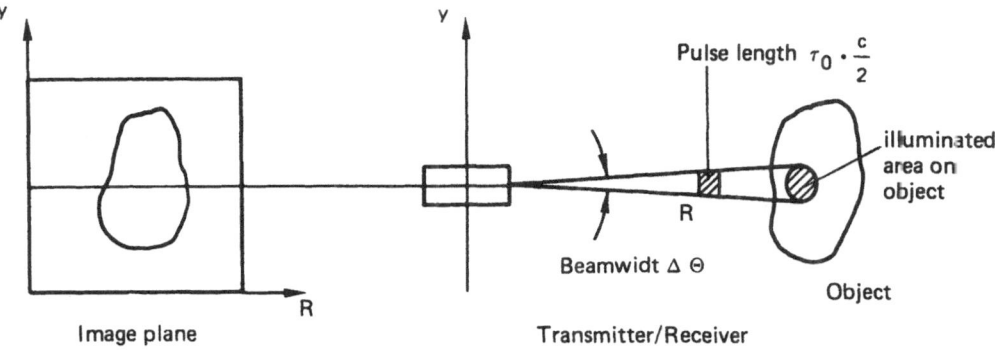

Fig. 2: B-scan system, the transitter/receiver scans the object while moving on the y-axis. An intensity display is generated on the image plane.

The principal features of a scanning system are briefly referred to as follows. The travel time τ of the echo gives the distance R

$$R = c\, \frac{\tau}{2} \qquad\qquad c = \text{sound speed} \qquad (3)$$

Range resolution depends on the pulse length τ_o, or more general on the bandwidth B of the transmitted pulse.

$$R = \frac{c\,\tau_o}{2} \;=\; c\, \frac{1}{2B} \qquad\qquad (4)$$

The lateral resolution of the transducer system is

$$\delta = \frac{\lambda}{D}\, R \qquad\qquad D = \text{dimension of the transducer array}$$

$$\Delta\Theta = 2\, \text{arc sin}\, \frac{\lambda}{2D} \qquad\qquad \lambda = \text{wave length}. \qquad (5)$$

The image formation is a sequential process. Correct spacial sampling of the field of view requires overlapping of sound beams at the 3 dB points. The number of scans therefore depends on the size of the field of view which has to be divided by the lateral or the angular resolution.

The acquisition time of an image is τ_{max} N where N is the number of scans and τ_{max} the echo travel time at maximum Range R_{max}.
To obtain a three dimensional image the transmitter or the receiver system has to be positioned on a two dimensional surface in order to scan the field of view which is a volume then.

The sequential image formation process requires a stable environment, during the scanning operation this may become a problem if objects in the field of view do move or if the platform itself carrying the system is in motion. This imposes restrictions on the application of scanning systems.

Holographic Acoustic Imaging

In holographic acoustic imaging the wave field distribution on the system input plane is sampled by a transducer array and is immediately converted to a stable set of dc values called a hologram. As in optical holography a reference signal is employed to demodulate the narrow band signal at the tranducer outputs.

In optical holography however, the reference signal is transmitted through the medium and the superposition of reference signal and wave field distribution gives the desired distribution of intensity on the system input plane.

In acoustical holography the reference signal essentially is a demodulation process after reception where amplitude and phase of the wave field are maintained. Either a real signal $A\cos \emptyset$ is obtained or a complex signal $A(\cos \emptyset + j\sin \emptyset) = A\exp(j\emptyset)$ is derived from the output of each transducer element by the demodulation process. Hence the acoustical hologram is an array of complex values $A_{n,m} \exp(j \emptyset_{n,m})$ evaluated at the positions of the transducers in the system input plane. These values are subject to image reconstruction.

The image reconstruction can be accomplished either by optical processing methods or by processing of the aperture data on a digital computer. Optical methods can be applied whenever it is possible or convenient to copy the acoustical hologram onto a transparent surface. Further processing is then identical to optical hologram reconstruction.

Reconstruction on a digital computer implies that prior to a spatial Fourier transform on the aperture data a phase correction compensates for wavefront curvature and is applied in a quadratic form in the Fresnel region ($D < R < D^2/\lambda$). It is called the Fresnel focus factor since it focuses the image at the desired range.

Digital computer processing can be very useful when all signal processing is accomplished in one single processor structure, thus ensuring uniformity in hardware and software.

Synthetic Aperture Imaging

The lateral resolution of an acoustic image can only be improved if the size of the receiver aperture (i.e. the systems input plane) is enlarged. One way to increase the aperture without filling it with all the required receiver elements is to synthesize the aperture either by motion of the object, the receiver, the transmitter or any of these. As the syntheses is a sequential process the scene to be analysed has to be sufficiently stable to allow for the necessary coherence of the wave field pattern.

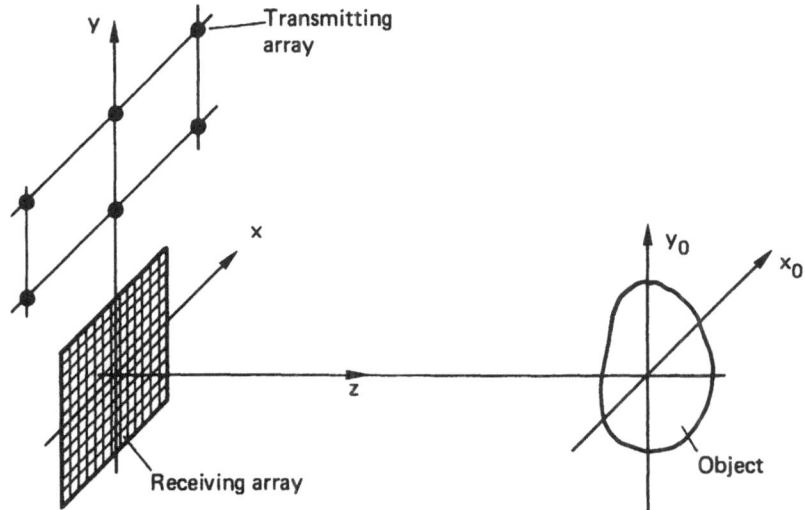

Fig. 3:   Synthetic aperture geometry using an array of transmitting elements and a receiving array of small size. /4/

From all the different approaches which have been realized or proposed
in acoustic imaging a solution is discussed here which seems to be particular attractive for robotic application. As shown in Fig. 3 the system
consists of a receiver array and several transducers which are arranged
in an array themselves. The synthetic aperture is accomplished when the
transmitters sequentially send off pulses and insonify the object from
different angles.
The sub-holograms have to be phase corrected before the combination of
all of them is spatially processed.

Tomographic Imaging

Computer tomography is often referred to in the context of x-ray diagnosis
to obtain tomographic images of cross sections of the human body. This
technique is now used for medical diagnosis in many hospitals.
X-rays are not the only kind of radiation for which tomographic processing
can be applied. Besides others also acoustic radiation can be used. As in
X-ray tomography transmission methods are applied in acoustic as well. A
special approach however is the acoustic reflection tomography where transmitter and receiver are essentially at the same position. This method is
explained in some more details.

As shown in Fig. 4 a transmitter/receiver insonifies the object and receives
a series of echos from the various reflective regions of the object. As
the beam width of the transducer is broad enough to insonify the whole object area all the reflective regions in the shaded area contribute to the
total echo energy received in one instant of time. Assume the transmitter/
receiver is rotating about the stationary body (alternatively the body is
rotating at a constant angular velocity about some centre of rotation) the
received echos exhibit a doppler which changes from high to low as the reflective regions rotate and change distance to the receiver. The maximal
doppler shift obtained depends on the distance of the reflective region
from the centre of rotation. In a computer doppler and range signals are
coherently evaluated and an image of the object is obtained which does not
contain shadow areas. The acoustic reflection tomography is a special mani-
festation of synthetic aperture technique described above, it requires

phase coherent processing of signals and as a consequence very precise control of the relative position of object and transmitter/receiver equipment

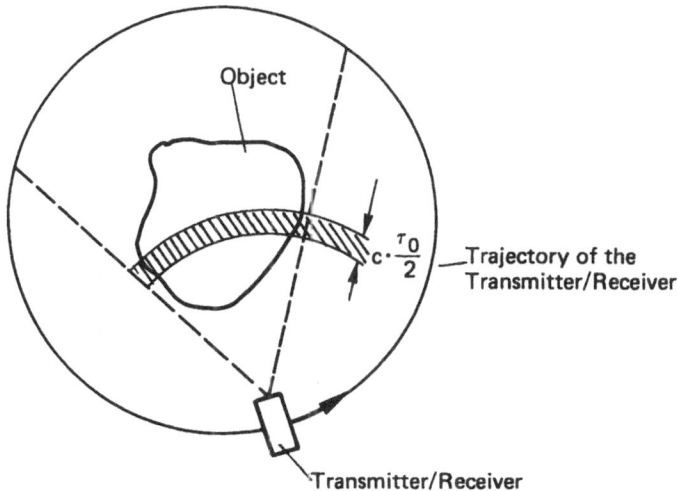

Fig. 4:   Reflection tomography, the transmitter/receiver rotates
           about the stationary object.

An uncoherent imaging technique employing a moving platform has been described by H.P. Movare and A. Elfes /6/. This technique was specially developed for an autonomous mobile robot. The area to be visualized is projected onto a rasterized two dimensional map, where somewhere occupied and mostly empty areas are finally displayed. The system is mounted on a mobile platform. It has horizontally wide angle transducer/transmitter and insonifies the area with short pulses. Range measurements of objects are projected onto the map. One single reflecting object occupies a ring-shaped area related to the illuminated area of the short pulse. Measurements taken from different points of view result in overlappings which are systematically integrated in the map. The map improves as more readings are taken. Cross sections of the ring-shaped areas are integrated and represent objects, non-overlapping sections show the empty space between objects.

The method is implemented as a probability profile of the area. In principle measurements of cross sections increase the probability of the related map elements. After a number of readings at different positions the probability distribution yields a useful estimate of the distribution of reflecting objects in the area. Compared to coherent signal processing methods the requirements on the sensor positioning system are much less. The accuracy is approximately a range cell whereas in coherent processing the position should be known to a fraction of the wave length.

3. PHYSICAL CONSTRAINTS AND DESIGN PARAMETERS

Acoustical radiation consists of mechanical waves which propagate in an elastic medium. Hence the elastic properties of the medium and the objects determine to a large extent the characteristics of the acoustic image.

Most systems use narrow bandwidth acoustic signals. These have the undesirable property of speckle. Speckle is caused by the interference of signals from different reflective regions on the object. They show up in the image as bright and dark spots and cannot be related to details on the object being imaged. As a consequence they are difficult to interpret and may cause confusion. To overcome speckle patterns broad band signals may be employed or the object may be illuminted from more than one direction thus causing the speckle patterns to change. Speckle patterns are also known in optical holography with lasers.

A characteristic of objects to be of interest is that they can be partly transparent. Sound penetrates material and may be reflected from inner structures such causing multiple echos. This effect exhibits information about the inner structure of objects which hardly can be obtained otherwise however it may be difficult to interpret these structures.
In addition most objects to be visualized have smooth surfaces with respect to the wavelength of sound. They produce specular or mirror-like reflections. The contrast of the image becomes very high, details of the object however remain hidden. Diffused images would be obtained if the object had a rough surface with respect to the wavelength. Specular reflections can be reduced if again the object is illuminated from different directions and many specular returns be integrated to form the final image.
Parameters concerning the performance of a sound ranging and imaging system

in global terms are:

- transmission loss (TL)
- source level of the projector (SL)
- target strength (i.e. the reflectivity of objects (TS) and
- noise level at the receiver input (NL).

The transmission loss is made up of the spherical spreading loss ($1/_R2$) and the attenuation. The attenuation of sound in air is due to absorption and depends mainly on the relative humidity, the temperature and the atmospheric pressure. In Fig. 5 the absorption coefficient versus frequency for various values of relative humidity and temperature is shown. At a frequency of 100 kHz for example the attenuation can vary between 1 dB/m and 4 dB/m. In logarithmic notation the transmission loss becomes

$$TL = 20 \lg (R/R_o) + R \cdot A \tag{6}$$

R_o is the reference distance and usually set to 1 m.

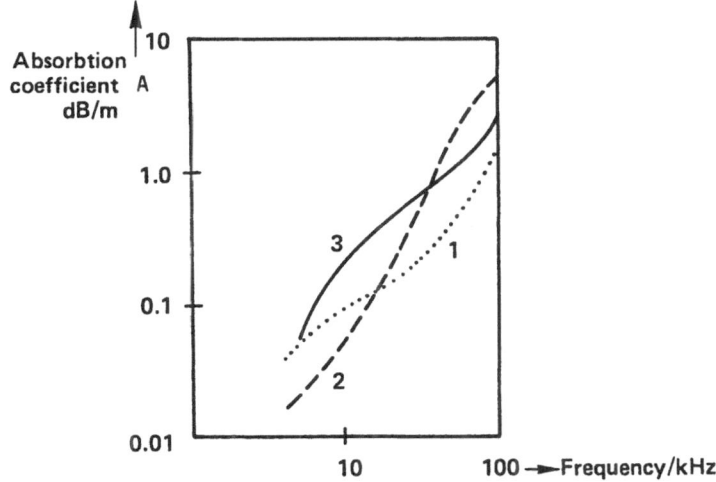

Fig. 5: Sound absorbtion in air at different temperatures and humidities /9/
   1: 266.5° K and 69.6% relative humidity
   2: 310.9° K and 50.0% relative humidity
   3: 288.7° K and 30.7% relative humidity

In the receiver of a sound ranging system some means have to be provided to compensate for the transmission loss, this is the automatic gain control device (AGC).

The radiated power of the projector is called the source level SL. It is defined at 1 m distance from the projector and is related in a simple way to the acoustic power it radiates and to its directivity index (i.e. the directionality of the projector). For a non-directional projector the radiated acoustical power output at distance r is:

$$P_r = 4 \pi r^2 \frac{p r^2}{\rho c} \tag{7}$$

p_r is the rms sound pressure, ρ the density of the medium and c the velocity of sound. Inserting the appropriate value for ρ and c (i.e. ρ = 1,189 $\frac{kg}{m^3}$ c = 343 $\frac{m}{s}$) and evaluating the power at r = 1 m distance gives

$$P_1 = 4 \pi \ 1 \frac{P_1^2}{c} = 0,0308 \left[\frac{p1}{\frac{N}{m^2}}\right]^2 \quad \text{Watt} \tag{8}$$

converting to decibels and taking 10 lg $\frac{P_1^2}{p_0^2}$ as the source level SL, with p_0 = 20 μ P_a as the reference sound level referring to a reference intensity $I_0 = 10^{-12}$ Watt/m^2, we have SL in decibel

$$SL = 109,1 + 10 \ \lg \left(\frac{P}{1 \, \text{Watt}}\right) \tag{9}$$

If the projector is directional with transmitting directivity DI we obtain

$$SL = 109,1 + 10 \ \lg \left(\frac{P}{1 \, \text{Watt}}\right) + DI \tag{10}$$

In sound ranging systems the parameter target strength TS is used as a measure of reflectivity. It refers to the intensity of sound returned at 1 m distance in relation to the incident intensity from a distant source. In logarithmic notation we have

$$TS = 10 \ \lg \left.\frac{Ir}{Ii}\right|_{r = 1 \, m} \tag{11}$$

The imaging quality can be severely effected by noise. In particular the noise level might become an important quantity if the acoustic imaging system operates in an industrial environment. Background noise is usually assumed to be isotropic and hence will be reduced by the directivity index of the receiving array. In practical applications however noise may be highly directional it has to be measured and the radiated source level has

to be adjusted if noise turns out being a limiting factor. The noise level
is defined in decibel

$$NL = 10 \lg \frac{I_{noise}}{I_o} \tag{12}$$

(The reference intensity in air I_o is taken as 10^{-12} Watt/m^2)

For sound ranging and imaging systems the active-sonar equation is a suitable mean to obtain an overall estimate of the system's performance. The
quantities used are in decibels as explained above.
The noise level is here assumed to be isotropic and hence is reduced by the
directivity of the receiving array. The echo-to-noise-ratio at the receiver
array gives the sonar equation as

$$10 \lg(\frac{S}{N}) = SL - 2 TL + TS - (NL - DI) \tag{13}$$

The sound level SL is reduced by the transmission loss TL whilst the sound
waves are travelling to the object. On reflection or scattering by the object of target strength TS, the reflected or back scattered intensity will
be reduced by TL. In travelling back towards the source this level is again
attenuated by the transmission loss.

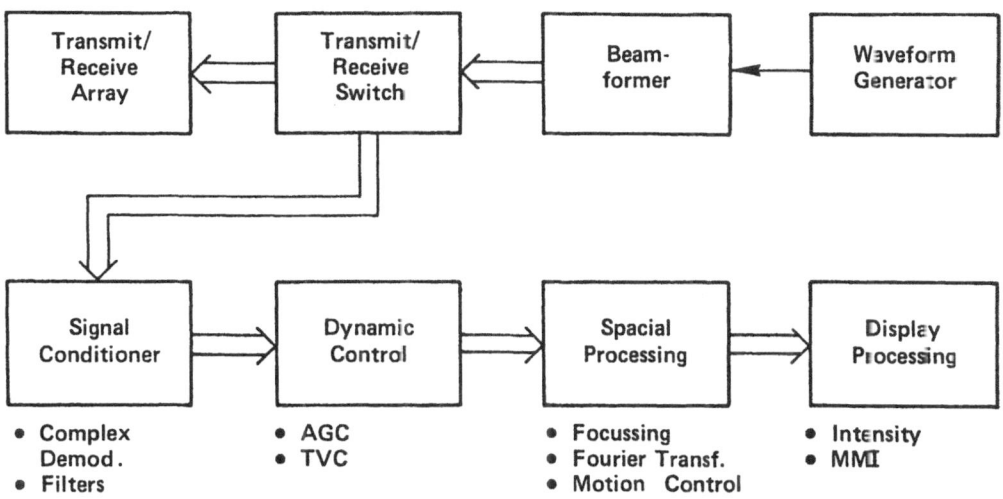

Fig. 6: Generic acoustic imaging system.

In Fig. 6 a generic acoustic imaging system with its major functional elements is shown. Transmission/reception can be accomplished either by the same transducer elements or by two seperate arrays. The transmit/receive switch is unnecessary in the latter case. The specific layout of the system depends on the application, mainly on the parameters as measuring range, lateral and radial resolution, the data acquisition time and the display facilities.

4. SUMMARY

A brief overview of the complex field of acoustic imaging was presented. Today application of acoustic imaging systems is mainly in the biomedical field, the non-destructive testing and in underwater. So far little work has been done in acoustic imaging in air. For robot application a combination of acoustic and optic imaging seems to be an attractive approach. Data fusion of image data of both systems has the potential to identify objects faster and more reliable.

Principal shortcomings of acoustical imaging systems have to be taken care of. Acoustic images are inferior to optical images in many respects, on the other side they give a three dimensional representation of the field of view which is difficult to obtain by visual methods.

REFERENCES

/1/ L.E. Kinsler, A.R. Frey, "Fundamentals of Acoustics", John Wiley & Sons Inc., Library of Congress Catalog No. 62-16151.

/2/ Ivar Veit, "Technische Akustik", Vogel-Verlag, Würzburg, ISBN3-8023-0063-7.

/3/ Jerry L. Sutton, "Underwater Acoustic Imaging", in Proc. IEEE, Vol. 67, No. 4, April 1979.

/4/ Patrick N. Keating, Takeo Sawatari, Gene Zilinskas, "Signal Processing in Acoustic Imaging" in Proc. IEEE, vol. 67, No. 4, April 1979.

/5/ Rolf K. Mueller, Mostafa Kaveh, Glen Wade, "Reconstructive Tomography and Applications to Ultrasonics", in Proc. IEEE, vol. 67, No. 4, April 1979.

/6/ Hans P. Roravec, Alberto Elfes, "High Resolution Maps from Wide Angle
    Sonar", in IEEE Internatinal Conference on Robotic and Automation,
    March 25-28, 1985, IEEE Computer Society Order No. 615.

/7/ E. Hundt, "Digitale Filterung für die Ultraschall-Abbildung", in NTZ
    Archiv Bd. 3 (1981) H. 9.

/8/ L. Kay, "Airborne Ultrasonic Imaging of a Robot Work Space", in Sensor
    Review, January 1985.

/9/ H.E. Bass, F.D. Shields, "Absorption of Sound in Air: High-Frequency
    Measurements", in J. Acoust. Soc. Am., Vol. 62, No. 3, Sept. 77.

ADAPTATIVE ULTRASONIC RANGE-FINDER FOR ROBOTICS

J.M. Martín, R. Ceres, J. No, L. Calderón
Instituto de Automática Industrial (CSIC)
Desvio Km 22.800, N-III, La Poveda
28500-ARGANDA DEL REY   Madrid   (Spain)

INTRODUCTION

The Instituto de Automática Industrial has carried cut re-
searches in the field of sensors since its very beginning.
During the last five years, the work on sensors has been, and
still is, going on within the general program of the Instituto
de Automática Industrial about Flexible Manufacturing
Integrated Systems.In this field of activities the research is
focused on two different topics: machining processes cutting
condition appraisement and industrial robots perception. On
the first one, and related with other works performed on
machine-tools at the Instituto, the pursued objectives are the
continuous appraisal of the cutting and tool wear down
conditions and its breakage prediction. Attainment of these
objectives would allow disposing of unmanned machining processes by means of adaptive control of machining parameters,
dynamic compensation of machining and replacement of tool at
the adequate moment.

As far as industrial robots perception systems, two
parallel complementary research subjects are under development
and are oriented towards detection, tracking and
discrimination of near by objects and to determine the
robot-object interaction efforts. A linear movement two
fingered jaw has been developed. It is activated by a step
motor and endowed with sensible structures with strain gauges

NATO ASI Series, Vol. F52
Sensor Devices and Systems for Robotics
Edited by A. Casals
© Springer-Verlag Berlin Heidelberg 1989

for measuring force components along the three coordinate axes. A computer at higher level will perform in real-time the effort spatial feedback operations in connection with the robot controller. This way it is expected that the problem concerning object grasping might be solved, and even, it may also be possible that one of the most important actual problems in the field of robotics and factories automation, as is the intelligent assembling of parts, which will allow limiting the precision of mechanical systems and to extend the operational universe.

Concerning near by object detection, we have used ultrasonic techniques based in measurement of fly time using the pulse-echo procedure. Two systems have been attached to the terminal organs of the robots, one on an experimental robot and the other on an EISA-IAI robot. These sensors incorporate specific processors that give it a determined adaptability to the environment, changing operational parameters (energy issued, prf, discriminating level, thermal compensation), as a function of the surrounding conditions at the moment, obtaining in this way greater trustworthiness and operativeness under industrial environments. Using ultrasonic techniques, other systems are being developed at present for objects characterization and speed measurement.

TRANSDUCER AND MEASUREMENT PRINCIPLE

The selection of the transducer must be related to the specific application. Within the technical possibilities of the generation and detection of ultrasound waves, the piezoelectricity offers better performance and it is more operational than other alternatives such as changes in the capacity of the transducer. Low acoustic impedance of the air, being our normal operating environment, imposes other conditions, solved by using piezoelectrical flexible ceramics, which require low voltage and pressure to deform, providing in this way a relatively high electroacoustic efficiency.

The selected transducers are Murata MA-40, with 40 KHz of

resonance frequency, which use pulse-echo as a measurement principle. They measure the time interval between wave emission and reception of the corresponding echo as reflected by the measured object.

The system uses two transducers. One as transmitter and the other as receiver to increase the measurement range as opposed to a single transducer acting alternately as transmitter and receiver. In this situation the ceramics inertia creates a dead time immediately after the emission where a measurement cannot be made. Thus, our measurement field extends from 20 to 2000 millimeters.

The emission lobe, (representation of emitted energy as a function of the angle),in these transducers is rather large, having a main lobe of 90^e and secondary lobes of $130°$. By installing a specifically designed horn, the secondary lobes can be eliminated and the directivity of the main lobe can be increased by up to 20^e. These changes allow for a higher angular resolution on the measurements.

Both transducers are placed in pairs at the horn block at a distance between centers of 28 millimeters. Each transducer has a 16 millimeter diameter and it is imbedded 26 millimeters in the structure.

GENERAL SYSTEM ARCHITECTURE

Fig. 1 shows a general diagram of the sensor system. An 8 bit microprocessor (Intel 8085 at 6 MHz), a minimum of 256 bytes RAM memory and an EPROM (2 Kbytes) memory make up a central microcomputer to control the peripheral units: temperature measurement device, counters for fly time measurement and emission pulse generation, external RS-232 asynchronous communications port and transmitting and receiving circuits.

RAM memory is mainly used to process the received data until a measurement is achieved which will be sent through the communications circuits to other external devices, in particular to the robot controller.

The EPROM memory stores the program which controls the sensor system and processes the obtained information.

The associated counter are used for different functions:

- to precisely determine the emission pulse width
- to measure the fly time with precision (the equivalent of 0.1 millimeters).
- to adjust the communication speed through the RS-232 port.

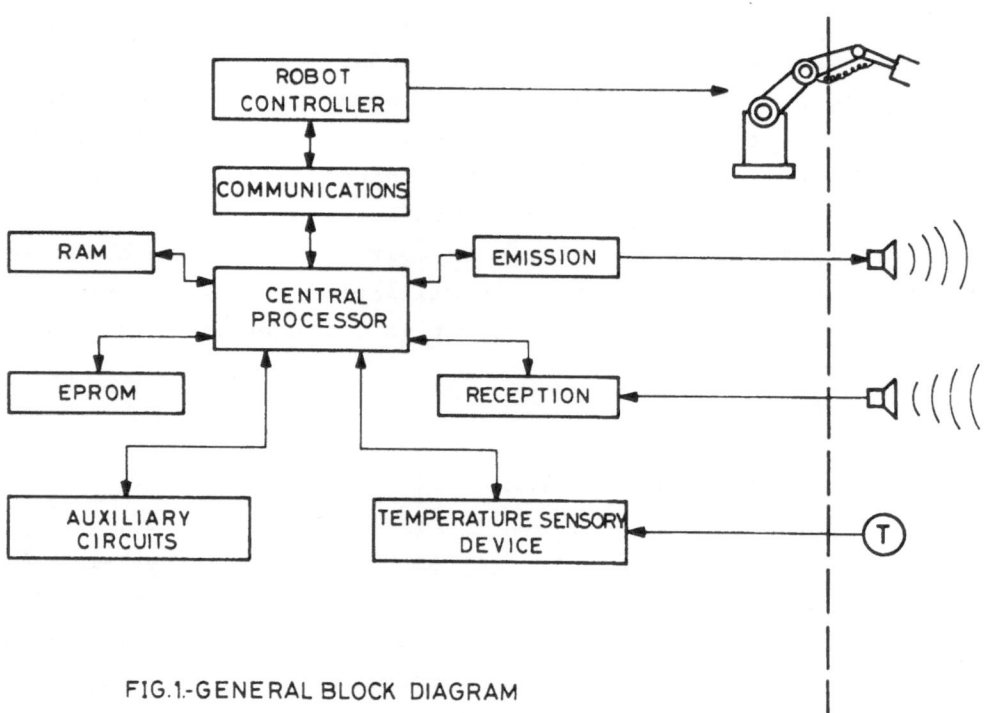

FIG.1.-GENERAL BLOCK DIAGRAM

To accomplish all the above, three 8-bit I/0 ports are used. The first one is connected to a D/A converter at the receiving circuit and it is used to generate the dynamic detection threshold. The second one is used to read from some microswitches the type of communications desired for the RS-232 port (speed, parity,... etc.). The third one (output), is connected to a set of visual indicators which show the operating status as well as diagnostics information.

Emission-Reception. Serial output from the microprocessor conveniently amplified in both voltage and current, is used to activate the emitting ceramics (24 v). Emission pulse with determination is done by a program with an auxiliary counter. The signal received from the transducer is amplified in two successive steps with an approximate gain of 10000. Afterwards the signal goes through a 40 KHz (Q=10) band-pass filter. The band in this filter has been set in such a way that suppressing a maximum of noise, the signal is not affected by the little differences in resonance frequency from some ceramics. Output from the filter is then compared with the A/D converter threshold controlled by the microprocessor and the result is passed on to a monostable and from there to a flip-flop. This avoids potential flickering in a low stability signal.

Temperature. A thermal solid state sensor (LM 335) detects the environment temperature, generating a voltage that is later amplified and converted into frequency by a specific circuit. This temperature is directly read by the microprocessor avoiding the use of the A/D converter. A table in the program allows us to convert the measured frequency into environment temperature with a 1 centigrade degree precision which is sufficient for our application.

Communications. This function is essentially performed by a USART which constitutes the nucleus of the circuit, thus allowing to translate the information obtained by the sensor system into the RS-232 norm, in such a way that this information can be read by any equipment which accepts this norm. In our case it is connected to our EI-25 robot controller, with the purpose of letting the robot detect, locate and grasp objects.

OPERATION MODE

If we observe the animal world, we realize that sensors, apart from the transducing systems (stimulus conversion into a signal which can be interpreted by the brain), exist together with an intermediate processing system which transforms the signal into understandable signs, obtaining maximum benefit from the detection system.

This function is performed in this sensor by the programming system that allows the sensor to adapt to the environment conditions, to improve the precision and reach a high reliability.

The information from the sensor system is a message or a measurement of the distance. These messages may be:

- No object detection within a fixed range
- Excessive dispersion between different measurements.
- Failure of the emission or reception systems
- Excessive environmental, electrical or acoustical noise to measure correctly.
- Environmental temperature out of range.

Fig. 2 shows a general flowchart of the program that controls sensor behavior. It begins with some tests to ensure proper operation of electronic components. When these tests are passed correctly, it goes to the next phase. Once the electronics is configured, the main operation parameters are automatically set allowing for working conditions to be predetermined and then the system tries to perform the first measurement.

If this measurement does not show any object within the detection range, then the system performs a control of correct operation of the emitting and receiving circuits (as well as of the associated transducers) taking advantage of the direct interaction through side lobes between transmitter and receiver. This control is very important since, if absence of

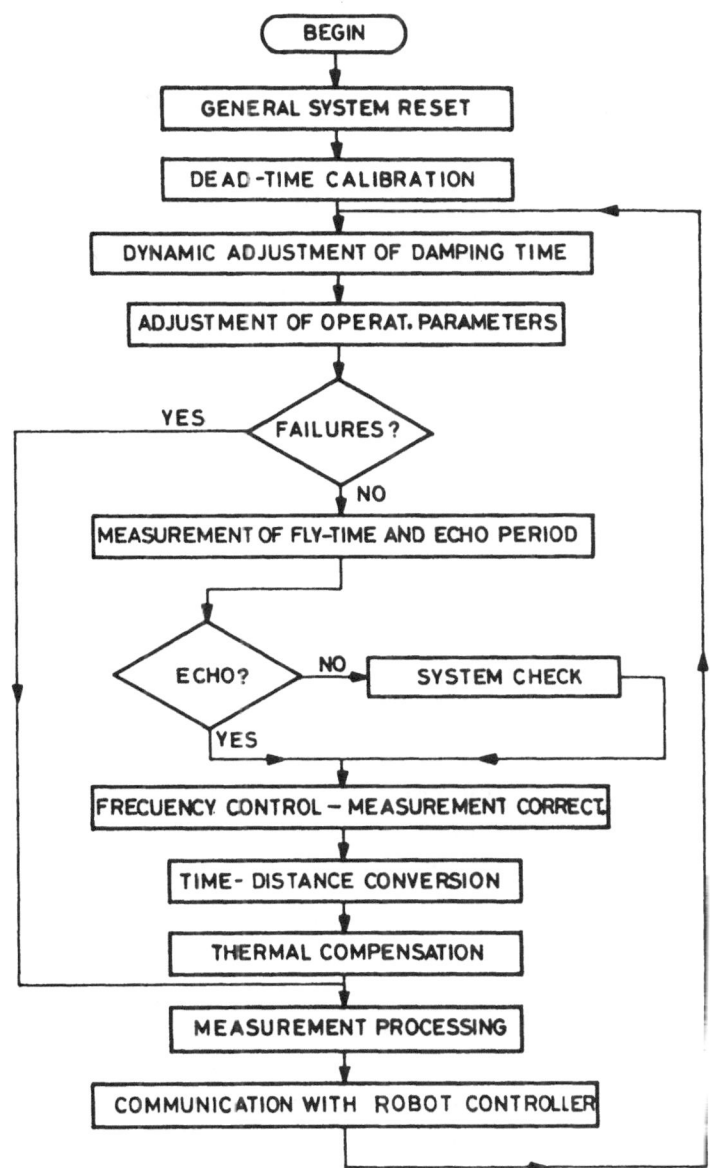

FIG. 2-GENERAL FLOW CHART

objects is erroneously indicated, the consequences for the robot system can be catastrophic.

Next, the environment temperature is measured and fly time measurement is corrected according to an empiric function, previously obtained, between environment temperature and the speed measurement performed. This correction, not only corrects the sound speed changes with temperature but it also balances out other possible variations in the fly time measurement circuits.

After the measurement is performed, other more distant echoes or echoes from the same object might arrive by indirect, and consequently longer paths. This prevents performing immediate measurements. On the other hand, if we wait as much time as in the worst case for these second echoes, the sampling frequency would decrease considerably. To solve this problem, the system remains listening during a time that corresponds to the largest reach. If during this time nothing is detected, then the system goes on to take the next measurement. Otherwise, the listening process is repeated. If this process does not end within a given time-frame, an excessive environment noise message is sent and it tries again. This technique provides the largest number of samples, ranging from 10 to 100 samples/second depending on the working environment.

SENSOR-ROBOT COUPLING AND EXPLORATION STRATEGIES

The sensor has been installed in a robot (EI-25) to perform and solve, in a near industrial environment, tasks such as search, location, object grasping and obstacle avoiding. This robot is anthropomorphous , having 5 degrees of freedom, a reaching range of 2 meters approximately and it is able to load a maximum of 25 Kg.

Although for many applications it will be sufficient that the transducers be placed in a fixed location external to the robot, it is evident that much more information could be obtained if they were located on the robot itself, using its moving ability to observe different fields or the same field from different positions. This robot-sensor symbiosis (the robot helps the sensor to provide more information) has been chosen because it seems to have more possibilities and it is more interesting from an experimental point of view. On the other hand, although the gripper of the robot its an interesting place to put the transducers, we choose the wrist (see fig. 3) to make it more universal, since the gripper are often interchanged in order to adapt the robot for different tasks.

After various trials, it was decided to use a self-calibrating system to determine the real position of the transducers with respect to the working coordinates of the robot. This method proves to be the most simple, universal and precise.

Tasks control and information processing is performed by the robots central controller, a 8086 microprocessor and an associated mathematical coprocessor 8087.

In our first stage we have worked on the location of cylindrical objects of different heights and a 4.5 cm diameter. After looking at different exploration strategies, we decided to use angular scanning since it is the simplest for the robot. It only requires the movement of one motor then computations are easier to make and position determination is more precise. On the other hand this scanning method provides a decrease of the apparent angular enlargement of the objects due to the lack of directivity of the transducers.

In the first test the simplest method was used performing a complete angular exploration of the working field assuming that the object was placed at the angle where the closest measurement is taken. The tests performed show an error of 5 degrees for objects placed at approximately 70 cm., which seemed unacceptable for our aims. In view of the results, other methods were tried which had been simulated in a computer using experimental data which had been previously filed. Among these methods it was decided to use the one to be described, which represents a good trade-off between computational simplicity and precision of results.

The method consists of searching for the angle which corresponds to the closest measured distance and storing a set of samples around this angle in such a way that they do not exceed a certain value (chosen 6 mm., although this value is not absolutely critical). In order to give more weight to closer distances in the angle estimation process, we found the maximum value of the distance (DM) in this interval and we estimated the angle as follows:

$$\alpha_e = \frac{\Sigma (DM-d_i)\ \alpha_i}{\Sigma (DM-d_i)} = \frac{DM.\Sigma\alpha_i - \Sigma\alpha_i d_i}{n.DM - \Sigma d_i}$$

where n is the number of measurements, d_i is the value of the distances within the interval and $_i$ is the corresponding angle.

Fig. 4a   Fig. 4b

This method reduces the error to a maximum of. 0.4 degrees
for different samples which allow the robot to grasp the
objects within any problem as the gripper width absorbs this
imprecision. These samples are taken every .0.5 degrees,, in
order to allow scanning at the highest possible speed
(approximately 3 sec. for 90 degrees scanned) and it can be
assumed that the precision can be improved at lower speeds.
Fig 4 shows a point distribution as it is received by the
robot close to the real position of a cylinder (Fig 4a) and a
prism (Fig 4b). In fig. 5 the values received by the robot

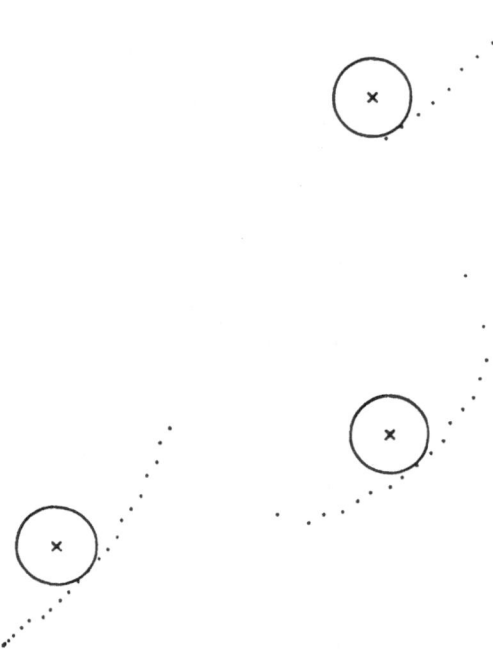

Fig. 5

when there are several cylinders in the working field can be
appreciated. This does not mean a deterioration of the method
as long as the distance between the cylinders is sufficient
for the gripper to penetrate and grasp them.

CONCLUSION

With the system just described and its installation in a
robot, the solution of the initially stated operations of
detection, location and grasping has been accomplished. In
this manner the sensor system, operating within a 2m range,
is able to measure with a 1mm precision, over objects of a
minimum effective surface of 1 cm^2, at a sampling speed of
between 10 and 100 sample/sec. Likewise, simple objects
(cylinders and prisms) discrimination operations have been
made in an experimental way, based on the measurement of some
of their geometric parameters (height and diameter), by means
of spatial explorations. On the other hand, this sensor has
been used to perform laboratory experiments of advancing and
returning the claws in mazes as well as the tracking of part
joints by detection of discontinuities on flat surfaces.
Currently, the described sensor has been upgraded with an
integrated module to measure speed using a phase shifting
technique. Finally, object characterization (size, curvature,
surface type) is being studied starting with the parametric
analysis of the previously digitized echo signal with
ultrasound waves of 40 and 200 Khz.
All this is complemented with other studies being made at
the Instituto de Automática Industrial, including design,
development and implementation of other sensor systems such as
force/torque and vision, together with some other AI works to
provide the robot with the ability to perform more complex
tasks in a more efficient manner.

REFERENCES

CALDERON, L. -"Sensor ultrasónico adaptativo de medida de distancias. Aplicación en el campo de la Robótica", Univ. Complutense Madrid, 1984.

CERES, R.; MARTIN, J.M.; CALDERON,L.- "Microprocessors in Measurements Processes an Application to a Range Finder". Proceedings of the ISMM International Symposium. June 1985. ISBM 7488-121-8.

K. BROWN,MICHAEL.- "Locating Object Surfaces with an Ultrasonic Ranging Sensor". LH2152-7/85/-1985-IEEE.

ESTOCHEN, ERICH,L.- "Application of acoustic sensors to robotic seam tracking". IEEE Transactions on Industrial Electronics, vol 31 n 3 , 1984.

BROWN,M.K. - "The Extraction of curved surface features with generic range sensors". The International Journal of Robotics Research. vol 5, n 1, Spring, 1986.

IV. OPTICAL SENSORS

TH 7864
AREA ARRAY CHARGE-COUPLED DEVICE (CCD) IMAGE SENSOR WITH BUILT-IN ANTIBLOOMING DEVICE

D. Herault, G. Boucharlat
Thomson-CSF Division Tubes Electroniques
38 rue Vauthier BP 305
92102 BOULOGNE-BILLANCOURT CEDEX   (France)

INTRODUCTION

The TH 7864 is a 2/3" format area array CCD image sensor that incorporates an antiblooming system. It delivers 576 lines of 550 pixels in the CCIR TV standard.

It is fabricated in three-level polysilicon buried-channel CCD technology. The p-doped substrate (3×10^{15} cm^{-3}) consists of an epitaxial layer so as to achieve good resolution in the near infrared.

Reflecting the company's latest improvements, the sensor achieves a dynamic range of 4000:1, with 30 dB signal/noise at 40 milli-lux sensor illumination. The contrast transfer function is about 80 % at 412 TV lines when using a BG38 filter.

Its built-in antiblooming protection is effective at up to 1000 x saturation exposure, enabling it to pick-up scenes presenting a wide illumination dynamic range without risk of image deterioration. The antiblooming structure also provides an electronic shutter function when using an integration time shorter than the readout time.

The TH 7864 shares all the advantages associated with CCDs : small size, high reliability, low-voltage drive and low light level operation without light bias. Moreover, its high overillumination resistance gives good adaptability to wide scene dynamics.

ORGANIZATION AND OPERATION

The TH 7864 has a frame-transfer organization comprising separate image and memory zones, a readout register and an output stage (see figure 1).

The geometrical characteristics are as follows :

- Image zone dimensions : 8.8 mm (H) x 6.6 mm (V)
- Pixel pitch : 16 μm (H) x 23 μm (V)
- Chip dimensions : 9900 μm (H) x 15,200 μm (V)

NATO ASI Series, Vol. F52
Sensor Devices and Systems for Robotics
Edited by A. Casals
© Springer-Verlag Berlin Heidelberg 1989

The sensor operates in the following manner :

. During the integration period, i.e. T_i = 20 ms in CCIR TV standard, the scene is sensed by the image zone.

. The charges accumulated in the image zone are subsequently transferred, during the field blanking period, into the memory zone.

. Integration then resumes in the image zone to gather the next field while the memory zone is emptied line by line into the readout register to be read by the output stage.

THE PHOTOSITE

Each photosite is a MOS element equipped with a horizontal antiblooming device formed by a gate (Φ_A) and a drain V_A (see figure 2).

Charge Integration

The image-forming photons cross the gate insulation structure before absorption in the substrate, where they create electron-hole pairs. The holes are evacuated by the substrate while the electrons are drawn into the potential wells formed by the MOS capacitors when their Φ_p gates are positively biased.

The Φ_p phases are inverted at alternate fields during the integration period (see figure 3). The charge collection centres are thus displaced vertically by half a pixel between odd and even fields.

Antiblooming Operation

When a photosite is overilluminated, the antiblooming device serves to limit the integrated charge to the "full well" value and evacuates excess electrons, thus preventing them from spilling over into neighboring photosites.

The antiblooming control gate (Φ_A) controls the height of the potential barrier separating the potential well where the charges are stored and the antiblooming drain. Thus, when a photosite is overilluminated, excess electrons spill from the potential well and into the drain (V_A) where they are evacuated (see figure 4).

Advantages of the Antiblooming Structure

In contrast with a "buried drain" structure, horizontal antiblooming allows to maintain good sensitivity in the red and near infrared. Resistance to overillumination is very high and is in fact only limited by smearing (smearing coefficient $S = 6 \times 10^{-5}$).

The TH 7864's antiblooming device can also be used as an electronic shutter. Indeed, during the integration period (T_i) (20 ms in CCIR TV standard) the image zone can be activated over any continuously variable time (t_i), with ti $\leqslant$ T_i, without altering the readout frequency (see figure 5).

The exposure reduction takes place during period I, when the channel under Φ_A is conducting and all photocharges are evacuated into the drain V_A (see figure 6)·

This operating mode enables the TH 7864 to be used over very large scene dynamics with conservation of information in strongly illuminated zones.

TH 7864 PERFORMANCE

Parameter	Typical value	Unit
Output sensitivity	410	e^-/mV
Saturation output voltage	800	mV
Saturation exposure	0.25	$\mu J/cm^2$
Responsivity (1)	4	$V/\mu J/cm^2$
	15	mV/lux
r.m.s. noise in darkness	0.25	mV
Dynamic range (relative to r.m.s. noise)	70	dB
Average dark signal at 50 °C	10	mV
Dark signal non-uniformity at 50 °C	2	mV
Horizontal resolution at 412 TV lines (2)	80	%
Vertical resolution at 244 TV lines (2)	70	%

(1) With BG38 filter and scene illumination measured before filter.
(2) With BG 38 filter.

CONCLUSION

The combined characteristics of the TH 7864 CCD image sensor make it well suited for applications requiring good resolution in the near-infrared. The TH 7864 may be used in modes other than CCIR TV standard. It has a built-in antiblooming control which serves both as an overilluminatior protection and an electronic shutter.

This sensor is also available in the US RS170 standard (ref. TH 7866).

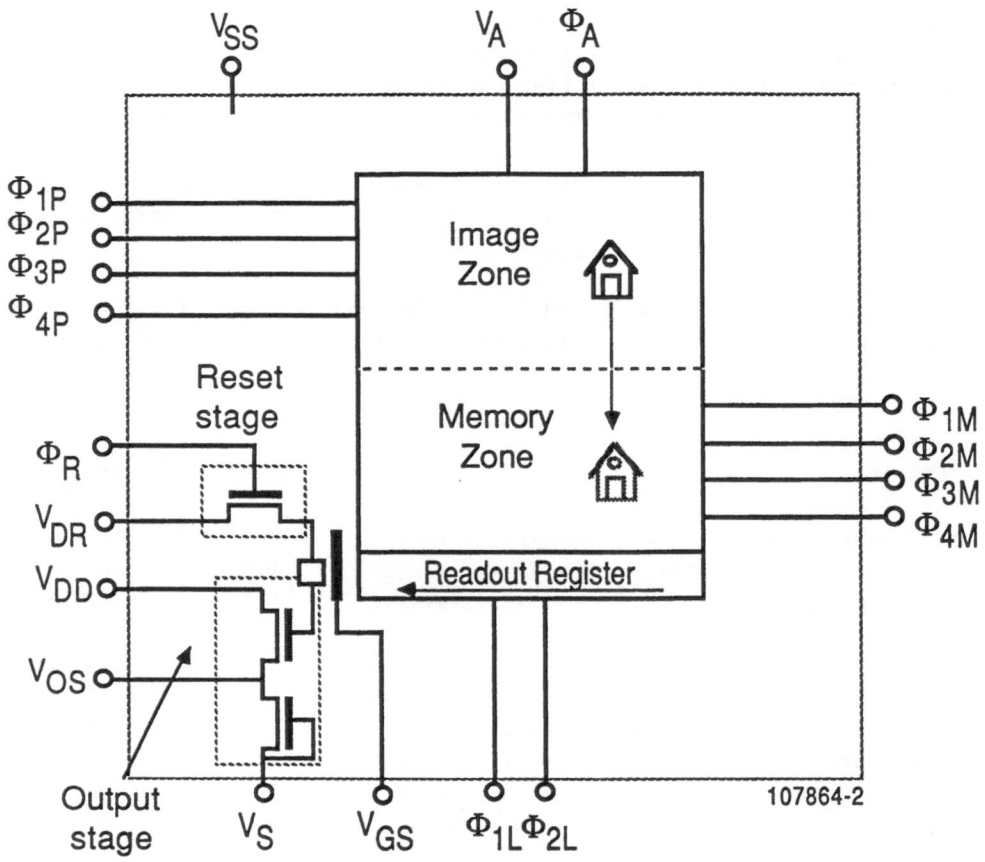

FIGURE 1 : TH 7864 ORGANIZATION

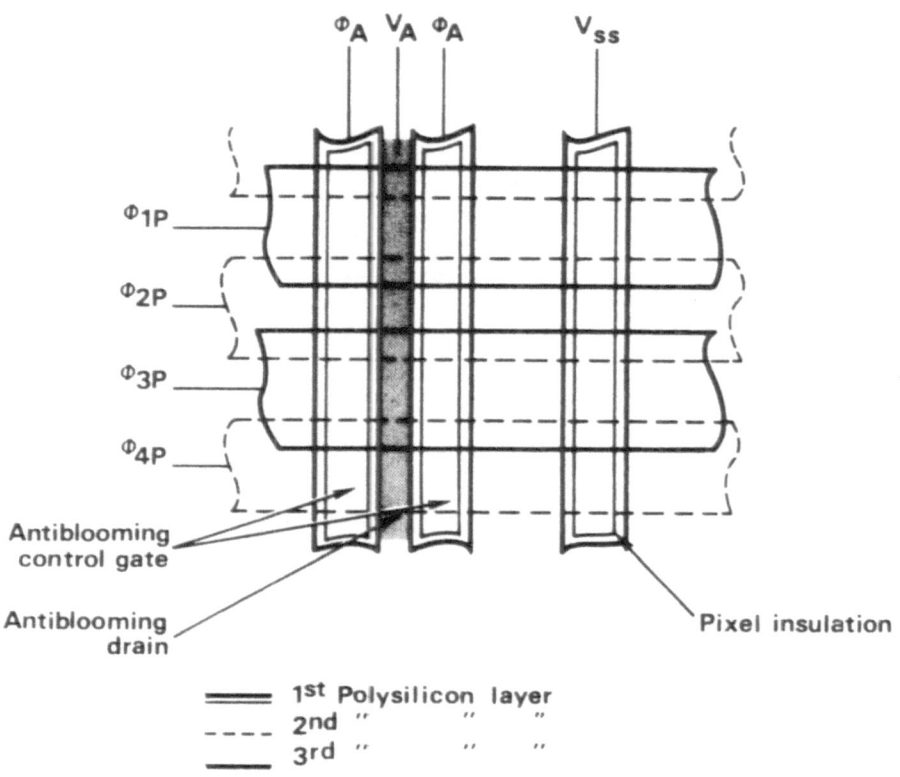

Φ_A V_A Φ_A V_{ss}

Φ_{1P}

Φ_{2P}

Φ_{3P}

Φ_{4P}

Antiblooming
control gate

Antiblooming
drain

Pixel insulation

─── 1st Polysilicon layer
──── 2nd " " "
─── 3rd " " "

FIGURE 2 : DETAIL OF A PHOTOSITE

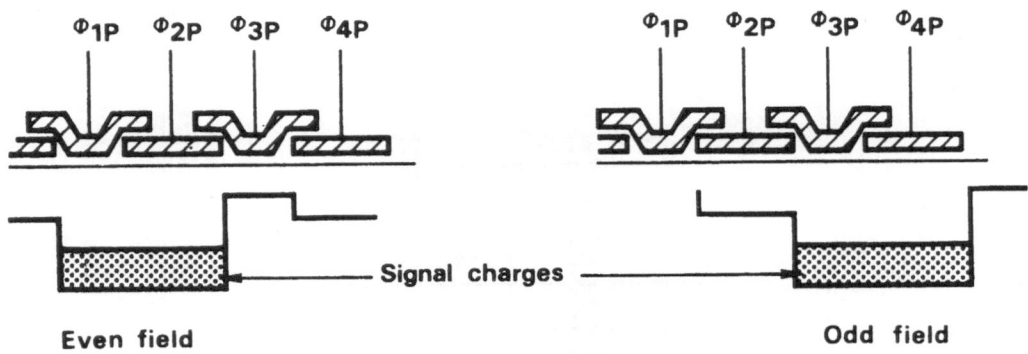

Φ_{1P} Φ_{2P} Φ_{3P} Φ_{4P} Φ_{1P} Φ_{2P} Φ_{3P} Φ_{4P}

Signal charges

Even field   Odd field

FIGURE 3 : INTEGRATION OF PHOTOCHARGES

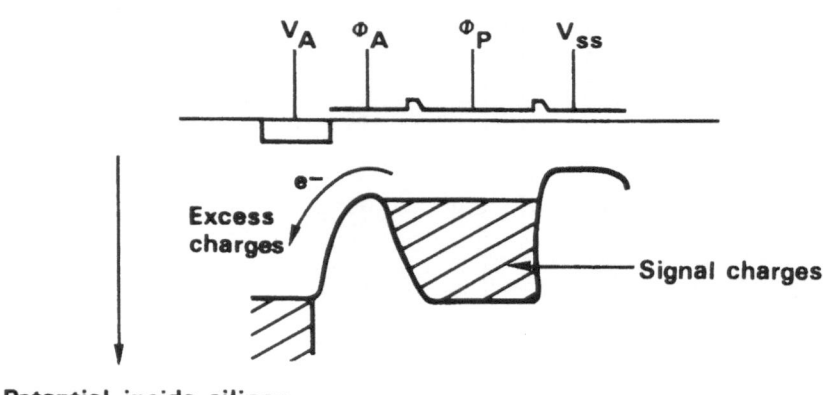

V_A Φ_A Φ_P V_{ss}

e⁻

Excess charges

Signal charges

Potential inside silicon

FIGURE 4 : PRINCIPLE OF ANTIBLOOMING

165

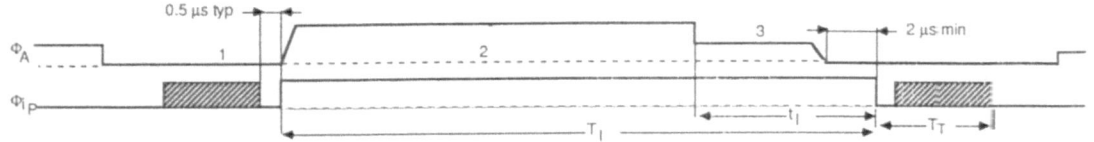

FIGURE 5 : CLOCK SIGNALS FOR ELECTRONIC SHUTTER FUNCTION

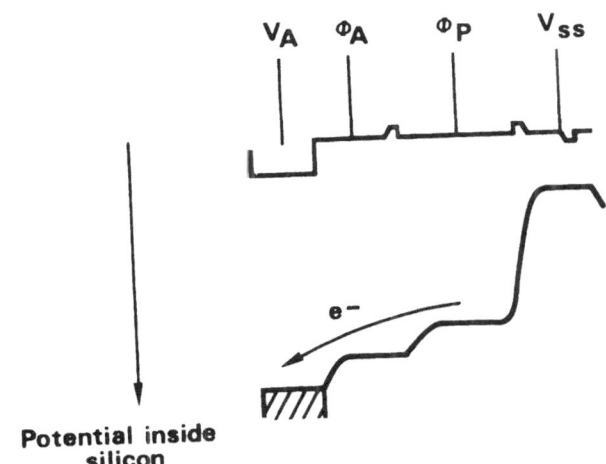

Potential inside
silicon

FIGURE 6 : ELECTRONS INJECTION IN THE ANTIBLOOMING DRAIN
FOR ELECTRONIC SHUTTER FUNCTION

REAL TIME HOLES LOCATION.
A STEP FORWARD IN BIN PICKING TASKS

Antonio B. Martínez, Vicenç Llario
Departament d´Enginyería de Sistemes, Automàtica i Informàtica Industrial
Facultat d'Informàtica de Barcelona
Universitat Politècnica de Catalunya
C/ Pau Gargallo, 5
08028-BARCELONA   (SPAIN)

Abstract

A special purpose hardware module developed to identify and locate holes in 3D is described. The system is based on the matching of the virtual points corresponding to the centers of the holes in a stereo pair. Location of the target regions on the image has been performed using two different approaches: the Radon transform and a modified version of the Hough transform.

Since the projection of a hole in the field of view on the image plane may be either a circle or an ellipse depending on the angle between the optical axis of the sensor and the base plane of each part in the scene, the system has been designed to identify and detect both, circular shapes and elliptical shapes.

Disparity analysis of the stereo pair is based on the determination of the coordinates on the image plane of the virtual points corresponding to the centers of the holes. The system supplies the 3D coordinates of the centers as well as the radii of the holes detected in the scene every 20 ms.

Although the system was initially thought to solve some bin picking applications it can be used to solve a great deal of applications not only in industrial environments but also in mobile robot guidance, traffic control, inspection and surveillance among others.

We present two real time implementations mostly based on a pipeline architecture wich allow real time performance. and as a consequence may be suitable for robotics applications.

NATO ASI Series, Vol. F52
Sensor Devices and Systems for Robotics
Edited by A. Casals
© Springer-Verlag Berlin Heidelberg 1989

1. Introduction

The acquisition of workpieces out of a bin or a conveyor is a common operation in a production plant. The problem of feeding parts which are in an unknown position and orientation inside bins has been adopted as a main research line in many research laboratories and industries because it is intrinsically an important generic problem in manufacturing environments. Oriented workpieces are required for proper operation of most manufacturing machines. The trade-off between productivity and cost has powered the incorporation of robots to deal with those tasks which in the past were performed by human operators. Although in most cases workpieces can be oriented by mechanical feeders to facilitate the job of robots, there are a lot of situations in which the use of a vision system brings more flexibility. Generally speaking, sensor based robots have the potential of dealing with less organized environments as well as unpredictable situations.

The main goal of the system designed is to look for specific features such as circular marks or holes and to locate them in the 3D space to allow a robot to pick them up out of a supply bin.

Since there are a great deal of workpieces in the manufacturing industry which have holes or circular marks, we thought that a good approach to solve a lot of bin picking tasks would be to look for such features. As a mather of fact, the system has obviously intrinsic drawbacks. A common problem of those tasks is that parts are organised in piles and, as a consequence, overlapped. Unfortunatelly, up to now, the system has been unable to deal with occlusions. Nevertheless, the basic research developed until now has emphasized the design of robust algorithms, easily implementable in hardware up to some extent, in order to guarantee real time performance of the whole process. We looked for a reasonable trade-off between hardware and software to shorten the response time.

We can assume that Robot Vision is just a special case of vision processing in which one can apply engineering solutions to solve what are, in general, difficult problems. For our purposes one of the most important advantages is that usually an industrial environment may be controlled someway, for instance arranging a suitable lighting system, avoiding, if possible, positional uncertainty of parts within the field of view, introducing limitations in the visual context of objects. All the a priori knowledge about a given task may be used to specify the capabilities and goals of the system as well as the characteristics of the environment in which it must perform.

The problem stated this way is not as simple as it might seem. In fact there are a lot of difficult inspection, acquisition and handling problems which exceed the abilities of current vision research.

Nevertheless, the principal motivation behind robot vision is increased flexibility and lower cost. That's why it is possible to find a great deal of commercially available vision systems suited to solve different kind of problems.

The purpose of the vision system developed is to produce a symbolic description of the feature regions on the objects being imaged. The level of description will depend on the kind of interaction between the robotic system and the environment, but in any case the system must be able to supply reliable data on time to be useful for the specific application feedback. In our context we have focused the identification and location of circular or elliptical shapes corresponding, in most cases, to holes.

We will try to concentrate on two different approaches which have been tested to detect such features in real time. First of all we will outline the general structure of a vision pre-processor based on an implementation of the Radon transform, and afterwards we will emphasize the modifications introduced in the Hough transform for the same purpose, showing a possible implementation. The reader will be able to find many references about this subject in the bibliography.

2. An Implementation of the Radon Transform to detect circular Shapes

We will outline at this point how we have used the Radon Transform to detect defined regions on the image in a rather easy way, better said, we would like to present a straightforward approach to solve the identification problem thinking about the suitability of a final hardware implementation able to perform in real time.

In the case of the Radon transform extended to circular shapes we can write:

$p = C(x,y;k)$ with $k = k_1, k_2, \ldots, k_n$

where C is the equation of a family of curves on the X,Y plane parameterized by p and the elements k_1, k_2,.... belonging to a vector k.

We supose $F(x,y) = a(x,y) X(c)$, where $a(x,y)$ is a probability density function along c and $X(c)$ is the function to be transformed. The generalized Radon transform will be given by the following expression:

$f(p,k) = \int\int a(x,y) X(c) (p-C(x,y;k)) \, dx \, dy$

$f(p,k)$ can be interpreted as the level of coincidence in the parametric space among the family of curves and the selected curve.

Since we want to detect circles, the family of curves chosen will be of the form:

$C(x,y;k) = \{ (x-a)^2 + (y-b)^2 \}^{1/2}$ $k = \{ a,b \}$

Since the detection module will work on contour images, we can assume some simplifications, for instance:

-The density function $a(x,y)$ will have only two values, 1 if the pixel belongs to the curve and 0 elsewhere. If we were looking for a circular shape of the form:

$X(c) = b' + \{ R^2 + (x'-a)^2 \}^{1/2}$ as shown in Figure 1

For every point (a_i, b_i) on the transformed space we will analise the level of coincidence. When this level will be above a threshold we will assume that a circle with radii R and center (a',b') has been detected.

In our case, the Radon transform will give us the lateral area of a cilinder with a height $h = 1$. So this value will correspond to the length of the base circle and

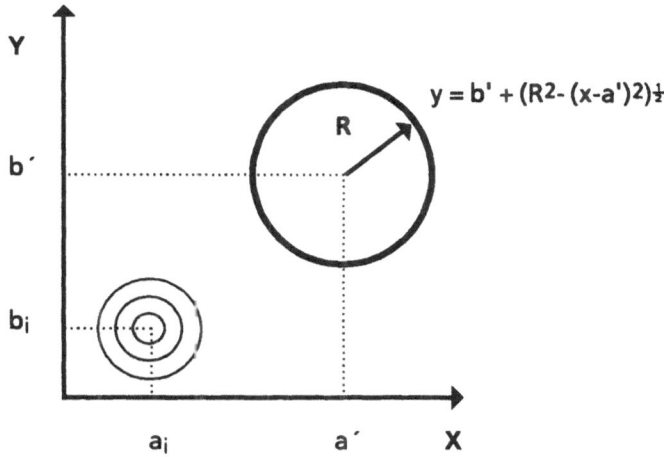

Figure 1.- Mode of operation of the Radon Transform

equal to $2\pi R$. Thus for every different radii we will get the transform as a discrete approximation of $2\pi R$. Figure 2.

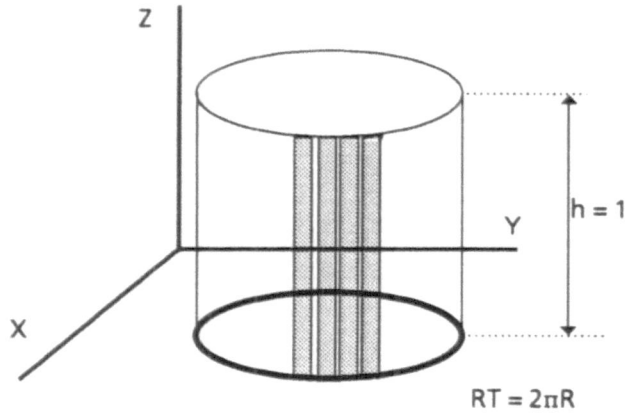

Figure 2.- Radon Transform over a circular geometry on a contour image

The discrete Radon transform may be implemented by sweeping the image space by means of a window of size nxn pixels. Locating the center of the window on every image point and assuming that each point is a potential candidate to be the center of a circle, we are mapping all the points on the parametric space. To exploit parallelism, we perform, at the same time, a circular scanning over the window. Figure 3.

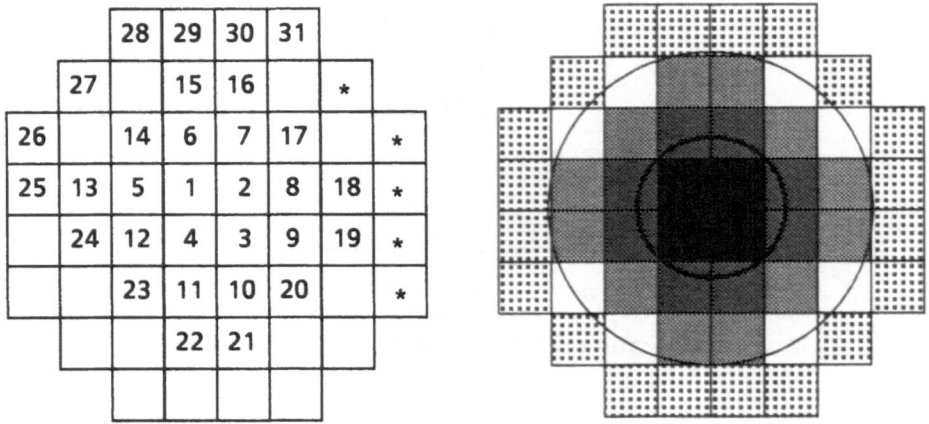

Figure 3.- Circular Scanning over the transformation window

This way we obtain for every scanned point (x_1, y_1) a longitudinal register ordered by radii as shown in Figure 4, containing all the radii $R \leq n/2$ and comparing for each one of them the level of coincidence with the curve $X(c)$.

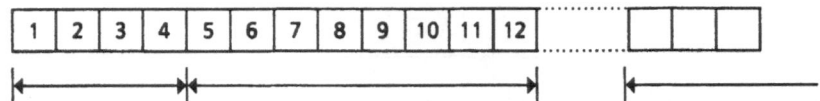

Figure 4.- Longitudinal register

Figure 5 shows a blocks diagramm of the whole system. In order to simplify the final implementation we introduced a new concept consisting on performing a radial sampling. This way we reduced the information concerning to every circle to 15 pixels. At the same time, the fact that we were working with discrete images powered the idea of introducing a band of confidence able to guarantee the detection independently of the discretization errors. Figure 6.

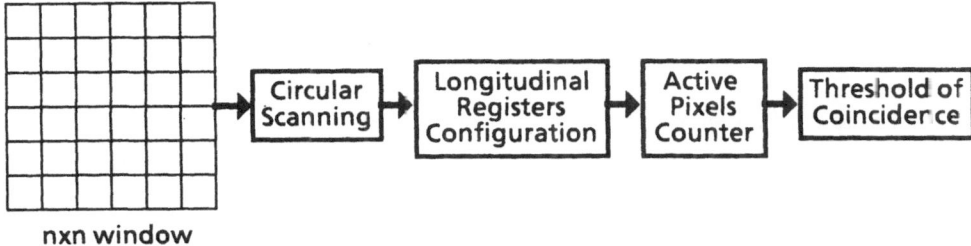

nxn window

Figure 5.- Blocks Diagramm of the whole system

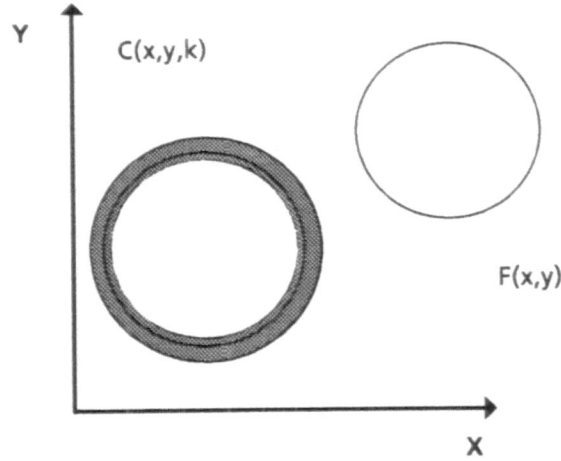

Figure 6.- Band of confidence for the matching process

The main drawbacks of this implementation of the Radon transform are, on one hand, the propagation time for the different stages of the piece of hardware developed which restricted the maximum number of radii to 20 and, on the other hand, the fact that the extension of the transform to be able to detect ellipses needed the generation of the family of all the possible ellipses for every point analised.

3. Real time Hough Transform to detect circular and elliptic geometries

A second method tested was based on a generalized Hough transform. Having a circle given by its parametric equation $R^2 = (x-a)^2 + (y-b)^2$, if we draw for every point (x_1,y_1), belonging to the contour, a circle of radii R on the parametric space (a,b) with its center (x_1,y_1). Figure 7. Then the point of intersection (a',b')

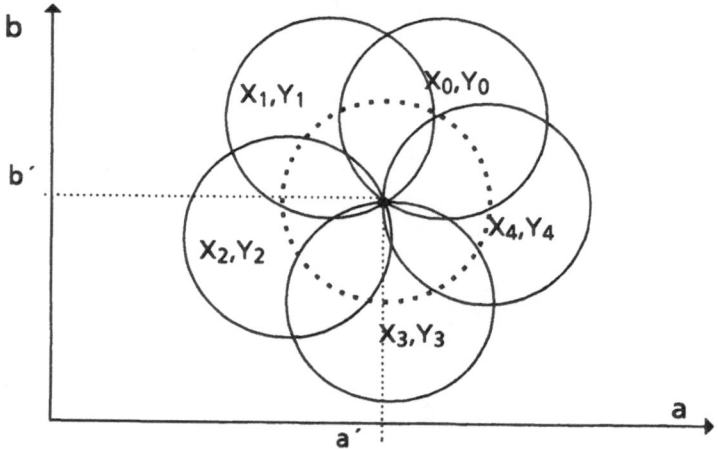

Figure 7.- Parametric space

of all those circles on the transformed space corresponds to the coordinates of the center of the circle on the image space. Figure 8 shows a blocks diagramm of

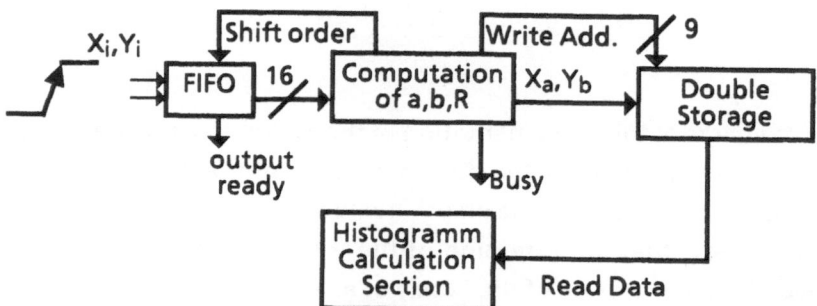

Figure 8.- Blocks diagramm of preprocessing stages (Hanara 88)

the system.

As a matter of fact, the system needs some FIFO memories since the time to process the information for every point exceeds the time between two consecutive pixels. Figure 9 shows a possible implementation of the calculations of x_a and y_b.

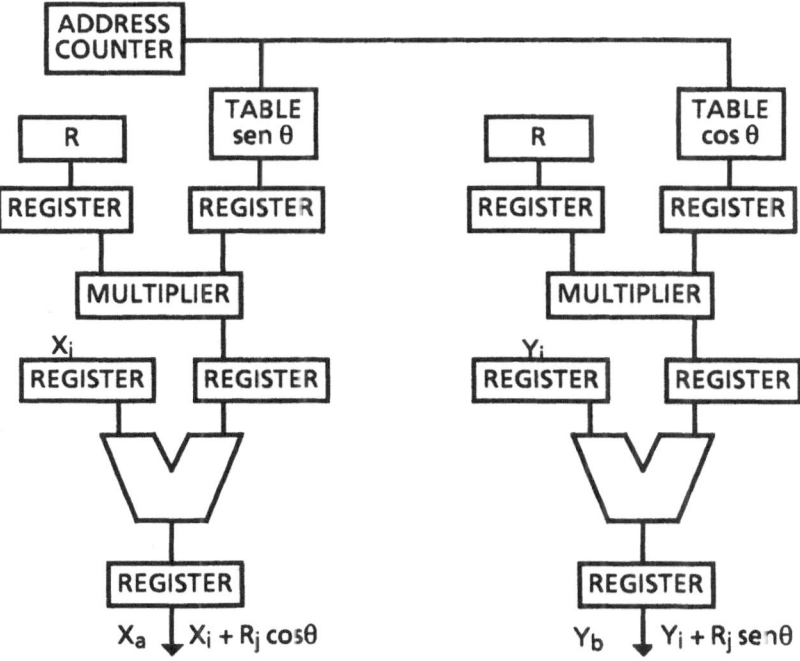

Figure 9.- Hardware computation of X_a and Y_b

The whole process may be decomposed in two parts: the transformation itself and the identification of the pattern. To deal with the transformation, the image space must be swept by a window able to perform the local analysis of inner pixels. In the ideal case, this window should have infinite segments to be able to supply the information corresponding to the angle with the X axis as well as the distance from the origin to the point of intersection with the curve.Thus any curve could be approximated by an infinite number of straight segments tangent to it and orthogonal to the radial ones.

Since the shape we want to detect is a circle and we assume that the transformation takes place from the center, the transformed space will correspond to the relationship between the radii and the swept angle. For the final implementation we used a 32x32 pixels window and we toke only 16 segments to sample the curve. To simplify and taking advantage of simetry we developed 4 subwindows with 4 segments each, which in total constitute the whole window as shown in Figure 10. Thus, every segment radially samples with

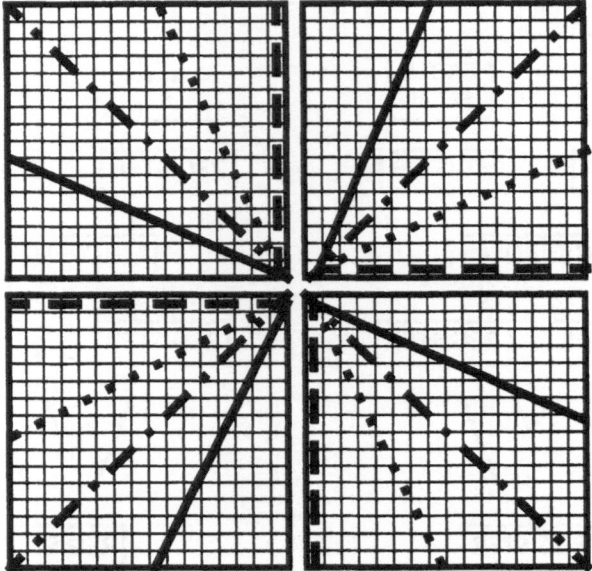

Figure 10.- Transformation window

its corresponding slope the intersected pixels within the window, providing the information of angle and distance from the center of the window. The way we can calculate those two parameters is by using priority encoders as shown in Figure 11.

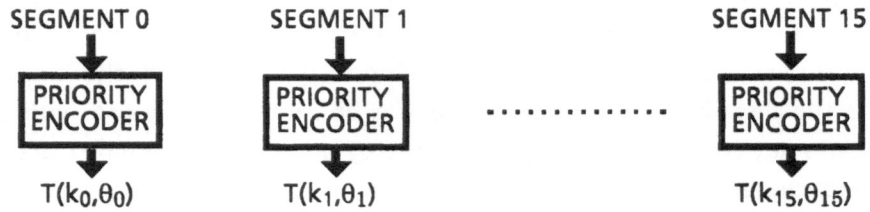

Figure 11.- Transformation stage

For every point of the image plane we will obtain a transformed space in which we will have to identify the pattern during one pixel time. One possible implementation of the Hough transform to detect circular patterns is shown in Figure 12. Since the transform is performed from the center of the window, the

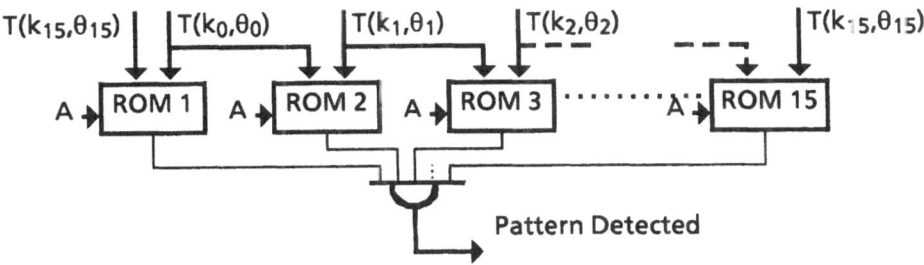

Figure 12.- Recognition Stage

distance must be the same for all the segments and, as a consequence, the bank of ROM memories calculates the difference between every pair of consecutive segments. Obviously the match between the pattern and the transformed space will happen when all the differences will be equal to zero.

As far as we are working with discrete images we can establish a level of confidence for the matching process. That is why we assume a band which can be selected by the user to search for the maximum coincidence between the pattern and the transformed space.Figure 13

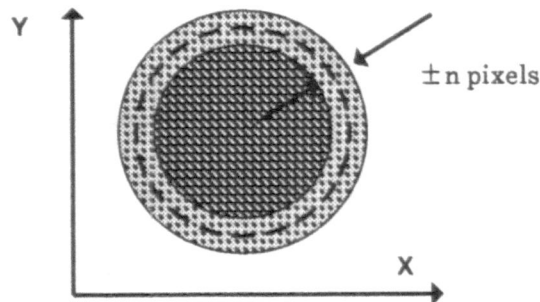

Figure 13.- Band of confidence for the matching process

The direct consequence of the searching band is that we will detect the circular geometry several times inside the band and this will supply us a cluster of points corresponding to the virtual centers. Figure 14 The last step will be then to

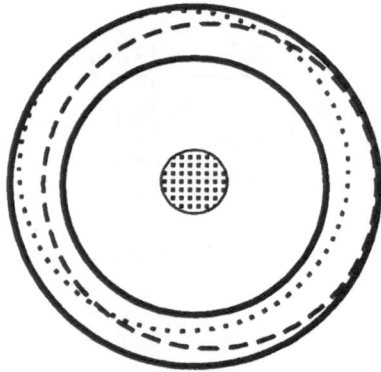

Figure 14.- Cluster of points corresponding to virtual centers

calculate the centroid for every cluster which will correspond to the estimated centers of the detected holes. Figure 15 shows one possible implementation of this filtering stage.

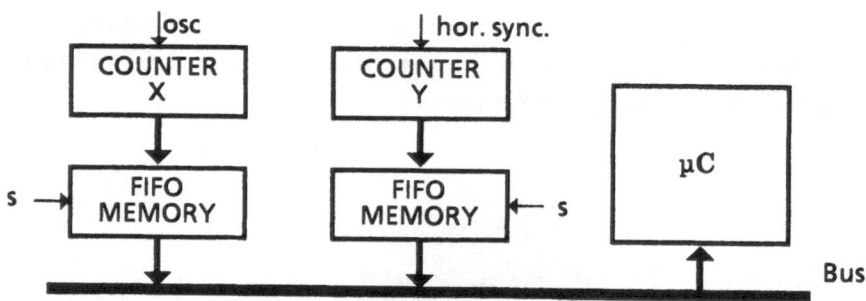

Figure 15.- Filtering stage for the cluster of virtual centers

4.From Ellipses detection to 3D Data

As far as we are working in a 3D space, the natural extension of the proposed method is to detect elliptical geometries which in general will correspond to the projection on the image plane of a circular geometry oriented with respect to the optical axis of the sensor. Knowing the ellipse parameters we will be able to recover or estimate the orientation of the circular surface.

Most of the authors have been working with the Hough transform although ellipses detection is a little bit more complicated since one must work on a 5 dimension space to take into account the five parameters defining a conic which in general will be given by an expression such as:

$$x^2 + 2Axy + By^2 + 2Cx + 2Dx + E = 0$$

If we try to implement an algorithm similar to the one used for circles detection for every point of the parametric space, the result will be disappointing since the needed processing will be time consuming and we won't have real time performance.

The proposed approach to solve this transformation consists of approximating the ellipse by means of tangent segments the same way we did for circles detection. If we get the Hough transform of each segment we will obtain a characteristic function on the transformed space, thus simplifying the identification process.

If the ellipse is given by its parametric equation

$$x^2/a^2 + y^2/b^2 = 1$$

Then we get the slope m of the tangent to the curve on (x,y) Figure 16

$$m = tg\ \beta = -cotg\ \phi = dy/dx = -(b^2/a^2)\ cotg\phi$$

the distance from the origin to the tangent will be

$$p = [mx-y]/\sqrt{m^2 + 1}$$

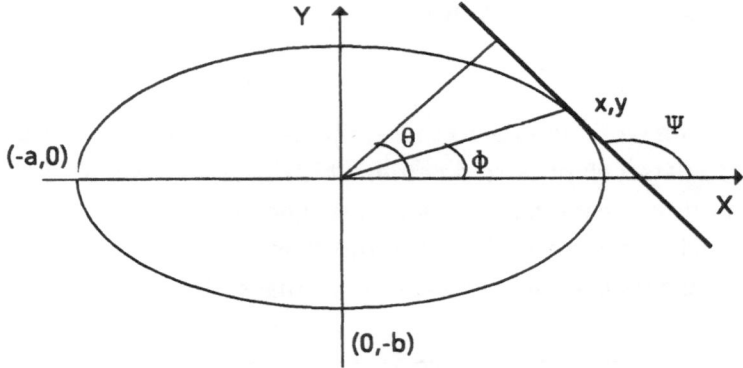

Figure 16.- Approximation of the Hough Transform

If we describe the ellipse with polar coordinates then

$$r^2 = 1/\cos\phi/a^2 + \sin\phi/b^2$$

The transformed space will be given by

$$p = HT(a,b,\phi) = \sqrt{a^2 \cos^2\phi + b^2 \sin^2\phi} \text{ (Casasent 87)}$$

As a consequence the transformed space will correspond to a sinusoid with a period equal to π. Figure 17

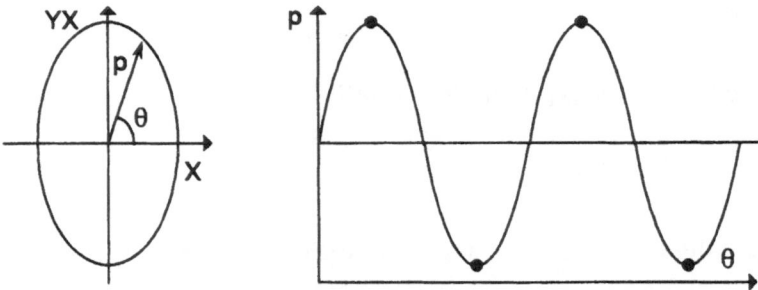

Figure 17.- Transformed space for an ellipse

To identify an ellipse on the transformed space we will need two parameters. The first one corresponds to the number of inflexions of the transformed function, Figure 17, and the second one to the maximum excentricity allowed.

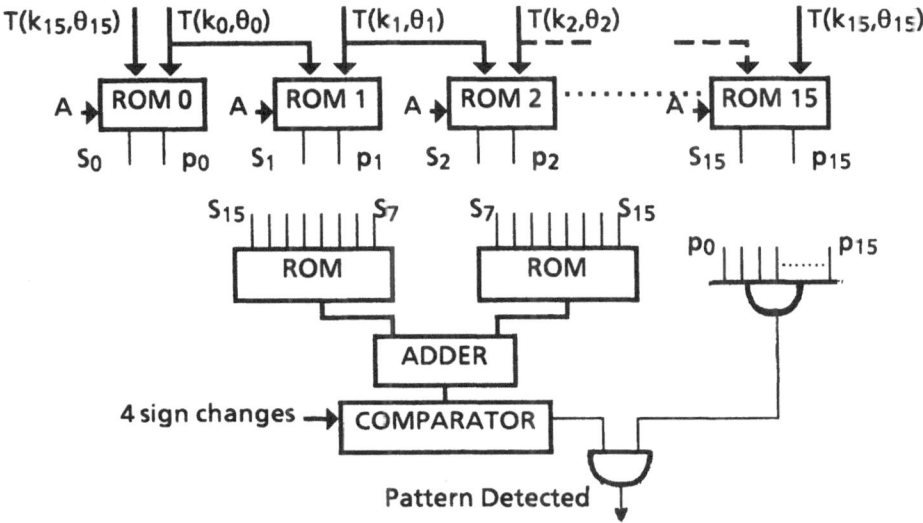

Figure 18.- Hardware implementation of Ellipses detection

Figure 18 shows the hardware implementation tested to detect elliptical shapes.

The role of ROM memories is double. On one hand they must supply the information corresponding to the slope on the 16 sampled pixels and on the other hand they must guarantee that the difference between two consecutive segments is within a defined threshold Ω.

A simple way to determine the slope consists of distinguishing for every consecutive pair of segments which one is longer assuming a sign and knowing whether we are rotating clockwise or counterclockwise. However the value of Ω is not only a function of the excentricity but also of the maximum diameter that can be detected and of the matching tolerance. Figure 19.

Once the ellipse has been located on the image plane, the next step consists of getting its spatial position and orientation. The orientation in the 3D space can be derived from a,b and β. The direction of the smallest axis directly provides

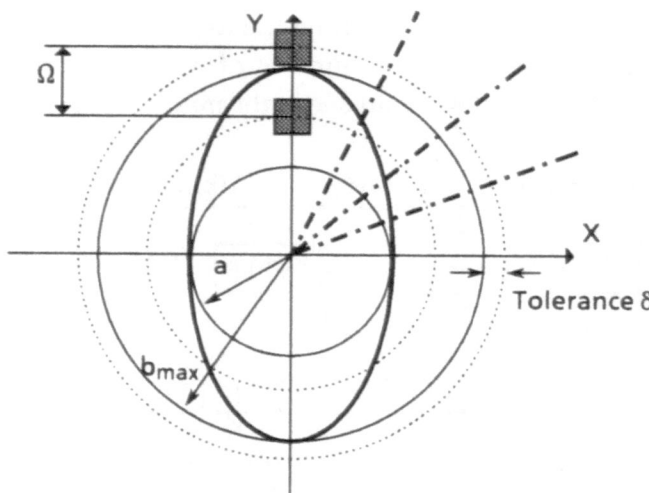

Figure 19.- Best match for Ω and δ inside the band of confidence

the angle β , while to get α we must know with which angle should be cut the elliptic cone to get a circular section.

To locate the surface in the 3D space we need only one point. The approach we have used has been to solve the correspondence problem for a stereo pair, matching the virtual points corresponding to the centers of the circles and ellipses. Assuming epipolar geometry, this means to look for correspondences along epipolar lines. As far as the system provides the coordinates (x,y) of the virtual centers once the clusters have been filtered, the application of constraints such as unicity, continuity and order will guarantee a right match of the corresponding points on the stereo pair. Afterwards we calculate disparity for every point and finally we obtain the depth map of virtual points.

5. Results and conclusions

The system developed has shown its competence to solve, in real time, the recognition and location of circular and elliptical geometries. Although the final implementation is hardwired and not easily reconfigurable, the trade-off between hardware and software allows a wide range of applications such as bin picking, qualitative and semiquantitative inspection of surfaces, traffic control.

Concernig the future work, we would like to face the occlusions problem which probably must be approached analysing gray scale images instead of contour images. The recognition process in noisy images will be also of our interest because of the great deal of potential applications in industrial environments where reliability is a principal requirement.

Acknowledgments

We would like to thank the research assistants of the Computers Technology Lab, J. Mª Asensio and Toni Perera, who ran and debugged some of the algorithms which later were hardwired.

Bibliography

- Amat, J.; Llario, V. (1985). A Vision System with 3D Capabilities. IEEE Conference on Robotics and Automation. Saint Louis.
- Asada, H.; Brady, M. (1986). The curvature primal sketch.IEEE Transactions on Pattern Analysis and Machine Intelligence. Vol PAMI 8 pp. 2-14.
- Basille, J. L.; Castan, S.; Latil, J. Y. (1984). Parallel Structures in Image Processing. Premier Colloque Image. CESTA. Biarritz.
- Bookstein, F. L. (1979). Fitting conic sections to Scattered Data. Computer Graphics and Image Processing.Vol 9 No 1 pp. 56-71.
- Danielsson, P. E.; Levialdi, S. (1981) Computer Architectures for Pictorical Information Processing. Computer. pp. 53-67.
- Davies, E. R. et al. (1985). Radial Histograms as an aidd in the inspection of Circular Objects. IEEE Proceedings, Vol 132 No. 4.
- Davies, E. R. (1987). A High speed algorithm for circular object location. Pattern Recognition Letters 6 pp. 323-333.
- Davies, E. R. (1988).A modified Hough scheme for General Circle Location. Pattern Recognition Letters 7. pp. 37-43.
- Deans, S. R. (1981). Hough Transform from Radon Transform. IEEE Transactions on Pattern Analysis and Machine Intelligence. Vol PAMI 3 no. 2.
- Casasent,D.; Krishnapuram, Raghuram. (1987) Curved Object location by Hough transformations and inversions. Pattern Recognition Vol 20 No 2 pp.181-188.
- Dodd, G. G.; Rossol, L.(eds) (1979) Computer Vision and Sensor based Robots. Plenum Press.
- Duff, M. J. B.; Levialdi, S; Preston, K.; Uhr, L.(eds) (1985). Computing Structures and Image Processing. Academic Press.
- Duff, M. J. B. (ed) (1985). SPIE Conference on Architectures and Algorithms for Digital Image Processing. Cannes.
- Eskenazi, R.; Wilft, J. M. (1979). Low Level Processing for Real Time Image Analysis. Proceedings of the COMPSAC Conference, Chicago. pp. 340-343.
- Foster, N.J.; Sanderson, A.C. (1984). Determining object orientation using ellipse fitting. SPIE Vol. 521 Intelligent Robots and Computer Vision.pp.34-42.
- Gerritsen, F. A.; Verbeek, P. W. (1984). Implementation of Cellular-Logic Operators using 3x3 Convolution and Table Look-up. Computer Vision, Graphics and Image Processing.Vol 27. pp. 115-123.
- Granlund, G. H. (1980). GOP : A Fast and Flexible Processor for Image Analysis. Proceedings of the 5th IEEE Conference on Pattern Recognition. pp. 489-492.
- Hanara, K. et al. (1988). A real time processor for the Hough Transform. IEEE Transactions on Pattern Analysis and Machine Intelligence. Vol PAMI 10 No. 1.
- Helgason, S. (1965). The Radon Transform on Euclidean Space, Compact two-points, Homogeneous space and Grassman Manifolds. Acta Mathematica 113.
- Horn, B. K. P. (1986). Robot Vision. MIT Press.
- Hough, P. V. C. (1962). Method and means for recognizing complex patterns.U. S. Patent 3069654.
- Hwang, K., Su, S. P. (1983). VLSI Architectures for Feature Extraction and Pattern Classification. Computer Vision, Graphics and Image Processing.Vol 24. pp. 215-218.
- Ianino, A.; Shaphiro, S. D. (1978). A survey of the Hough Transform and its Extensions for Curve Detection.IEEE Proceedings.
- Kung, H. T.; Picard, R. L. (1981). Hardware Pipelines for Multi-Dimensional Convolution and Resampling. IEEE Computer Society Workshop on Computer Architectures for Pattern Analysis and Image Database Management. Hot Springs, Va, pp. 273-277.

- Kung, H. T.; Song, S. W. (1982). A Systolic 2-D convolution chip. Multicomputers for Image Processing: Algorithms and Programs. Academic Press. New York.
- Lea, R. M. (1985). SCAPE: A single chip array processing element for signal and image processing. IEE Proceedings, Vol 133.
- Llario, V. (1987).Hardware-Software trade-offs in Robot Vision. NATO Advanced Research Workshop on Real Time object and environment measurement and classification. Maratea, Italy.
- Ludwig, D. (1966). The Radon Transform on Euclidean Space. Communications on Pure and Applied Mathematics Vol 19 pp. 49-81.
- Mori, K. I.; Kinode, M.; Shinoda, H.; Asada, A. (1978). Design of Local Pattern Processor for Image Processing. National Computer Conference. pp. 1025-1031.
- Nagao, M; Nakajima,S. (1987). On the relation between the Hough Transformation and the project on Curves of a rectangular window. Pattern Recognition Letters 6 pp. 185-188.
- Nagata, T. et al. (1985). Detection of an Ellipse by use of a recursive least-square estimator. Journal of Robotics Systems. 2 pp. 163-177.
- Poggio, T. (1985). Early Vision: From Computational Structure to Algorithms and Parallel Hardware. Computer Vision, Graphics and Image Processing 31. pp. 145-151.
- Overington, I.; Greenway, P. (1986). Accurate local form and motion extraction by a composite 1st and 2nd difference processor based on an interpretation of human vision. SPIE Conference on Advances in Intelligent Robotics Systems. Cambridge.
- Pugh, A. (ed) (1983). Robot Vision. IFS Publications. Springer Verlag.
- Reeves, A. P. (1980). The Anatomy of VLSI Binary Array Processors. Workshop on New Computer Architectures and Image Processing. Ischia.
- Reeves, A. P. (1984). Parallel Computer Architectureds for Image Processing. Computer Vision, Graphics and Image Processing.Vol 25. pp. 68-88.
- Sheen, J. A.; Hatch, M. (1987). The VISIVE Demonstrator. Current especifications and Future Enhancements. Report No JS 10549. British Aerospace.
- Sherdell, D. (1980). A Low level Architecture for a Real Time Computer Vision System. Proceedings of the 5th IEEE Conference on Pattern Recognition. pp. 290-295.
- Shiray, Y. (1985). Robot Vision. North Holland FGCS .
- Snyder, W. E., Husson, C.; Benz,H. F. (1979). Satellite Pattern Classification Using Charge Transfer Devices. Conference on Pattern Recognition and Image Processing. pp. 246-249.
- Sternberg, S. R. (1981). Architectures for Neighborhood Processing. Proceedings of the IEEE Conference on Pattern Recognition and Image Processing. pp. 374-380.
- Stockman, G. C.; Agrawala, A. K. (1987). Equivalence of Hough Curve Detection to Template matching. Communications of the ACM Vol 20 No. 11.
- Swartzlander, E. E.; Gilbert, B. K.; Reed, I. S. (1978). Inner Product Computers. IEEE Transactions on Computers. C-27. pp.21-23.
- Thissen, F. L. A. M.; Rosier,W. J. (1986). PAPS: A Picture Acquisition and Processing System . PAPS Symposium.
- Tsuji,S.; Matsumoto,F. (1978). Detection of ellipses by a modified Hough Transformation. IEEE Transactions on Computers. Vol. C.27 No. 8. pp.777-781.
- Venema, W. J.; Sterken, H. P. M. (1986). IdentiVision System. Machine V sion Symposium.
- Willet, T. J., Tisdale, G. (1978). Hardware Implementation of a Smart Sensor: A Preview. ARPA Image Understanding Workshop. pp. 1-8.
- Ye,Q.(1986). Apreprocessing method for Hough Transform to detect Circles.1986 IEEE Conference on Robotics and Automation. San Francisco.pp651-653.

COMBINED 2-D AND 3-D ROBOT VISION SYSTEM

P. Levi, L. Vajtá*
Forschungszentrum Informatik, Haid- und Neu-Str. 10-14
D-7500 Karlsruhe   (F.R. Germany)

INTRODUCTION

The automation of assembly and handling of tools or workpieces can be performed by the integration of sensory systems only. Sensors have a fundamental importance in this process. Sensors used in manufacturing systems can be classified into two groups:

- Obtaining internal data from robots (e.g. joint position, angular velocity, etc.)
- Obtaining external data from the environment to detect the presence, type, orientation, surface or other characteristics of objects in order to perform different operations with them.

The majority of robots installed in industry uptil now are used in less complicated working processes. The control of the necessary robot-movements in these simple cases are usually performed without external sensors. Handling of tools and workpieces without any control or supervision is possible only with significant increase of costs due to the employment of additional accessories, for example, in the form of precision made conveyors. Though such a complicated and rather expensive investement eliminates the problem of sophisticated robot sensor systems, it has a negative effect on production, viz., neither complex manufacturing tasks nor flexible production can be accomplished.

The solution to this problem is to use an adequate sensory system which supplies precise information from the environment to the robot controller. Such a system is difficult to imagine without image processing unit. Early vision systems were designed on the basis of intensity sensing. Various methods have been developed to evaluate such pictures to recognise objects. Intensity picture however depends on the effects of environment, they are difficult to evaluate and also the volumetric properties are suppressed. On the other hand, object recognition is not the only task to be solved in robot vision. One needs to measure position, orientation and distances for adequate robot control. For this purpose range sensing techniques offer a practical solution. The more is the

This work has been supported by the Deutsche Forschungsgemeinschaft
*Technical University of Budapest Department of Process Control
Müegyetem rkp.9. 1111 Budapest Hungary

complexity of the sensor system, better is the information achieved. The application of combined 2-D (intensity) and 3-D (range) sensors seems are a good choice to solve these sensing problems.

SENSORS FOR THREE DIMENSIONAL VISION

In general, vision systems are composed of an illuminating wave and detector system. 3-D systems can be classified as active or passive ones depending on the application of the illuminate wave.

In passive systems the illumination is independent on the measuring system. Target area is obtained simultaneously by two cameras situated in a strictly defined geometrical arrangement and distance data are calculated from the corresponding points of the two images. This method needs large amount of memory because of the problem of correlation. Considering the present technological level we come to the conclusion that such a system cannot be used in industry yet.

The problem of correlation is solved by controlling the light waves in active systems. There are several systems where structured light is employed: punctual (with deflection), split beam, raster blend, etc. The use of more complex arrangement of light sources makes the evaluation method simpler and also the number of necessary measurements can be reduced when calculating a distance image.

Disadvantages worth mentioning in this method is the error occurred due to the use of external light waves. Light sources with high intensity are used in a special spectral domain to ensure the necessary contrast. On the basis of its phisical characteristics laser beam is predestinated for this task.

Systems mentioned above were based on the triangulation measuring method. The application of laser beam offers another possibility as well by measuring the so called propagation time instead of the position of a light point (LIDAR) or the phase shift of a laser beam modulated by an analog signal.

As small distances increase problems in connection with the measurement of propagation time and phase shift, the applciation of structured light sources in the near future seems to be a better choice. Uniqueness is guaranteed only in the case of punctual projection. Consequently this solution offers the simplest construciton and fastet data acquisition time.

LASER SCANNER BASED ON THE TRIANGULATION THEOREM

At the Research Center for Informatik in Karlsruhe, in cooperation with the Technical University of Budapest (Hungary), we are developing a triangulation laser scanner for combined evaluation of intensity and range data.

Figure 1. illustrates the principle of triangulation. Light beam comes from a laser source. The reflected signal is detected by a sensor (a position sensitive diode in our case) situated at a distance of d (base line). The source, the reflection point of the object (P) and the sensor form a triangle. Knowing the side lengths or the angles of this triangle the distance between any object point and the source of the laser beam or the origin of the coordinate system can be determined.

The source of the laser beam is situated in the origin of the coordinate system (x,y,z). The object point P is projected to the B pixel of the picture plane (x-y coordinate system) which is at a distance of f (focus) from the center of the detector model. Distance data can be calculated if we know either the horizontal or the vertical deviation angle of the emitted beam and the calibration parameters d and f by measuring both the pixel coordinates x and y. In this case, x and y values correspond to the horizontal and vertical deviation angle respectively.

If we use a parallel projection instead of the central one, and calculate the distance of the object point P from the base line (h) instead that of P (fig. 2), two angles out of the four (e.g. α and β) are sufficient to determine h. The parallel projection is advantegeous in volumetric calculations (centre of gravity, mass) and 3-D representation as well. From fig. 2 distance h can be written in the following form:

$$h = \frac{d}{\cot \delta + x/f_0}$$

The main parts of a laser scanner based 3-D vision system are the laser source, the deflection unit and the detector part.

The *laser source* must have an optical power large enough to produce a detectable light-spot. Due to the wide variation in surface quality the received optical power is mostly not more than 0.001% of the laser power. On the other hand specular reflection directed toward the detector will produce a signal level which is higher in several orders of magnitude. Variation of surface quality in the area of the light-spot causes an error by the spot position (x) detection, i.e. small spot diameter is needed.

A large number of methods exist for *deflection* of laser light. Solid-state acusto-optic cells, polygon mirrors, resonant scanners or holographic devices offer capability for scanning at high rates. For our purposes of combining 2-D and 3-D data a random scanner is needed and therefore we use galvano-mirrors in our system.

The most advanced *detection method* for light-spot position makes use of position sensitive detector called a lateral effect photodiode. The continuous position detectors locate the centroid of the light spot making them somewhat insensitive to focusing. The position sensitivity is directly proportional to the amount of light received. Therefore one must be aware of decreasing performance when viewing dark surfaces.

Fig. 3 shows the structure of the triangulation laser scanner schematically. The beam of a 24 mW diode laser is directed towards the galvano-mirrors. The laser beam is intensity modulated by use of TTL control signal. The position of the galvano-mirrors is servo-feedback stabilized providing random scanning capability with increased speed. The reflected light is measured by a position sensitive camera. A commonly used CCD camera is placed in the middle of the optical axis ot the system. Its task is to pick-up the intensity picture. The laser scanner is controlled by a multiprocessor system based on the microprocessor Motorola 68000. Fig. 4 shows the structure of the multiprocessor system.

COMBINED EVALUATION OF INTENSITY AND RANGE DATA

The current signals of the PSD are digitized and filtered. The sum of the currents is proportional to the reflected light intensity, and the difference of them divided by the intensity corresponds to the actual range.

There are three different kinds of range data produced by the laser scanner:

- Normal range data corrected on the basis of the intensity values measured by the PSD (intensity values over the noise limit and below the saturation)
- Pixels belonging to shadow-areas. The laser light cannot be projected upon these areas or they are invisible for the detector (intensity values of nearly zero)
- Background data (there is no difference in range value belonging to the same position in background and active picture)

Pixels belonging to shadows or background are marked in the range matrix.

Due to the large variation of surface quality on volumetric contours of object their range data are mostly not considered. On the other hand both the textural and volumetric contours produce fast intensity changes. These intensity variations are easy to detect. Therefore the position of contours is defined on the basis of the intensity data in our system. The volumetric contours divide the scene into elementary surface regions. The characteristics of surface segments are defined by the use of range data.

As shadow areas are without ambiguity marked in the range matrix, it is possible (due to the random scan capability of the scanner) to drive the laser light along the boundary of the shadow area. The laser light causes a saturation in the CCD camera. There are the shadow regions of the laser scanner marked in the CCD camera picture exposed in this case. The evaluation of these regions in the camera picture makes the correction of range data and the solving of ambiguity problems on the recognition level possible.

EXPERIMENTAL RESULTS

The resolution of distance using this system was designed to be 1.5 mm for an object distance of 1.5 m. The resolution depends on the base length (which was 0.3 m in our case). The corresponding resolution can be increased up to 0.25 mm using a deflection compensated optics, or using larger base length. The increase of base length however has disadvantage viz., shade effect becomes more significant.

Another problem arises with highly reflective surfaces due to which the sensor would become "blind", as it can be seen on figure 5-6. Fig. 5. shows a laser intensity picture of a tea-cup. The scanning method is very sensitive for textural variations which makes a surface quality control possible with the same sensor system. Fig. 6 shows a background compensated range picture with the three kinds of range data.

The exposure time for a 256^2 image is in the order of few seconds.

CONCLUSIONS

Structured light methods seem to be a promising approach to 3-D vision. For industrial application it would be necessary to increase the speed of the system. The shade effects can be decreased by combining 2 and 3-D pictures. This arrangement seems to be useful aiming to extract textural properties too.

A new system is under development based on these experiences. The system will contain among others two galvano-mirrors to deflect the laser light, and a deflection compensator optics for the PSD providing random tracking with an increased speed. The system will be in operation in 1988.

REFERENCES

1. J.L. Mundy, G.B. Porter: A Three-Dimensional Sensor Based on Structured Light - in Three Dimensional Machine Vision (ed. by Takeo Kanade), Kluwer Academic Publishers, Boston/Dordrecht/Lancaster, 1987
2. D.J. Svetkoff: Towards a High Resolution, Video Rate, 3-D Sensor for Machine Vision, SPIE Vol. 728, Optics, Illumination, and Image Sensing for Machine Vision, 1986
3. J. Jalkio, R. Kim, S. Case: Three Dimensional Inspection Using Multi-Stripe Structured Light, Optical Engineering, Vol. 24, No. 6, Nov. 1985
4. F. Blais, M. Rioux: "BIRIS": a Simple 3-D Sensor, SPIE Conference on Advances in Intelligent Robot Systems, Cambridge, MA, Oct. 1986
5. M. Rioux: Laser Range Finder Based Upon Synchronized Scanners, Applied Optics, Vol. 23, No. 21, Nov. 1984
6. P. Levi, E. Weirich: laser-gestützte Qualitätskontrolle mit Synthetischen Bildern, 5. DAGM Symposium (Mustererkennung), Karlsruhe, VDE Fachberichte 35, 101-106, 1983

192

7. P. Levi, E. Weirich: Differential Reflectance Functions and their Use for Surface Identification, Proc. of the 13th Int. Symp. on Industrial Robots, Chicago, 17.61-17.77, 1983
8. M. Hebert, T. Kanade: Outdoor Scene Analysis Using Range Data, in IEEE Conference on Robotics and Automation, 1426-1432, 1986

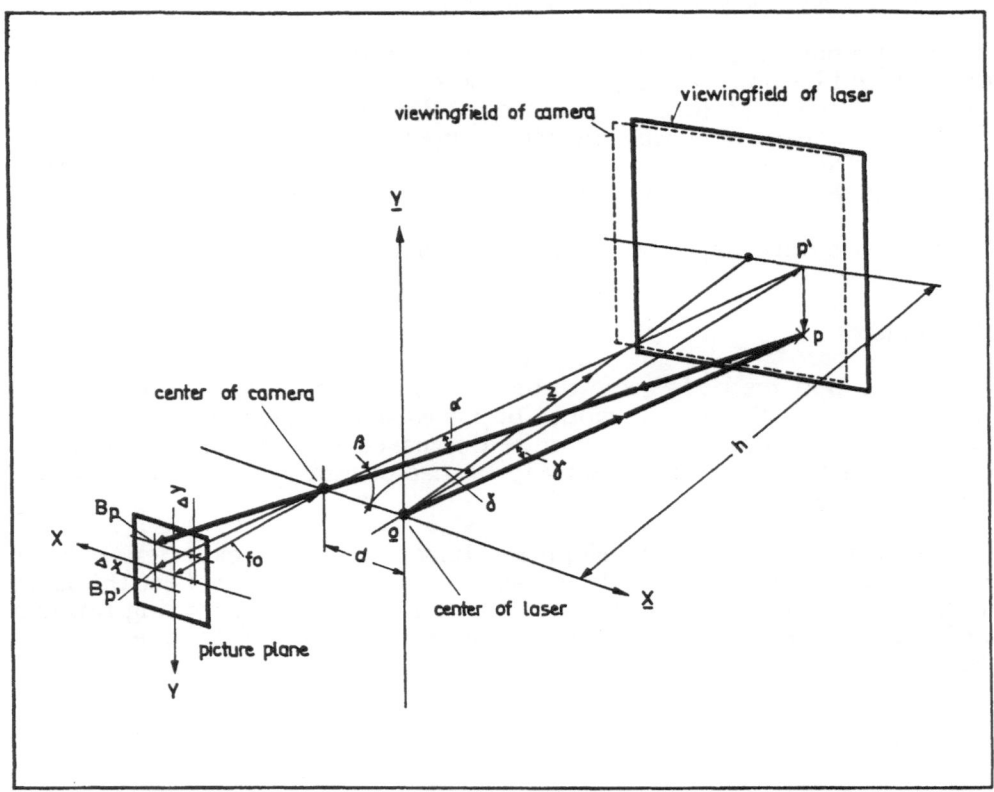

Figure 1: Principle of triangulation

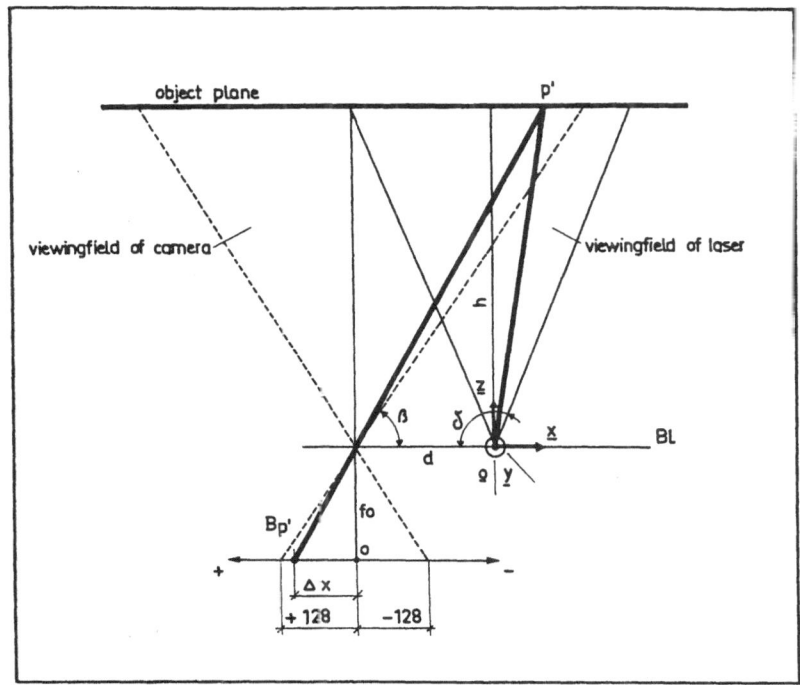

Figure 2: Principle of parallel projection

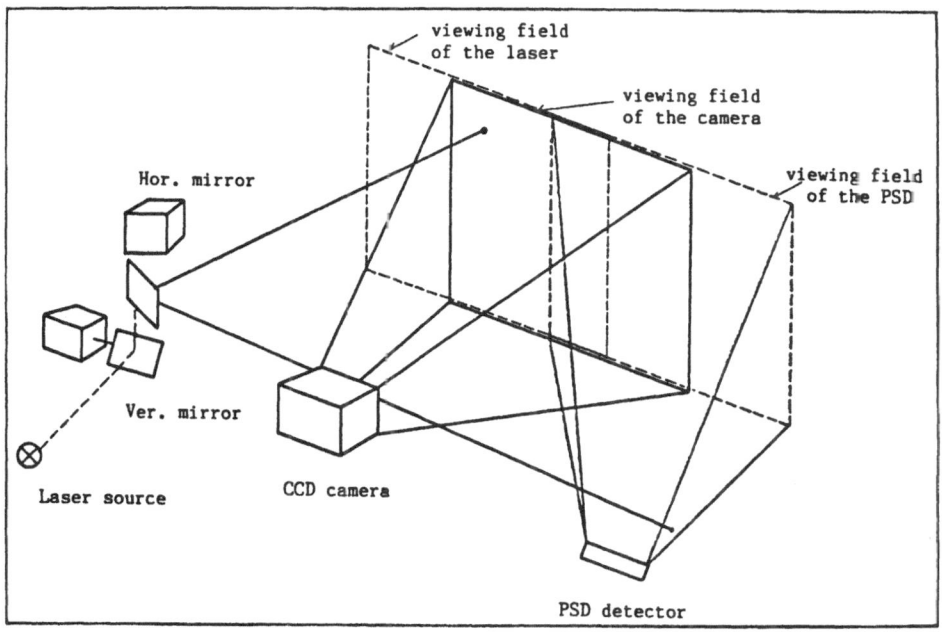

Figure 3: Schematical structure of the triangulation scanner

194

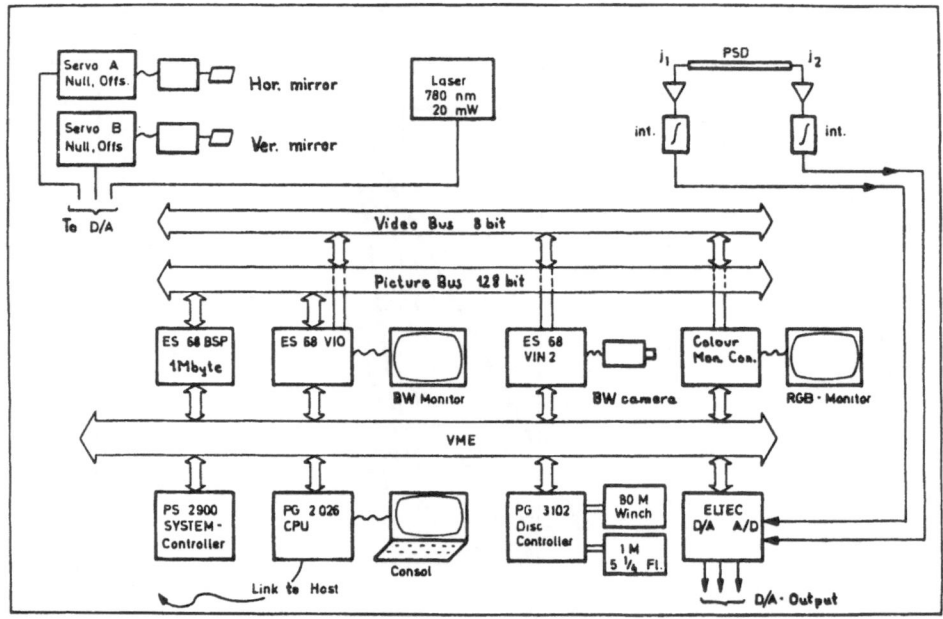

Figure 4: Structure of the multi-processor system

Figure 5: Intensity image of
a coffee cup

Figure 6: Distance image of
a coffee cup

THE CALIBRATION PROBLEM FOR STEREOSCOPIC VISION

O.D. Faugueras, G. Toscani
INRIA
78153 Le Chesnay   (France)

Abstract.

The problem of calibrating a stereo system is extremely important in practical applications. We describe in this paper our approach for coming up with an efficient and accurate solution. We first review the pinhole camera model that is used and analyze its relationship with respect to the internal camera parameters and its position in space. We then study its behavior with respect to changes of coordinate systems.
This yields a constraint which is used in the meansquare solution of the calibration problem that we propose. Since an estimation of the uncertainty is also important, we suggest another solution based on Kalman filtering.
We show a number of experimental results and compare them with those obtained by Tsai [8]. We finish with two practical applications of our calibration technique: reconstructing 3D points and computing the epipolar geometry of a stereo system.

I Introduction.

The problem of calibrating a number of cameras is becoming extremely important in the field of Computer Vision to make 3D Perception possible. For Vision Systems used in industrial applications the *accuracy* of 3D reconstruction relies on it.
Furthermore it can prove important [2] for an intelligent vision system to have a measure of the *uncertainty* related to the reconstructed 3D points for the definition of a sensing strategy: where to look and how to increase the accuracy on the location of an object. Again this uncertainty depends, among other factors, on the calibration.
Even though this problem has been often considered as minor by many people as compared to the more noble problem of Stereo, or Motion, we think that this attitude is wrong for two reasons.
First, in Stereo, the epipolar constraint (detailed in the last Section of this paper) plays a very important role since it reduces the search for correspondences from a two-dimensional space to a one-dimensional one. Accurate knowledge of the epipolar geometry is therefore a prerequisite of all stereo algorithms and can be achieved only by optical calibration if we want to avoid complicated mechanical setups. Due to the lack of easily usable calibration procedures, this has lead most workers in the field to assume in their algorithms a very special camera geometry yielding a simple epipolar geometry with epipolar lines parallel to the scanlines, hoping either that this configuration could be achieved by careful mechanical alignment (not a very practical assumption), or that a more complicated geometry could be compensated for by performing the so-called epipolar transformation on both images in order to make the epipolar lines parallel to the scanlines. Unfortunately, the epipolar transform can only be derived through the very same calibration procedure which is discussed here, and applying it to the images implies interpolating them on a new grid and is therefore bound to be

                i) computationally expensive
  and/or
                ii) inaccurate.

We show in the last Section that the computation of the epipolar geometry is extremely simple from the results of the calibration and that therefore rectifying the stereo pair should only be necessary when numerical correlation techniques are used in the stereo matching. Moreover, if one wants to use more than

NATO ASI Series, Vol. F52
Sensor Devices and Systems for Robotics
Edited by A. Casals
© Springer-Verlag Berlin Heidelberg 1989

This work was partially supported by ESPRIT grant P940
two cameras the notion of epipolar transformation vanishes as fog in the sun.

The second reason is that after working for a while with calibration it becomes clear that the problem is fundamentally the same as that of estimating the motion (or more precisely the displacement) of a camera. Indeed, the knowledge of the displacement from the first camera to the second, i.e rotation and direction of translation, plus the knowledge of some parameters internal to each camera (to be defined later as the intrinsic parameters), completely defines the epipolar geometry. Thus, calibration is equivalent to motion analysis or more precisely, since usually the displacement between the two cameras is not small, can be attacked by token matching techniques [1,3]. The results obtained by the corresponding techniques which are not detailed in this paper but in the two previous references seem to indicate the possibility of self-calibration without the use of a any special pattern.
Our main concern is to use off-the-shelf cameras. Standard cameras and lenses have the advantage of being cheap and easily available, and the disadvantage of usually presenting high symmetric and asymmetric lens distortion, poor definition, together with an absence or poor reliability of information concerning intrinsic parameters.
Here, we analyse the pinhole model of camera from two points of view:
        i) first, the relationship between the perspective transformation performed by such a model and the physical parameters attached to the camera.
        ii) second, the properties of the transformation with respect to changes of coordinates.
In particular, we show that the position and orientation of the camera (extrinsic parameters), and some internal (intrinsic) parameters can be recovered from the perspective transformation matrix while guaranteeing that the computed displacement is indeed a displacement, i.e the rotation matrix is orthogonal which is usually not guaranteed by most of the other techniques.

In order to compute the perspective transformation matrix, it has to satisfy at least one constraint. We show that the most commonly assumed constraint (for its simplicity), yields to a solution which is absurd since the intrinsic parameters depend upon the choice of the world coordinate system. We then propose another constraint, based on the analysis of the properties of the perspective transformation matrix with respect to changes of coordinate systems, which does not suffer from this drawback.

We discuss how the matrix can be estimated both by meansquare and by linear filtering techniques (Kalman Filter) and present a number of ways of measuring the accuracy of the technique and comparing it with the predicted accuracy obtained by the Kalman Filter. One advantage of using the Kalman Filter is that it allows us to incorporate measurement noise at several levels : pixel noise, 3D reference points coordinates noise. We give experimental results and compare them to those obtained by a method reported recently by Tsai [8].

We finish by showing how the results of calibration can be used to reconstruct 3D points and compute the epipolar geometry.

Notations.
Let us define a few notations that will be heavily used in the remaining of this paper:
- bold characters are used for row or column vectors: l, m.
- outline characters are used for matrices: $\mathbb{M}$, $\mathbb{R}$.
- lm denotes the matrix product of row vector l by column vector m.
- l × m stands for the cross product of vectors l and m.
- $\mathbb{R}$m denotes the matrix $\mathbb{R}$ applied to the column vector m.

II The camera model.

In this section we describe the camera model which is used in the sequel.

II.1 The perspective transformation model.
We assume that the camera performs a perfect perspective transformation with center O_1 (the camera optic center) at a distance f (the focal length) of the retina plane (figure 1). Let us consider an ideal image plane parallel to the retina plane at a distance 1 from the optic center.

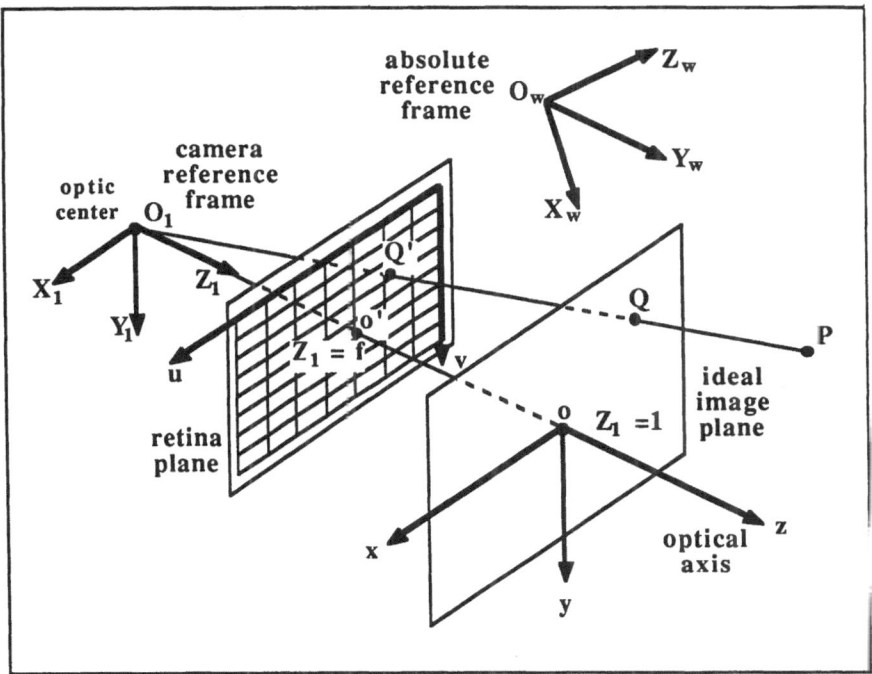

Figure 1. The camera model.

With the camera 3-D coordinate system $(O_1X_1Y_1Z_1)$ and the ideal image plane coordinate system (oxy) as in figure 1 we have the following relationship:

$$\frac{1}{Z_1} = \frac{x}{X_1} = \frac{y}{Y_1} \tag{1}$$

Equation (1) can be rewritten as a linear relationship in the homogeneous coordinate system:

$$\begin{bmatrix} sx \\ sy \\ s \end{bmatrix} = \begin{bmatrix} 1 & 0 & 0 & 0 \\ 0 & 1 & 0 & 0 \\ 0 & 0 & 1 & 0 \end{bmatrix} \begin{bmatrix} X_1 \\ Y_1 \\ Z_1 \\ 1 \end{bmatrix} \tag{2}$$

or, on matrix form:

$$U = \mathbb{M}\, X \tag{3}$$

where matrix $\mathbb{M}$, defined up to a scale factor, is also referred to in the literature [4,6,8] as the perspective transformation matrix. Note that s=0 corresponds to a point at infinity in the camera plane. Indeed, this is equivalent to Z_1=0, i.e. to the object point being in the camera focal plane.
The system (3) can be considered in many ways: if X and $\mathbb{M}$ are known, it yields the image coordinates of the point P; if U and $\mathbb{M}$ are known, it permits the 3-D reconstruction (note that in this case (3) defines a line). The case where U and X are known is used in the sequel to recover the perspective transformation matrix $\mathbb{M}$.

II.2 Effect of rigid displacements.

If we rotate and translate the (O_1, X_1, Y_1, Z_1) coordinate system, as in figure 1:

$$\begin{bmatrix} X_1 \\ Y_1 \\ Z_1 \end{bmatrix} = \mathbb{R} \begin{bmatrix} X_w \\ Y_w \\ Z_w \end{bmatrix} + t$$

where $\mathbb{R}$ is the rotation matrix and t the translation vector, matrix $\mathbb{M}$ becomes:

$$\mathbb{M} = \begin{bmatrix} \mathbb{R} & t \end{bmatrix}$$

II.3 The retina model.

The transformation from retina coordinates (u, v) (in pixels) into image coordinates (x, y) (in meters) is modelled in the following manner (figure 2):

$$x' = a' + b'u + c'v$$
$$y' = d' + e'v \tag{4}$$

with $e' > 0$ and $b' < 0$ or $b' > 0$ depending on the computer acquisition. This model allows two different scale factors along the x and y axes (scanning and sampling) and takes into account a possible non perpendicularity of the u and v axes. If we drop this last assuption then $c'=0$. This means that we assume that the u and v axes are perpendicular, even if they are not. We show in section III that this may cause problems when we recover matrix $\mathbb{R}$.

Letting: $a=a'/f$, $b=b'/f$, $c=c'/f$, $d=d'/f$, $e=e'/f$, and $g = (cd-ae)/be$, from (4) we have:

$$x = x'/f = a + bu + cv$$
$$y = y'/f = d + ev \tag{5}$$

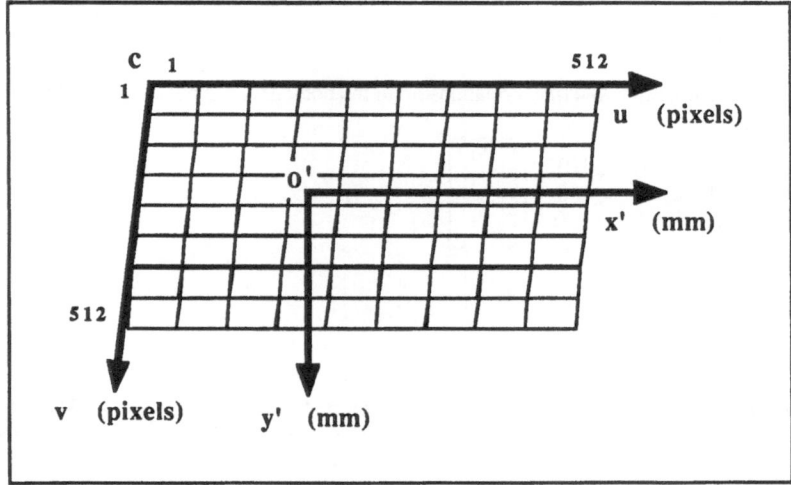

Figure 2. The retina model.

We can rewrite relationship (2) as follows:

$$\begin{bmatrix} su \\ sv \\ s \end{bmatrix} = \begin{bmatrix} 1/b & -c/be & g & 0 \\ 0 & 1/e & -d/e & 0 \\ 0 & 0 & 1 & 0 \end{bmatrix} \begin{bmatrix} X_1 \\ Y_1 \\ Z_1 \\ 1 \end{bmatrix}$$

After rotation and translation of the absolute reference frame, matrix $\mathbb{M}$ becomes:

$$\mathbb{M} = \begin{bmatrix} 1/br_1 -c/ber_2+gr_3 & 1/b\,t_x -c/be\,t_y+gt_z \\ 1/er_2 -d/er_3 & 1/e\,t_y -d/\acute{e}\,t_z \\ r_3 & t_z \end{bmatrix} = \begin{bmatrix} l_1 & l_{14} \\ l_2 & l_{24} \\ l_3 & l_{34} \end{bmatrix} \quad (6)$$

where r_i is the ith (1x3) row vector of the (3x3) rotation matrix $\mathbb{R}$, and l_i the ith (1x3) row vector of matrix $\mathbb{M}$ without the last coordinate. Matrix $\mathbb{M}$, which relates object coordinates in an absolute reference frame to image coordinates in pixels, is the actual matrix to be calibrated.

II.4 Image distortion.

There are many mathematical models [10,11,12,13] which have been developed to describe image distortion. The various models are built using the collinearity equation (1) in the form:

$$u = \frac{l_1 X + l_{14}}{l_3 X + l_{34}} + \Delta u + \varepsilon_u$$

$$v = \frac{l_2 X + l_{24}}{l_3 X + l_{34}} + \Delta v + \varepsilon_v$$

where $X = (X_w, Y_w, Z_w)$ are the object coordinates in the world reference frame, (u, v) are the image coordinates (in pixels), $\Delta u, \Delta v$ are systematic errors (e.g. due to lens distortion) and $\varepsilon_u, \varepsilon_v$ are random errors. In the case of a perfectly centered lens, the distortion is symmetrical about the optical axis and may be represented, according to Franke [14], by an odd polynomial:

$$\Delta r = K_1 r^3 + K_2 r^5 + K_3 r^7 + \dots$$

or, separated in the components along the u and v axes:

$$\Delta u_s = \bar{u} (K_1 r^2 + K_2 r^4 + K_3 r^6 + \dots)$$
$$\Delta v_s = \bar{v} (K_1 r^2 + K_2 r^4 + K_3 r^6 + \dots)$$

where:

$$r = \sqrt{ \bar{u}^2 + \bar{v}^2 }$$
$$\bar{u} = u - u_0 \qquad \bar{v} = v - v_0$$

(u, v) are the distorted image coordinates
(u_0, v_0) are the coordinates of the point of symmetry

Tsai [8] proposed a linear technique in the case of a symmetric (radial) error. However, the asymmetric component of lens distortion found in standard cameras and lenses is often non negligible. This can be seen in figure 4, which represents the distortion of the calibration pattern of figure 6 in the following manner: least-squares lines are computed from the distorted image lines, as shown in figure 3, and the residual error, multiplied by a factor of 15, is plotted in figure 4.

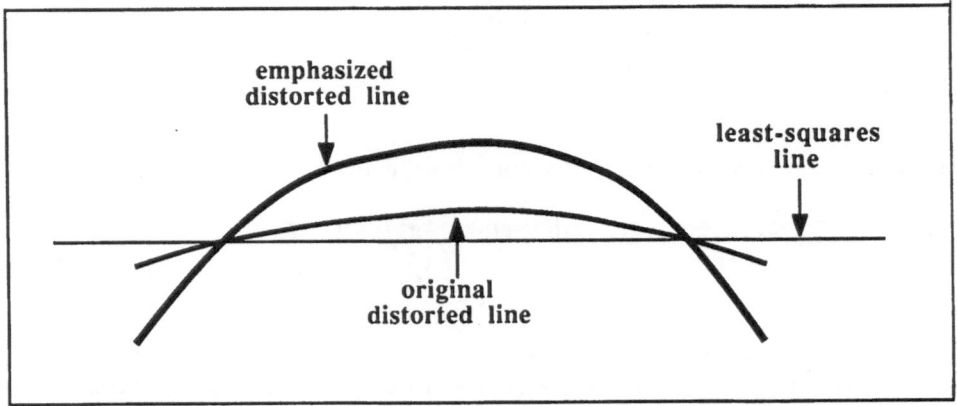

Figure 3. Representing the distortion of straight lines.

According to Conrady [15], the error due to decentering of lens elements is:

$$\Delta u_a = [\ P_1(\ r^2 + 2\ \overline{u}^2\ ) + 2P_2\ \overline{u}\ \overline{v}]\ (\ 1 + P_3 r^2 + P_4 r^4 + \ldots\ )$$
$$\Delta v_a = [\ P_2(\ r^2 + 2\ \overline{v}^2\ ) + 2P_1\ \overline{u}\ \overline{v}]\ (\ 1 + P_3 r^2 + P_4 r^4 + \ldots\ )$$

The mathematical model of the global distortion is then given by:

$$\Delta u = \Delta u_s + \Delta u_a$$
$$\Delta v = \Delta v_s + \Delta v_a$$

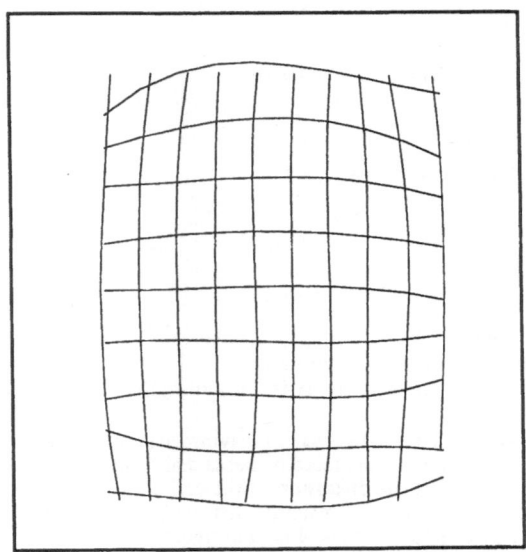

Figure 4. Distortion of the calibration pattern.

Brown [11] proposed a technique for correcting radial and decentering distortion based on this model using a calibration test pattern formed by straight lines. The idea is to correct the distorted coordinates so that they conform to a straight line.

Our approach consists in computing the perspective transformation matrix first, and then correct for the image distortion.

II.5 Parameters to be calibrated.

We consider the camera as a black box with no a-priori knowledge of the intrinsic parameters, such as scale factors (b, e), focal length (f), perpendicularity error (c), or image center coordinates (-a/b, -d/e). This is the case, in particular, when normal vidicon type cameras are used, compared to discrete array cameras where scale factors can be obtained using information supplied by the manufacturer. Since there is no way with such model to separate focal length from scale factors, there are 5 intrinsic parameters to be calibrated: a, b, c, d, e.

The position of the camera 3-D coordinate system with respect to the world coordinate system yields 6 extrinsic parameters to be calibrated: 3 for the rotation and 3 for the translation.

III Computing the parameters.

Matrix $\mathbb{M}$ having been measured by a method described later, a first question is, how can we recover the parameters a, b, c, d, e, $\mathbb{R}$ and t. According to (6), the fact that $\mathbb{R}$ is a rotation matrix and $\mathbb{M}$ is defined up to a scale factor k, we can recover all the parameters if we know the sign of k. Since from (6) $l_{34} = k\, t_z$, we can see that this is equivalent to know whether the origin of the world coordinate system O_w is in front or behind the camera ($t_z > 0$ or $t_z < 0$).

The parameters are recovered as follows:

$$k = \sqrt{l_3 l_3^T}$$

$$t_z = l_{34} / k$$

if $t_z < 0$ then $t_z = -t_z$ and k = -k (since we have choosen in this way the absolute reference frame)

$$r_3 = l_3 / k$$

$$g = l_1 l_3^T / k^2$$

$$d/e = -l_2 l_3^T / k^2$$

$$e = \frac{1}{\sqrt{l_2 l_2^T / k^2 - d^2/e^2}} \qquad e > 0 \text{ by definition}$$

$$d = (d/e)\, e$$

$$r_2 = e/k\, l_2 + d/k\, l_3$$

$$t_y = (e/k)\, l_{24} + (d/k)\, l_{34}$$

$$c/b = -e^2 l_1 l_2^T / k^2 - gde$$

$$b = \frac{1}{\sqrt{l_1 l_1^T / k^2 - c^2/(b^2 e^2) - g^2}}$$

b > 0 or b < 0 (depending on the orientation of the v-axis of the retina: b>0 in all our setups)

$$c = (c/b)\, b$$

$$r_1 = b/k\, l_1 + c/k\, l_2 + a/k\, l_3$$

$$t_x = (b/k)\, l_{14} + (c/k)\, l_{24} + (a/k)\, l_{34}$$

$$a = (cd - gbe) / e$$

It can be easily verified that the rotation matrix thus obtained is orthogonal. This is not true in general, when c=0. Indeed, assuming without loss of generality, that k = 1, it easy to show that:

$$r_1 = (l_1 - (l_1 l_3^T) l_3) / A$$
$$r_2 = (l_2 - (l_2 l_3^T) l_3) / B$$

where $A = \sqrt{l_1 l_1^T - (l_1 l_3^T)^2}$ and $B = \sqrt{l_2 l_2^T - (l_2 l_3^T)^2}$. Therefore:

$$r_1 r_2^T = (l_1 l_2^T - (l_1 l_3^T)(l_2 l_3^T)) / AB$$

which is not zero in general.

IV Estimation of the perspective transformation matrix: two mean-square techniques.

We assume that we are given the 3-D coordinates $X_i = [x_i, y_i, z_i]^T$, and the 2-D coordinates (u_i, v_i) of N (N>5) non coplanar reference points P_i. We are going to use (3) to estimate M. For every reference point P_i, (3) yields two linear equations in the unknowns l_i and l_{ij}. By multiplying the 3rd equation of (3) by u (resp. v) and subtracting it from the 1st one (resp. 2nd), we can rewrite (3) as follows:

$$l_1 X_i - u_i l_3 X_i + l_{14} - u_i l_{34} = 0$$
$$l_2 X_i - v_i l_3 X_i + l_{24} - v_i l_{34} = 0 \qquad (7)$$

For N points we obtain an homogeneous linear system of 2xN equations:

$$A L = 0_{2N} \qquad (8)$$

where A is a (2Nx12) matrix depending on the reference points 3-D and 2-D coordinates, 0_{2N} is the (2Nx1) null vector and L is the vector:

$$L = [l_1, l_{14}, l_2, l_{24}, l_3, l_{34}]^T \qquad (9)$$

It can be proved that, for a large number (N>>5) of non-coplanar calibration points, the rank of A is 11. System (8) is thus overdetermined and the problem is to find the vector L that best satisfies equation (8), i.e. that minimizes the criterion $C = \| A L \|$.

IV.1 Constraint $l_{34} = 1$.
Equation (8) has been solved so far [4,9] using standard least-squares approach by setting the constraint $l_{34}=1$. This yields the solution:

$$L = - (C^T C)^{-1} C^T B \qquad (10)$$

where C is the (2Nx11) matrix formed by the first 11 columns of A and B is the 12th column.

IV.2 Changing coordinate system.
It is now useful to ask ourselves how a perspective transformation matrix M transforms itself if we decide to change the current coordinate system.

Using the same notation as before, is is easily shown that:

$$M' = \begin{bmatrix} l_1 R & l_1 t + l_{14} \\ l_2 R & l_2 t + l_{24} \\ l_3 R & l_3 t + l_{34} \end{bmatrix} = M \begin{bmatrix} R & t \\ 0_3{}^T & 1 \end{bmatrix} \qquad (11)$$

where R and t are the (3x3) rotation matrix and (3x1) translation vector from the new world coordinate system, E, to the old world coordinate system, E', defined by:

$$X_{old} = R \, X_{new} + t$$

Let:

$$M' = M \, U \qquad (12)$$

Equation (7) becomes:

$$A' L = 0_{2N} \qquad (13)$$

where A' depends on the new world coordinates of the reference points.
Letting $X_i' = [x_i', y_i', z_i', 1]^T$, the coordinate vector in the new reference frame E', we have:

$$X_i' = U^{-1} X_i$$

and:

$$A_{2i}' = [X_i^T (U^{-1})^T, 0_4{}^T, -u_i X_i^T (U^{-1})^T]$$
$$A_{2i+1}' = [0_4{}^T, X_i^T (U^{-1})^T, -v_i X_i^T (U^{-1})^T]$$

or:

$$A' = A \, (W^{-1})^T$$

where we have set:

$$W = \begin{bmatrix} U & 0 & 0 \\ 0 & U & 0 \\ 0 & 0 & U \end{bmatrix}$$

with 0 the (4x4) null matrix. Let K be the perspective transformation matrix derived from the solution of (8) and K' the one corresponding to the solution of (13). From (11) we expect them to verify :

$$K' = K \, U \qquad (14)$$

This is not the case if we use the constraint $l_{34}=1$, as can be seen from equation (11). In particular, the intrinsic parameters will depend on the choice of the world coordinate system.

IV.3 Constraint $\| l_3 \|^2 = 1$.
It is interesting to note from (6) and (11) that the relation $l_3 l_3{}^T = 1$ is left invariant by the group of rigid motions.

Setting $X_9 = [l_1, l_{14}, l_2, l_{24}, l_{34}]^T$ and $X_3 = l_3^T$ equation (8) can be rewritten as follows:

$$\mathbb{B}\,X_9 + \mathbb{C}\,X_3 = 0_{2N}$$

Setting the constraint $\| X_3 \|^2 = 1$, the criterion to be minimized becomes:

$$C = \| \mathbb{B}\,X_9 + \mathbb{C}\,X_3 \|^2$$

or, using the method of Lagrange multipliers:

$$C = \| \mathbb{B}\,X_9 + \mathbb{C}\,X_3 \|^2 + \lambda\,(1 - \| X_3 \|^2)$$

for all real number λ. We have:

$$
\begin{aligned}
C &= (\mathbb{B}\,X_9 + \mathbb{C}\,X_3)^T (\mathbb{B}X_9 + \mathbb{C}\,X_3) + \\
&+ \lambda\,(1 - X_3^T X_3) = X_9^T \mathbb{B}^T \mathbb{B} X_9 + X_3^T \mathbb{C}^T \mathbb{C} X_3 + \\
&+ X_9^T \mathbb{B}^T \mathbb{C} X_3 + X_3^T \mathbb{C}^T \mathbb{B} X_9 + \lambda\,(1 - X_3^T X_3)
\end{aligned}
\tag{15}
$$

Differentiating C with respect to X_9 and X_3 and setting the derivatives to 0, we find:

$$(1/2)\,\partial C/\partial X_9 = \mathbb{B}^T \mathbb{B} X_9 + \mathbb{B}^T \mathbb{C} X_3 = 0_9 \tag{16}$$

$$(1/2)\,\partial C/\partial X_3 = \mathbb{C}^T \mathbb{C} X_3 + \mathbb{C}^T \mathbb{B} X_9 - \lambda X_3 = 0_3 \tag{17}$$

which yields:

$$X_9 = -(\mathbb{B}^T \mathbb{B})^{-1}\,\mathbb{B}^T \mathbb{C}\,X_3$$

$$(\mathbb{C}^T \mathbb{C} - \mathbb{C}^T \mathbb{B}\,(\mathbb{B}^T \mathbb{B})^{-1} \mathbb{B}^T \mathbb{C})\,X_3 = \lambda\,X_3$$

Let us define the (3x3) matrix $\mathbb{D}$ and (2Nx2N) matrix $\mathbb{E}$ as follows:

$$\mathbb{D} = \mathbb{C}^T \mathbb{C} - \mathbb{C}^T \mathbb{B}(\mathbb{B}^T \mathbb{B})^{-1}\mathbb{B}^T \mathbb{C} = \mathbb{C}^T(\mathbb{I}_{2N} - \mathbb{B}(\mathbb{B}^T \mathbb{B})^{-1}\mathbb{B}^T)\mathbb{C} = \mathbb{C}^T \mathbb{E}\,\mathbb{C}$$

where $\mathbb{I}_{2N}$ is the (2Nx2N) identity matrix. $\mathbb{D}$ and $\mathbb{E}$ are symmetric and we prove that they are positive.

Let Y be an eigenvector of $\mathbb{E}$:

$$Y - \mathbb{B}(\mathbb{B}^T \mathbb{B})^{-1}\mathbb{B}^T Y = \lambda Y \tag{18}$$

If $Y \notin \mathrm{Ker}(\mathbb{B}^T)$, multiplying both members of (18) by $\mathbb{B}^T$:

$$\lambda \mathbb{B}^T Y = 0_9$$

which yields $\lambda = 0$.

If $Y \in \mathrm{Ker}(\mathbb{B}^T)$, (18) yields $\lambda = 1$. The only eigenvalues of $\mathbb{E}$ being 0 and 1, $\mathbb{E}$ is positive, which in turn yields $\mathbb{D}$ positive.

If X_3' is a unit eigenvector coresponding to the eigenvalue λ', (15), (16) and (17) yield $C = \lambda'$.

The solution is thus obtained by taking a unit eigenvector of $\mathbb{D}$ corresponding to the smallest eigenvalue: this yields X_9 and $\mathbb{M}$.

We now prove that the matrix $\mathbb{M}$, obtained using the constraint $\| l_3 \|^2 = 1$, verifies (14). In this paragraph, $(L)_3$ will denote the vector l_3 defined in equation (6).

Let L_1 (resp. L_1') be the vector that minimizes the criterion $C(L) = \|\mathbb{A}L\|$ (resp. $C'(L') = \|\mathbb{A}'L'\|$) subject to the constraint $\| (L)_3 \| = 1$ (resp. $| (L')_3 \| = 1$).

We prove that $L_1 = (\mathbb{W}^{-1})^T L_1'$.

Let : $L_2 = (\mathbb{W}^{-1})^T L_1'$ and $L_2' = (\mathbb{W})^T L_1$. We show that:

$$L_2 = L_1 \tag{19}$$

It is easy to verify that (19) is equivalent to (14). We have :

$$(L_2)_3 = \mathbb{R} \, (L_1')_3 \quad \text{and} \quad (L_2')_3 = \mathbb{R}^T \, (L_1)_3$$

which yields : $\| (L_2)_3 \| = 1$ and $\| (L_2')_3 \| = 1$.
Furthermore from the definitions:

$$C'(L_1') = \| \mathbb{A}'L_1' \|^2 = \| \mathbb{A}(\mathbb{W}^{-1})^T L_1' \|^2 = C (L_2)$$

and similarly:

$$C'(L_2') = C (L_1)$$

We cannot have $C (L_2) < C (L_1)$ since it is inconsistent with the hypothesis that L_1 minimises the criterion $C (L)$.

We cannot have $C (L_2) > C (L_1)$ either, since $C (L_2) = C' (L_1') > C (L_1) = C' (L_2')$ would be inconsistent with the hypothesis that L_1' minimises the criterion $C' (L)$. We must then have $C (L_2) = C (L_1)$. The criterion is the smallest eigenvalue of matrix $\mathbb{D}$ of the previous paragraph. The solution is unique up to the sign if the smallest eigenvalue is unique. As shown in the section on computing parameters, the sign is unique. Therefore in that case $L_1 = L_2$.

IV.4   Constraint $l_1 l_2^T - (l_1 l_3^T) (l_2 l_3^T) = 0$.

We have seen previously that if we assume the u and v axes to be orthogonal, the matrix $\mathbb{R}$ computed from matrix $\mathbb{M}$ may satisfy $r_1 r_2^T \neq 0$.

To insure that $r_1 r_2^T = 0$ we must add to $| l_3 \|^2 = 1$ the constraint $l_1 l_2^T - (l_1 l_3^T) (l_2 l_3^T) = 0$, when minimizing criterion C. It can be verified that this constraint is left invariant by the group of rigid motions. Unfortunately, unlike the constraint $\| l_3 \|^2 = 1$, this constraint is a polynomial of degree 4 in the coefficients of matrix $\mathbb{M}$ and a closed form solution cannot be found. Numerical techniques can also be used, but we have not pursued in this direction since relaxing the orthogonality assumption of u and v does the job.

V   Estimation of the perspective transformation matrix: Kalman filtering.

Kalman filtering [2,7] estimates a vector a, of dimension n, called the state vector, of an invariant linear system, using a measured vector y, of dimension p, in presence of measurement noise u, independent of a. The measurement equation is:

$$y = \mathbb{M}\, a + u \tag{20}$$

with $\mathbb{M}$ a known (pxn) matrix and u's second order statistics known.

If we have N measurements $y_1,...,y_N$ satisfying (20) and we start with an initial estimate $\hat{a}_0$ of a and its associated covariance matrix $\mathbb{S}_0 = E\,(\,(\hat{a}_0 - a)(\hat{a}_0 - a)^T)$ the Kalman filter approach deduce recursively an estimate $\hat{a}_N$ of a and its covariance matrix $\mathbb{S}_N = E\,(\,(\hat{a}_N - a)(\hat{a}_N - a)^T)$ after taking into account N observations.

The key assumption is that the parameters to be calibrated can be modeled by a state vector and that the measurements y are corrupted with an additive zero mean Gaussian noise:

$$y = y' + \varepsilon$$

V.1 Classical Kalman filtering.

One way to keep the linear approach of (20) is to use the constraint $l_{34} = 1$. The function to consider is the one formed by the system (7): y is the vector $[u, v, X, Y, Z]^T$ of the 2-D and 3-D observations of a reference point; the state vector a is formed by the first 11 elements of vector L defined in (9). Note that $\mathbb{M} = \mathbb{C}$ and $y = B$, where $\mathbb{C}$ and B are the one used for (10).

VI Accuracy Analysis.

VI.1 Four types of measures.

Many ways of measuring the calibration accuracy have been proposed in the literature; we discuss four of them:

Measure of type I: Angle from inverse projection.
As shown in Fig. 5, the calibration process tries to find camera model parameters such that the ray starting from the optical center O, passing through the image point Q_t will eventually pass through the three-dimensional test point P_t. Due to various errors, the ray will not exactly pass through P_t. One measure of the camera calibration accuracy is the extent of the angle $P_t O Q_t$. The practical computation of this angle is detailed below.

Measure of type II: Three-dimensional reconstruction.
Let (u_1, v_1) and (u_2, v_2) be the measured image points on two image planes of an objet point $P_t = (x_t, y_t, z_t)$. The rays starting from the optical centers O_1 and O_2 define a point $P_c = (x_c, y_c, z_c)$ which is the solution of the system (eventually obtained using a least-squares technique):

$$\begin{bmatrix} su_1 \\ sv_1 \\ s \end{bmatrix} = \mathbb{M}_1 \begin{bmatrix} x_c \\ y_c \\ z_c \\ 1 \end{bmatrix} \qquad \begin{bmatrix} su_2 \\ sv_2 \\ s \end{bmatrix} = \mathbb{M}_2 \begin{bmatrix} x_c \\ y_c \\ z_c \\ 1 \end{bmatrix} \tag{23}$$

where $\mathbb{M}_1$ and $\mathbb{M}_2$ are the calibration matrices associated to the two cameras. A measure of the calibration accuracy of the stereo pair is the vector $P_c P_t$ i.e. the difference between the measured 3D coordinates of the test point and the coordinates computed in (23).

Measure of type III: Stability of intrinsic parameters.
Another measure of the calibration quality is the stability of the camera intrinsic parameters over a number of calibrations obtained by using different sets of three-dimensional objet points. Two calibrations yield two matrices $\mathbb{M}$ and $\mathbb{M}'$. From these perspective transformation matrices we derive the intrinsic parameters: a, b, c, d, e. The difference between the parameters obtained from the two calibrations is then a measure of the calibration quality.

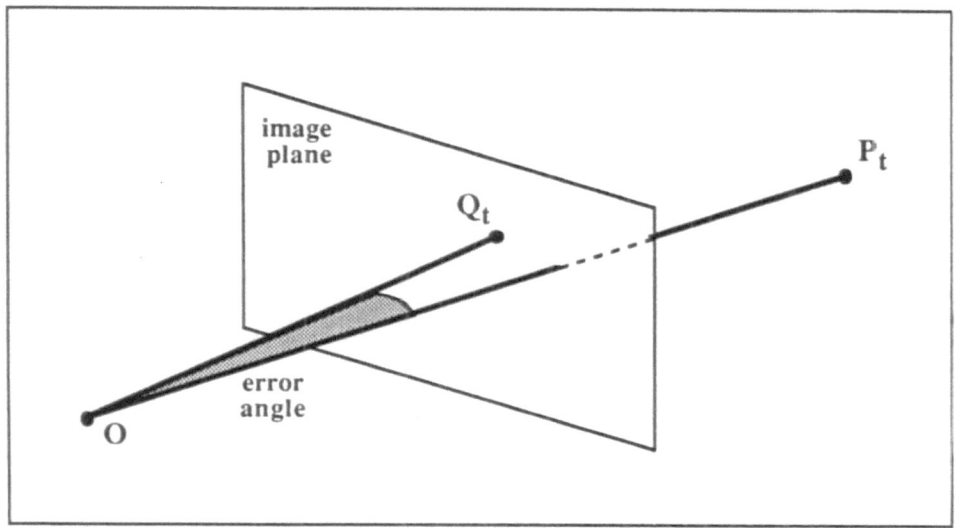

Figure 5. Accuracy measure of type I.

Measure of type IV: Dimensional measurements.
It is sometimes useful to determine how well the camera can perform relative instead of absolute measurements.

Measure of type II vs. measure of type III.
The objective of the calibration is the three-dimensional reconstruction. The knowledge of the camera parameters accuracy (measure of type III) does not tell us anything more about our primary concern: 3D accuracy. Nevertheless it could be useful to consider it in some applications, such as the manipulation of a camera held by a robot arm.

Measure of type I vs. measure of type II.
The accuracy of stereo reconstruction (measure of type II) depends on the accuracy of the calibration of each camera of the stereo pair (measure of type I) and on their relative position. Given the accuracy of the calibration of each camera, we are always able to determine the accuracy of the 3D reconstruction, but not the opposite. So the measure of type I is the information that determines completely the accuracy of the calibration and is the only type that will be considered in the sequel.

VI.2 Computing the error angle.
To determine the angle $P_t O Q_t$ the procedure is as follows:
Step 1). Calibrate the camera using a set of non coplanar points $(P_m)_{m=1,M}$. This step yields the perspective transformation matrix $\mathbf{M}$.
Step 2). Compute the optic center coordinates in the world reference frame (see section VIII.2)
Step 3). Given a three-dimensional test point $P_t = (x_t, y_t, z_t)$, not used for calibration in step 1, let $Q_t = (u_t, v_t)$ be its corresponding image point. The direction of the line OQ_t in the world reference frame is obtained using equation (24) (see section VII). We can then compute the angle $P_t O Q_t$.

VI.3 Computing the uncertainty ellipses.
For every test point P_t let R_p and R_q be the intersection of the segments OP_t and OQ_t respectively with the ideal image plane. Given a set of test points $(P_{t_m})_{m=1,M}$, over a certain region of space, we consider the covariance matrix of the vectors $(R_{p_m} R_{q_m})_{m=1,M}$, which can be visualized by drawing the ellipse associated to it. The score reported on table I is the inverse of the average standard deviation.

VII   Experimental results.

VII.1   Experimental procedure.

The calibration test pattern we use is a grid of lines created with a plotter, as shown in figure 6. The center of the intersection of any two lines is a measurement point. We used a plane containing 113 points. The first image was taken with three cameras at approximately 3 meters. Other three images were taken at 50, 100 and 150 cm from the first position, giving a total of 452 calibration points. Two planes out of the four were used for the calibration and the other two for the test. Two techniques proposed by TSAI [8] have been implemented and the test results for the calibration of one of the three cameras are shown in Table I.

Tsai monoview single plane technique.

The intrinsic parameters used were the one supplied by the manufacturer of the camera. The calibration planes used formed a small angle (between 0.3 and 5 degrees) with the image plane. This could be a reason of the poor performance, since the algorithm fails in the case of a plane parallel to the image plane.

Tsai multiple plane technique.

The intrinsic parameters used were again the one supplied by the manufacturer of the camera. Tsai technique is based upon the hypothesis that the parameters given by the manufacturer are accurate enough and that asymmetric distortion is negligible: this seems not to be the case with the cameras we used. Note that for vidicon cameras this techique is not applicable since such parameters are not available.

VII.2   Test pattern: lines versus circles.

The perspective projection of a circle, situated on a plane not parallel to the image plane, is an ellipse. The center of the circle does not project onto the center of the ellipse. If we take the center of the ellipse in the image plane as the calibration point we make an error that depends on the angle between the plane of the circle and the image plane and it is proportional to the image size of the ellipse. Furthermore, the precision in determining the center of the ellipse depends on the number of pixels on its contour, (i.e. image size): the bigger the image size and the less ellipses can fit into the image, the less calibration points are available. Those disadvantages are solved by using as calibration points the intersections of the centers of lines at right angle. The number of pixels used to determined an image calibration point depends only on the size of the image. The thickness of a line can be reduced to a few pixels, leaving a negligible error in the determination of the medium point.

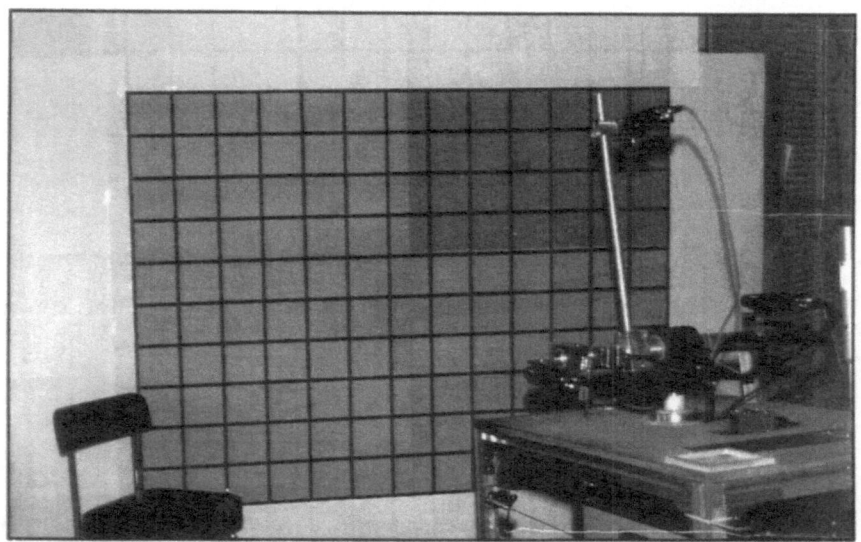

Figure 6. Set-up for camera calibration.

VII.3 Distortion correction.

In order to compensate for the geometric distortions present in Figure 4, we have decomposed our images into a number of square buckets (about a hundred) in which we assumed that the distortion was homogeneous and could be modelled by an affine transformation:

$$u' = \alpha + \beta u + \gamma v + \delta uv$$
$$v' = \varepsilon + \eta u + \vartheta v + \chi uv$$

(u,v) are the measured pixel coordinates for a given test point, (u',v') are the corrected pixel coordinates; the coefficients $\alpha, \beta, \gamma, \delta, \varepsilon, \eta, \vartheta, \chi$ are found by minimizing the square distances between the corrected pixels and the pixels obtained by applying the matrix $\mathbf{M}$ to the test points, while preserving the continuity at the border of the squares. This situation is depicted in Figure 7.

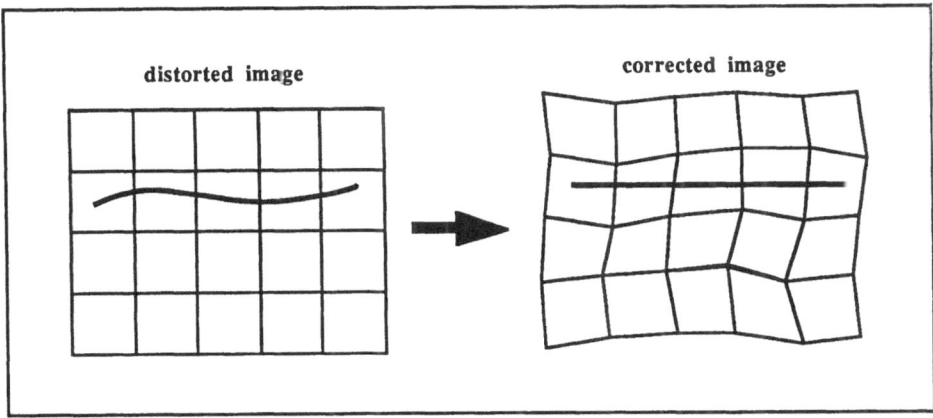

Figure 7. Geometrical distortion correction.

VII.4 Statistical uncertainty ellipses.

A way to characterize the reconstruction error is to compute the covariance, within each bucket, of the error vectors between the measured pixel coordinates and the projected pixel coordinates for each test point whose projection falls in the bucket. The results are shown in figure 8.

VII.5 Error prediction using Kalman filtering.

We can predict the error made in reconstruction from the results of the Kalman filtering tehcnique described in section V.1. Indeed, after calibration we have the covariance matrix on the coefficients of matrix $\mathbf{M}$. Given an image point $Q = [u, v]^T$, we consider the point P, intersection of the line OQ with the ideal image plane. This point has coordinates x and y. Equations (7) can be considered as measurements equations with $y = L$ and $a = [x, y]^T$. Therefore the Kalman filter provides us with an estimate of x and y as well as an estimate of their covariance matrix which we represent as an ellipse in figure 9. These have to be compared with the experimentally measured ellipses of figure 8. It can be seen that the agreement is fairly good, indicating that our overall modelling is accurate enough.

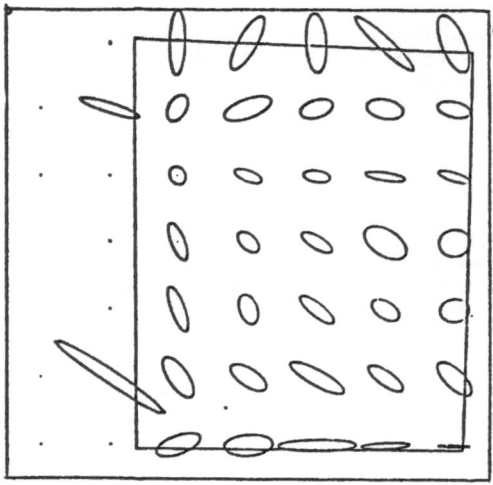

| Figure 8. Statistical uncertainty ellipses associated to measure type I. | Figure 9. Uncertainty ellipses predicted by using the Kalman filter. |

VIII Applications.

We now show two simple applications of our calibration procedure to the problem of Stereo.

VIII.1 3D reconstruction of points.

The first one is the reconstruction of 3D points from their image coordinates $(u_i, v_i)_{i=1,\ldots,n}$ in n camera planes, given for each camera the corresponding matrix M_i obtained by calibration. The case of standard stereo corresponds to n=2, but we can have n>2. Let $X=[x, y, z]^T$ the coordinate vector representing the 3D point to be reconstructed. We rewrite equations (7) as:

$$l_1{}^i X - u_i l_3{}^i X + l_{14}{}^i - u_i l_{34}{}^i = 0$$
$$l_2{}^i X - v_i l_3{}^i X + l_{24}{}^i - v_i l_{34}{}^i = 0$$

where the upper index i denotes the camera number. We have therefore a system of 2n linear equations in the three unknowns x, y, z which in the exact case (an unrealistic assumption) has a unique solution, and in the real case can be solved either by meansquare, or by Kalman Filtering which allows to take into account both the noise on the image coordinates (u_i, v_i), and on the result of calibration, i.e on the matrixes M_i (see Section V), and yields a best estimation of X and of its covariance matrix which can then be used in further processing [3].

VIII.2 Computing the epipolar geometry.

We describe briefly the epipolar geometry of a stereo pair. Let us make a simple geometric remark, helping ourselves with figure 10. Indeed, in this figure we can see that given m_1 in retina plane 1, all possible physical points M which may have produced m_1 are on the infinite half-line $O_1 m_1$. As a direct consequence, all possible correspondents m_2 of m_1 in plane 2 are located on the image, through the second imaging system, of this infinite half-line. This image is also an infinite half-line (ep_2) starting at point E_2, intersection of the line going through O_1 and O_2 and the plane 2. E_2 is called the Epipole of plane 2 with respect to plane 1, and the line (ep_2) is called the epipolar line of point m_1 in the plane 2. The corresponding constraint is that for a given point m_1 in plane 1, its possible correspondents in plane 2 all lie on a half line in plane 2 (see figure 10).

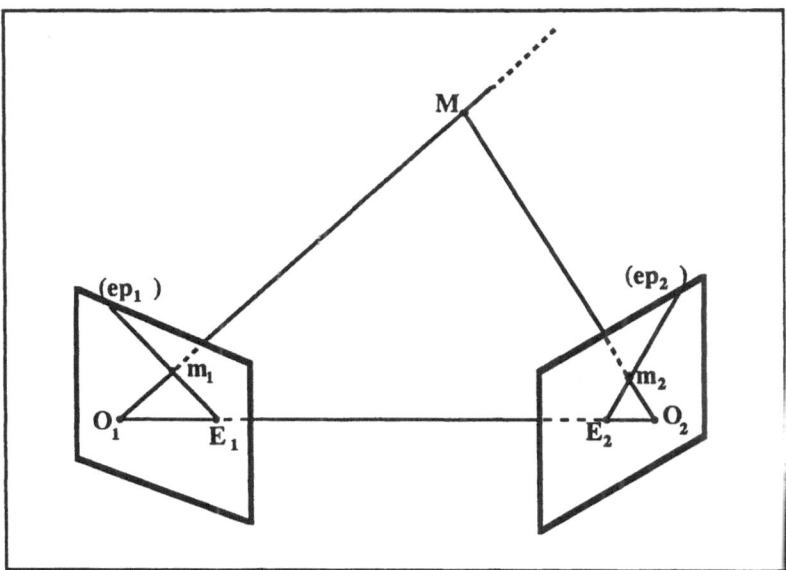

Figure 10. The epipolar geometry.

Therefore we have reduced the dimension of our search space from 2 to 1 dimension. The epipolar constraint is of course symmetric and for a given point m_2 in plane 2, its possible correspondents in plane 1 all lie on a half line (ep_1) starting at the epipole E_1, intersection of the line O_1O_2 with plane 1. (ep_1) and (ep_2) are the intersections of the plane O_1MO_2, called the epipolar plane, with planes 1 and 2, respectively. Of course, when plane 1 or plane 2, or both, are parallel to this line, one (or both) epipoles go to infinity and epipolar lines in one plane (or both) become parallel. The situation where both planes are parallel to the line O_1O_2 is often assumed because of its simplicity. But, in practice, it is difficult to align precisely and in a stable way the two optical systems and we show next that the problem of computing the epipolar lines is just as simple in the general case.

The coordinates of O_i ($i=1,2$), the two optical centers in the world reference frame are obtained by solving the following systems of linear equations :

$$\mathbb{M}_i O_i = 0_3 \qquad\qquad i = 1, 2$$

where $O_i = [x_i, y_i, z_i, 1]^T$ is the coordinate vector of O_i, $i=1, 2$ (see figure 10) and 0_3 is the (3×1) null vector. We also use $o_i = [x_i, y_i, z_i]^T$.

Since each epipole E_i is the image by the ith camera of the other camera's optical center O_j ($j \neq i$), the image coordinates of the epipoles E_i are obtained by applying matrices $\mathbb{M}_i$ to the vectors O_j ($i, j = 1, 2, i \neq j$):

$$e_i = \mathbb{M}_i O_j = [U_{ie}, V_{ie}, S_{ie}]^T$$

Let us now show how, for a given point m_1 in plane 1, its corresponding epipolar line can be easily computed from its coordinates. Using again equations (7) for camera 1, slightly modified to deal only with column vectors, we have:

$$l_1^T X - u_1 l_3^T X + l_{14} - u_1 l_{34} = 0$$
$$l_2^T X - v_1 l_3^T X + l_{24} - v_1 l_{34} = 0$$

These are the equations of two planes whose intersection is the line O_1m_1. A vector n parallel to the line is given by the cross product of the two normal vectors $l_1 - u_1l_3$ and $l_2 - v_1l_3$:

$$n = u_1 l_2 \times l_3 + v_1 l_3 \times l_1 + l_1 \times l_2 = u_1 g + v_1 h + k \tag{24}$$

The line $O_1 m_1$ is described by the vectors $o_1 + \lambda n$ where λ varies from $-\infty$ to $+\infty$. The epipolar line of m_1 is the image of that line through camera 2, therefore, from equation (3), the projective coordinates (U_2, V_2, S_2) of a point on that epipolar line are given by :

$$u_2 = \mathbb{M}_2 \begin{bmatrix} o_1 + \lambda n \\ 1 \end{bmatrix} = e_2 + \lambda\, \mathbb{M}_2' n$$

where $\mathbb{M}_2'$ is the (3x3) matrix obtained by dropping the last column of $\mathbb{M}_2$. Letting $a = [a_1, a_2, a_3]^T = \mathbb{M}_2' n$, we see that both the epipole E_2 represented by e_2 and the point represented by a belong to the epipolar line.

Letting $G = \mathbb{M}_2' g$, $H = \mathbb{M}_2' h$, and $K = \mathbb{M}_2' k$, we have :

$$a = u_1 G + v_1 H + K$$

and therefore :

$$a = F\,u$$

where F is the 3x3 matrix $[G, H, K]$ and $u = [u_1, v_1, 1]^T$.

The projective equation of the epipolar line attached to the point m_1 is given by the determinant:

$$\begin{vmatrix} U & U_{2e} & a_1 \\ V & V_{2e} & a_2 \\ S & S_{2e} & a_3 \end{vmatrix} = 0$$

Therefore, the equation of the line can be written as :

$$(V_{2e}a_3 - S_{2e}a_2)U + (S_{2e}a_1 - U_{2e}a_3)V + (U_{2e}a_2 - V_{2e}a_1)S = 0$$

The cartesian equation is of course :

$$(V_{2e}a_3 - S_{2e}a_2)u + (S_{2e}a_1 - U_{2e}a_3)v + U_{2e}a_2 - V_{2e}a_1 = 0$$

Notice that this encompasses the case where the epipole E_2 is at infinity ($S_{2e} = 0$) or at a finite distance ($S_{2e} \neq 0$). The coefficients of the equation of the line are affine functions of the coordinates u_1, v_1 of point m_1, with coefficients depending only on matrixes M_1 and M_2, and are therefore easily computed.

References

[1] O.D. FAUGERAS, G. TOSCANI, 1986, "The Calibration Problem for Stereo", Proc. IEEE Computer Vision and Pattern Recognition, Miami, pp. 15-20.

[2] O.D. FAUGERAS, N. AYACHE, and B. FAVERJON, 1986, "Building visual maps by combining noisy stereo measurements", Proc. IEEE Robotics and Automation, San Francisco.

[3] O.D. FAUGERAS, F. LUSTMAN, and G. TOSCANI, 1987, "Motion and Structure from Motion from Point and Line Matches", Proc. ICVV, London, U.K., June 8-11, 1987, pp. 25-34.

[4]   S. GANAPATHY, 1984, "Decomposition of Transformation Matrices for Robot Vision", Proceedings of Int. Conf. on Robotics and Automation, pp 130-139.

[5]   O.D. FAUGERAS, J.D. BOISSONNAT, 1987, "The Delaunay Triangulation and Passive Stereo", INRIA, in preparation.

[6]   R.M. HARALICK, 1980, "Using perspective Transformations in Scene Analysis", Computer graphics and Image Processing, 13, p 191-221.

[7]   A.P. SAGE, J.L. MELSEA, 1971, "Estimation theory with applications to communications and control", McGraw-Hill, NY, pp 89-90.

[8]   R.Y. TSAI, 1985,"A versatile Camera Calibration Technique for High Accuracy 3D Machine Vision Metrology using Off-the-Shelf TV Cameras and Lenses", IBM Research Report RC 11413.

[9]   Y. YAKIMOVSKY, R. CUNNINGHAM, 1978, "A System for Extracting Three-Dimensional Measurements from a Stereo Pair of TV Cameras", Computer Graphics and Image Processing, 7, pp 195-210.

[10] Y.I. ABDEL-AZIZ, H.M. KARARA, 1974, "Photogrammetric Potentials of Non-Metric Cameras", Civil Engineering Studies, University of Illinois, Urbana, Illinois.

[11] D. C. BROWN, 1971, "Close-Range Camera Calibration", Photogrammetric Engineering, Vol. 37 N° 8, pp 855-866.

[12] H. M. KARARA, Editor, 1979, "Handbook of Non-Topographic Photogrammetry", American Society of Photogrammetry.

[13] D. C. BROWN, 1966, "Decentering Distortion of Lenses", Photogrammetric Engineering, Vol. 32, N° 3, May 1966.

[14] G. FRANKE, 1966, "Physical Optics in Photography", London; New York, Focal Press, 1966.

[15] A. CONRADY, 1919, "Decentering Lens Systems", Monthly Notice of Royal Astronomical Society, Vol. 39, N° 9, September 1919.

TABLE I

Planes used for test	Score (inverse of the standard deviation)					
	1,2	1,3	1,4	2,3	2,4	3,4
Tsai - single plane	612	604	589	613	607	595
Tsai - multiple plane	688	688	678	697	702	685
Invariant perspective	932	912	943	917	934	937
Radial distortion correction	1103	1089	1140	1076	1145	1037
General distortion correction	2122	2014	2046	2089	2204	2114

Camera type: I2S IS400   Resolution: 288 lines x 384 columns

Number of points used for calibration: 226 Number of points used for test: 113-339

NON HEURISTIC ESTIMATION OF OBJECT SHAPES
FROM PARTIAL INFORMATION

J.D. Boissonnat, O. Monga
INRIA
Avenue Emile Hugues
06565 VALBONNE   (France)

Abstract

This paper considers the problem of reconstructing shapes of objects from sparse measures such as points on the boundary of an object. In most situations, the points are the end points of a curve, called a "ray", which does not cross the objects. For example, if the sensor is an optical device, the ray is the straight line (the optical ray) joining the camera center to the point. We show that the information provided by rays is crucial when determining the shapes of objects and describe non heuristic reconstruction methods in 2D and 3D space.

I- INTRODUCTION

The aim of the present paper is to show that for sensor data analysis, not only the data but also the way they have been measured is of interest. This is illustrated in the problem of reconstructing objects shapes from points measured by sensors. The set of measured points is not sufficient to reconstruct properly the shape; indeed, there are many possible ways of joining the points and thus additional knowledge is necessary. Usually heuristics are used but they are restricted to specific cases. Herein we show that the measures themselves carry the complementary informations which are needed. Moreover taking this complementary informations into account yields efficient reconstruction algorithms.

To be more concrete, let us consider the case of an optical device such as a laser range finder. The sensor measures coordinates of points on the boundaries of objects. However the optical ray, the straight line segment joining the sensor to the measured point, cannot pass, for obvious physical reasons, through an object. This is the kind of additional knowledge we will take into account. Similar situations hold for tactile sensors at the tip of a robot arm (the ray is the arm itself), for mobile robot (the trajectory of the robot is a ray) and for many others. We will show that considering the rays constrain very much the problem and yields a unique solution in the 2D case.

Section II is devoted to a more precise description of the problem.

Section III considers the 2D case and gives an efficient algorithm to compute the unique polygonal approximation associated to the data.

Section IV investigates the above problem in 3D space and shows how the results for the 2D case can be applied.

Section V presents an application of this 3D reconstruction method to stereo data. We use a set of 3D segments provided by a stereo vision algorithm. We compute a polyhedral surface containing, as edges, the measured segments. We present results obtained on real data [1].

II- INFERRING RAYS FROM SENSORS MEASURES

Let us consider a set of points in the plane belonging to the boundary of an object. Solving the reconstruction problem is to find the order of the points on the boundary of the object or equivalently to compute a polygonal approximation of the shape of the object whose vertices are the measured points. In general, this problem admits many solutions as is shown in Figure 1.

NATO ASI Series, Vol. F52
Sensor Devices and Systems for Robotics
Edited by A. Casals
© Springer-Verlag Berlin Heidelberg 1989

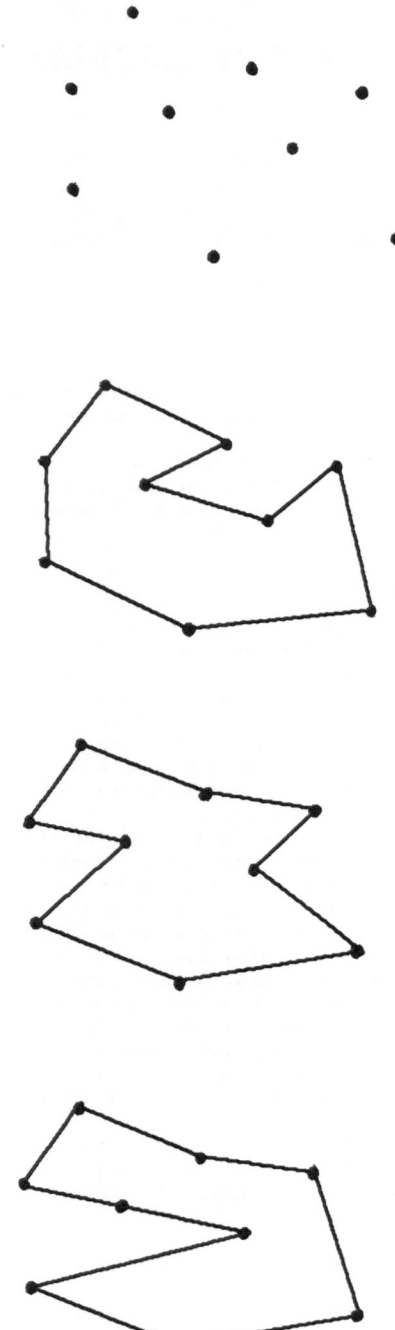

Figure 1 : points measured and some possible solutions for the contour of the object

However, as is shown in Figure 2, we can, in general, associate to each point a ray (a robot arm, an optical ray,...).

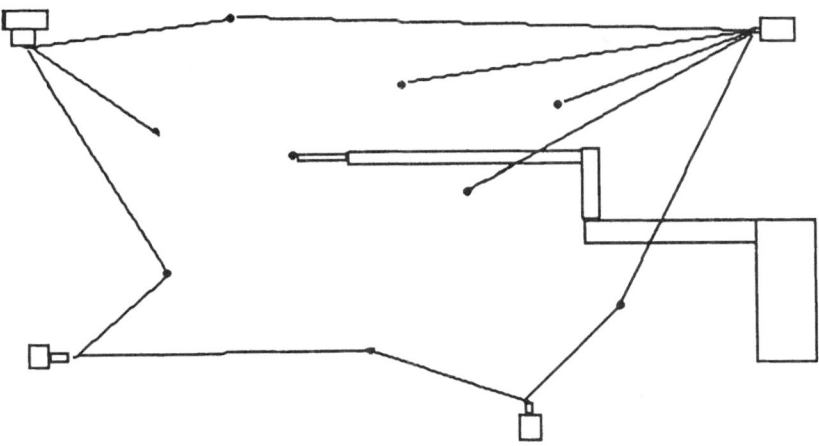

Figure 2 : rays attached to the points of figure 1

Because these rays cannot cross the object, it is clear that the object lies in the textured area of figure 3.

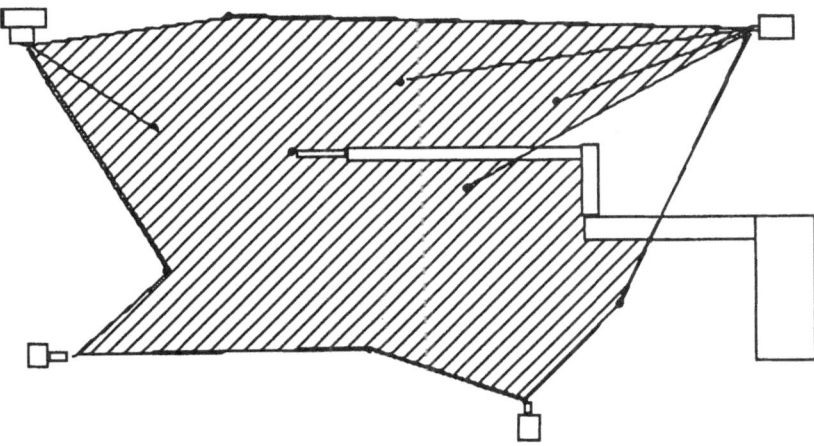

Figure 3 : the object lies in the marked area

Moreover, there exists only one polygonal approximation of the object contour (see figure 4). This simple example illustrates the importance of rays. In the next sections we show how rays can be used to reconstruct efficiently the contours of one or several objects in 2D or 3D space.

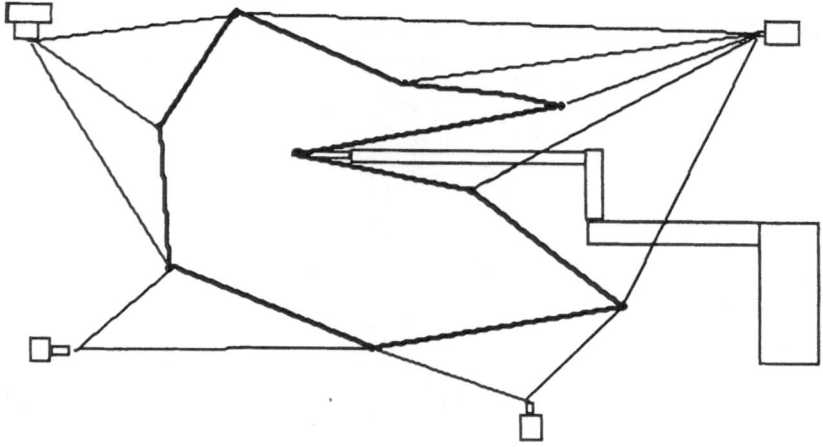

Figure 4 : contour of the object

III- RECONSTRUCTION OF CONTOURS IN 2D SPACE

III.1- CASE OF ONE OBJECT

Let $P=\{P_1,P_2,\ldots P_n\}$ be the set of the measured points and $L=\{L_1,L_2,\ldots L_n\}$ be the set of the corresponding rays. We assume L_i to be a semi infinite curve that originates at point P_i. The problem is to join the points of P by line segments which do not intersect the rays of L or, equivalently, to find a polygonal approximation of the object $(P_{i1},P_{i2},\ldots\ldots P_{in})$. We have to notice that such a polygon does not always exist; figures 7 and 8 show two typical cases.

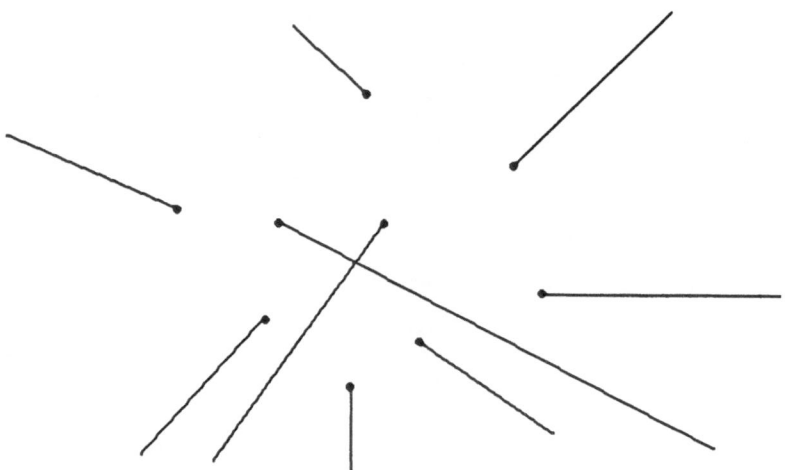

Figure 5 : the points belong to more than one object

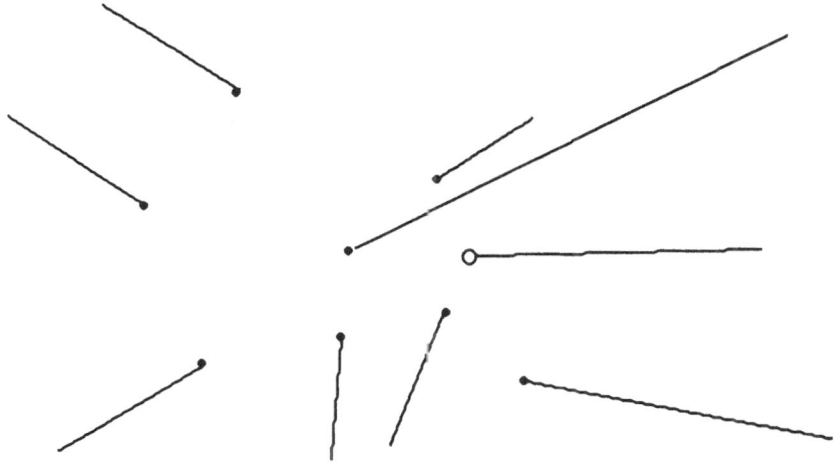

Figure 6 : there is not enough data ; we have to add a new measure (for instance the encircled point)

Alevizos, Boissonnat and Yvinec (5) have proved that if there exists a solution then it is unique ; moreover they propose an $O(nlog(n))$ algorithm which enables to find it. The key point of their approach is to reduce the search of the polygon to sorting the rays. Their algorithm is based on the following theorems :

Theorem 1 : *Assume without loss of generality that the same vertex has been chosen as an origin on both the convex hull of the measured points and on the contour. Then the vertices of the convex hull appear in the same order in counterclockwise marches along the polygonal contour solution to the problem (i.e. the searched polygonal approximation of the object) and along the convex hull.*

Theorem 2 : *Let a_i and a_{i+1} be two consecutive vertices of the convex hull. Then any point C whose ray intersects the edge a_ia_{i+1} of the convex hull is to be encountered between a_i and a_{i+1} in a counterclockwise march along the contour solution to the problem,*

The reconstruction problem is therefore reduced to sorting a class of points whose rays intersect the same edge of the convex hull **CH**. This can be done thanks to the following comparison rule (see figure 7) [5] :

Comparison rule : *c_1 is before c_2 ($c_1 < c_2$) if one of the two following sentences is true :*

    1 : i_{c1} is before i_{c2} on the oriented edge a_ia_{i+1} and the number of intersections between the clipped rays C_1 and C_2 (i.e. the portion of the rays between the points and edge a_ia_{i+1}) is even.

    2 : i_{c2} is before i_{c1} on the oriented edge a_ia_{i+1} and the number of intersections between the clipped rays C_1 and C_2 is odd.

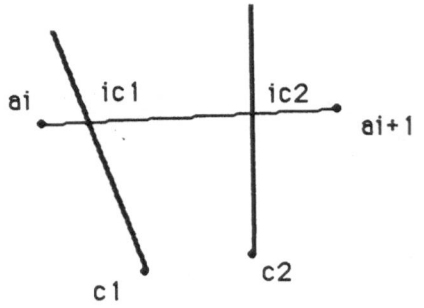 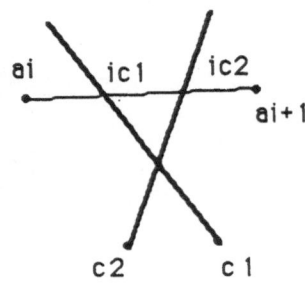

$$c_2 \text{-->} c_1 \qquad\qquad c_1 \text{-->} c_2$$

 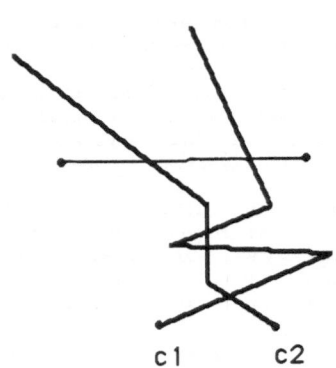

$$c_2 \text{-->} c_1 \qquad\qquad c_2 \text{-->} c_1$$

Figure 7 : illustration of the comparison rule; the correct order is written under the different examples.

From these results, we deduce the following algorithm which solves the reconstruction problem [5] :

ALGORITHM :

1- Compute the convex hull **CH** of set **P**

2- For each vertex **c** of **P-CH** find the first intersection of the ray **C** measuring **c** with the convex hull and form the equivalence classes corresponding to all the edges of the convex hull

3- Sort each equivalence class using the classical heapsort algorithm and the comparison rule to compare 2 points of a given class.

We have implemented this algorithm on a Sun 3 mini-computer. Figures 8 and 9 present some results we have obtained.

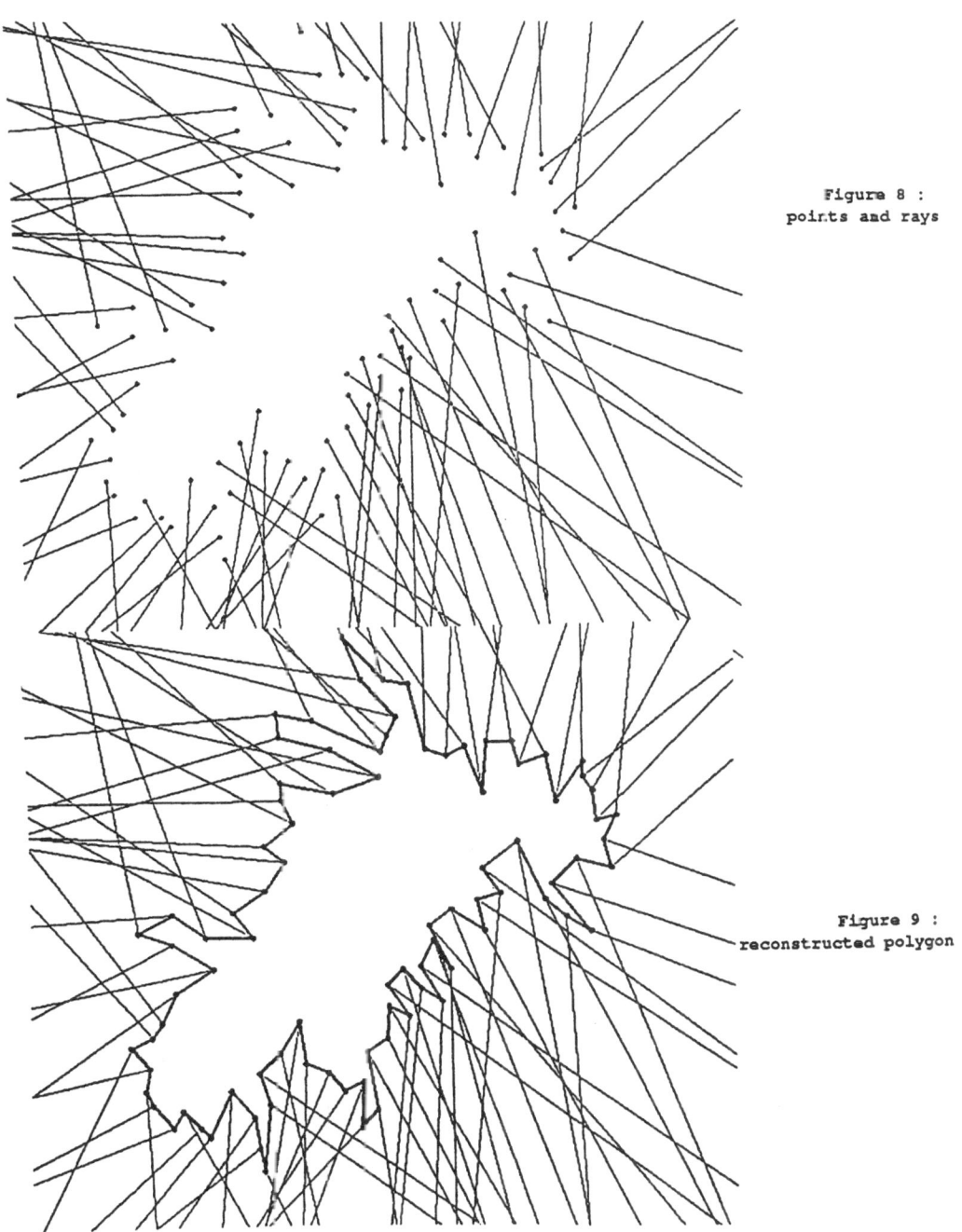

Figure 8 :
points and rays

Figure 9 :
reconstructed polygon

It is shown in [5] that the time complexity of the algorithm is **O(n.log n)** and that it is optimal. Moreover this algorithm can be implemented in order to support insertions of additional points; an insertion can be done in **O(logn)** time which is also optimal. More details and results can be found in [5].

III.2- CASE OF SEVERAL OBJECTS

If the measured points belong to more than one object, we have to search several simple polygons which represent polygonal approximations of the contours of the objects. Hence the algorithm described in the previous section does not work. In this case we solve the reconstruction problem by marching along the rays and turning left each time we meet an intersection betwwen two rays or between a ray and a rectangular box C which is taken to be large enough to enclose all the points (see figure 10). Our algorithm is roughly described below :

1 -   Compute the planar graph drawn by the rays and the enclosing box C;

2 -   Start at a non yet marked point; mark it and march along its ray until an intersection is reached; then turn left and repeat this process until another measured point is reached; goto 2.

The first stage can be performed in **O((n+k).log(n))**, **k** being the number of intersection points, thanks to a sweeping line algorithm [13]. The second stage costs **O(k)**. Therefore the worst case complexity of the whole task is **O(n^2.log(n))**. But in many practical cases **k = O(n)** and therefore the actual complexity is **O(n.log(n))**. We may notice that the march on the rays determines several regions, each one surrounding an object. Each region is the smallest region which surely contains the corresponding object. This may be useful if we are interested in determining free space. Figures 11 and 12 presents some results. The computing time ir a SUN 3 workstation is a few seconds.

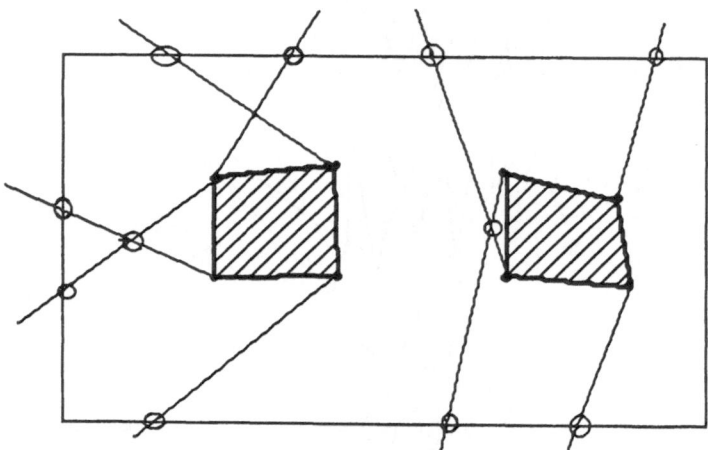

Figure 10 : the intersection points are encircled

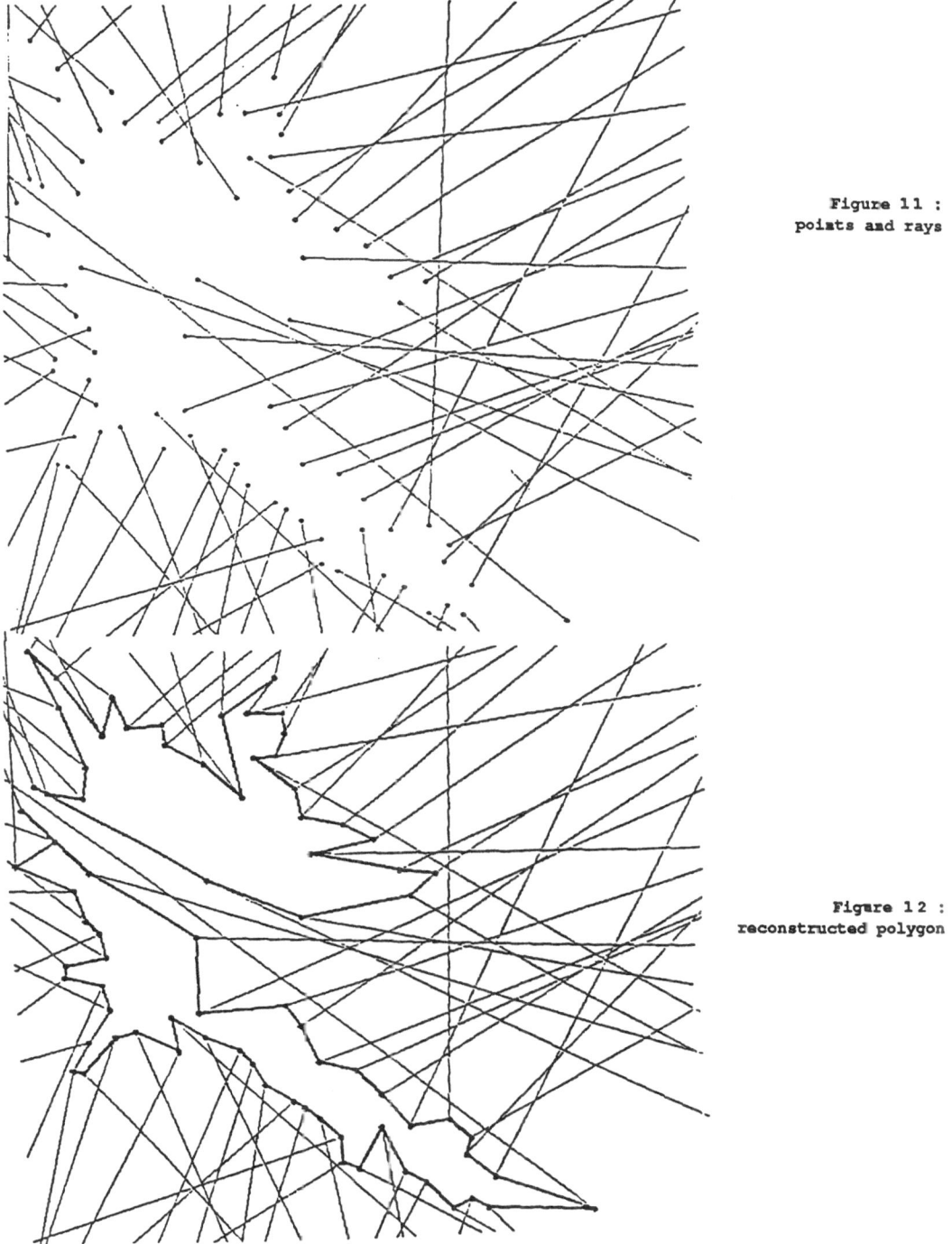

Figure 11 :
poiats aad rays

Figure 12 :
reconstructed polygon

IV- RECONSTRUCTION IN 3D SPACE

In 3D space the reconstruction problem is the following one : Given a set of points and the corresponding set of rays, we look for a polyhedral surface having the measured points as vertices and not crossed by any ray. In the case where a solution exists it is generally not unique. Moreover, for chosing the right solution, we have no means other than a priori informations. Therefore, we have to make some restrictive assumptions about the shape of the objects and the measures. We can derive from the above 2D reconstruction algorithm a 3D reconstruction method which can work if the following conditions are satisfied :

(A) The object is supposed to be a polyhedral surface whose edges have been all measured.

(B) There exists a set of planes P_i such as :

α- The cross sections of the object with the P_i do not intersect each other

β- Each vertex of the object is contained in a plane P_i for some i

γ- Each ray is contained in a plane P_i for some i.

These restrictive conditions are more precisely discussed in [14]. In practice (A) and (B) are only satisfied approximately. An example of such a situation is presented in the next section where experimental results are shown. Our approach can be split in three main stages :

1- Determination of a set of planes.

2- Reconstruction of polygons on each planar cross section by using rays.

3- Reconstruction of surfaces by connecting polygons obtained on two successive cross sections.

Step 1 consists in determining a set of planes P_i. Condition (A) implies that a plane will contain all the vertices of the polygons defining the section of the object. If (A) is not satisfied, we could have not enough information in each plane for reconstructing the section of the object.

Step (2) consists in reconstructing the contours of the objects in each P_i. According to the fact that a plane contains all the rays intersecting it (condition B.γ), the 2D reconstruction algorithm described in III can be used. Thus we find on each plane the polygons corresponding to the section of the object. We therefore determine in each plane the correct section of the object.

The condition (B.α) allows to order the cross sections. Thus we can achieve the reconstruction by linking the sections of the objects obtained in two successive planes (step (3)). It is important to notice that because (A) and (B) are satisfied this task is trivial. But generally (A) and (B) are not strictly satisfied and thus we need to use some heuristics for computing the portion of the surface between two adjacent cross sections. In the case of one simple object we have chosen an optimal triangulation method introduced by Fuchs and al. [8]. We look for the triangulation which minimizes the sum of the length of the edges [6,7]. This is performed by searching an optimal path in the triangulation graph. In the case of many objects we use a reconstruction method based on the Delaunay triangulation [3]. The aim of this approach is to compute a Delaunay triangulation of the points lying in the two sections and to remove the tetraedra making the object to have singularities. In this specific case the Delaunay triangulation can be determined efficiently by means of only 2D operators [3].

In the next section we deal with a 3D reconstruction problem where the above method can be applied.

V- APPLICATION TO SURFACE RECONSTRUCTION FROM STEREO DATA

In this section, we deal with the reconstruction of surfaces from 3D segments obtained by stereo vision [14]. The 3D segments are computed by matching edge segments issued from the two images of a stereo pair [1]. We can therefore assume that they correspond to the edges of a polyhedral surface. Thus we can apply the method described in the precedent section. In this specific case the reconstruction problem can be formalized as follows :

E : 3D space

$[A,B]$: segment whose endpoints are A and B

$[A,B[$: semi-infinite line defined by point A and direction AB

$C_1, C_2, \ldots\ldots C_n$: optic centers of cameras

$S = \{ s_1, s_2, \ldots\ldots s_n \}$: set of 3D segments obtained by stereo vision. At each segment s_i we attach an optic center C_{si} from which it can be seen.

$s_i = [M_{i1}, M_{i2}]$: a segment is caracterized by its two endpoints.

M_{jk} : points of E

$R = \{ [M,C[\ / \ \exists\, s \in S, \ (M \in s) \ \& \ (C = C_s) \}$

R is the set of semi-infinite lines defining rays attached to S. In this case rays are semi-infinite lines which join a point of a segment and the corresponding optic center (see figure 13)

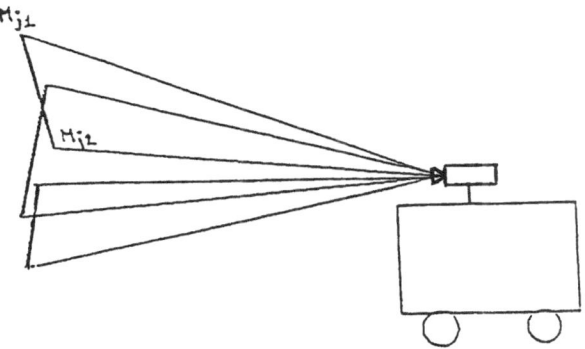

Figure 13 : A robot which see some segments

The problem is to find a polyhedral surface T where vertices belong to S and which satisfies the two following properties :

(P_1) T contains all points of P

All segments issued from the stereo vision process are included in T

(P_2) T is not intersected by any ray

We come up with a solution to this 3D reconstruction problem by means of the algorithm described in IV. We use a somehow different method in the cases of one or many view points.

. Case of one view point

In this case we can assume that the polyhedral surface we look for is connected. Our algorithm can be split in three main stages :

(a) Generation of a pencil of planes containing all the endpoints of the segments from the optic center of a camera (see figure 14)

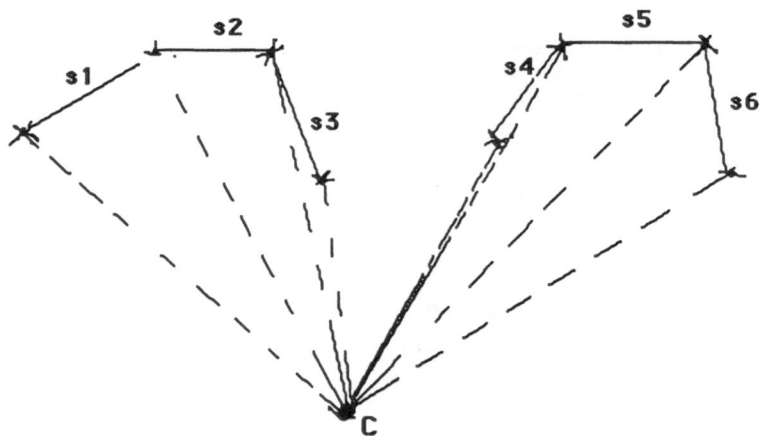

figure 14 : Cross section corresponding to Figure 13, the optical center of the camera is C

(b) Reconstruction of an open polygon in each cross section by sorting rays

In the specific case where all the rays originate from the same point, the 2D reconstruction method (see III) consists only in sorting the rays with respect to their orientations (see figure 15). We then obtain an open polygon on each planar cross section.

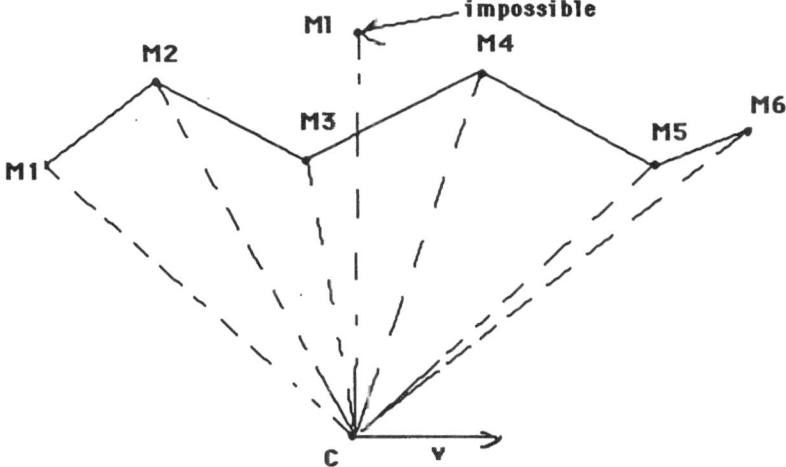

Figure 15

(c) Connection of the two polygons obtained in two successive planar cross sections by means of Fuchs and al. method (see IV). We added the constraint which consist in joining two points included in a same segment.

. Case of many view points

(a) Generation of a set of parallel cross sections containing all the endpoints of the segments, such that rays can be approximated by their projection on the planes.

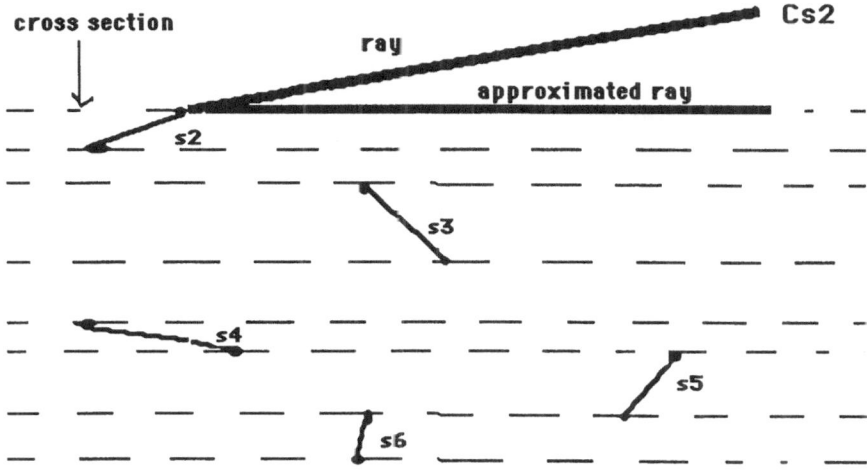

Figure 16

(b) Reconstruction of polygons in each planar cross section.
For this purpose we use the 2D reconstruction method for many objects described in III.

(c) Connection of the polygons obtained on two successive sections by means of Delaunay triangulation (see IV).

Our reconstruction method therefore provides a polyhedral surface defined by a set of triangular faces. Figures 17 and 18 present some results we have obtained on real data provided by a trinocular stereo vision algorithm [1]. The result of the reconstruction process is a set of triangles defining a polyhedral surface. In order to match the original image with the surface, we coloured each triangle with the mean grey level of its projection on the original image. We compute these mean luminances thanks to the calibration of the cameras [10].

Figure 17 : original image

Figure 18 : perspective view of the reconstructed surface

VI- CONCLUSION

The present paper illustrates how useful can be the information provided by rays when determining the shapes of objects from sensors measures. In many cases they lead to non heuristic and low computational time reconstruction methods which do not require any a priori information. We have presented some examples of such algorithms in 2D and in 3D space. We actually investigate how such methods can cope with uncertainties and errors in the mrasurements.

References :

[1] **N. Ayache et F. Lustman** : *"Fast and reliable passive stereovision using three cameras"*, dans Proceedings International Workshop of Industrial Applications of Machine Vision and Machine Intelligence, Tokyo, IEEE,Février 1987.

[2] **J.D. Boissonnat, O.D. Faugeras, E. Lebras** : *"Représentation des données stéréo par la triangulation de Delaunay"*, Congrès AFCET-RFIA, November 1987, Artibes, France (in french).

[3] **J.D. Boissonnat** : *"Geometric structures for three dimensional shape representation"*, ACM trans. on Graphics, Octobre 1984.

[4] **Williams** : *"Algorithm 232 heapsort"*, commun ACM 7,6, Juin 1984, pp. 347-348.

[5] **P. Alevizos, J.D. Boissonnat, M. Yvinec** : *"An optimal O(nlog(n)) algorithm for contour reconstruction from rays"*, ACM Symposium on Computational Geometry, Waterloo, Juin 1987.

[6] **E. Pauchon** : *"Numérisation et modélisation automatique d'objets 3D"*, These de troisième cycle de l'Université d'Orsay, 1983.

[7] **E. Keppel** : *"Approximating complex surfaces by triangulation of contcurs lines"*, IBM J. Res. Develop. 19 (Janvier 1975), 2-11.

[8] **H. Fuchs, Z.M. Kedem et S.P. Uselton** : *"Optimal surface reconstructior from planar contours"*, Graphic and Image Processing Communication of the ACM, Vol. 20, n° 10, Octobre 1977.

[9] **M. Gondran, M. Minoux** : *"Graphes et algorithmes"*, Editions Eyrolles, pp. 31-43, 1979.

[10] **O.D. Faugeras, G. Toscani** : *"The calibration problem for stereo"*, Proceedings CVPR 1986, Miami Beach, Florida, pages 15-20, IEEE, 1986.

[11] **J.D. Boissonat** : *"Shape reconstruction from planar cross sect:on"*, **INRIA** research report June 1986.

[12] **J.D. Boissonnat, F. Preparata** : *"On the external boundary of an union of rays"*, June 1986.

[13] **F. Preparata, Shamos** : *"computational geometry : an introduction"*, Springer Verlag 1985.

[14] **J.D. Boissonnat, O. Monga** : *"Surface reconstruction from stereo data by planar cross sections"* , Proposed for IEEE Robotics and Automation 1988, Philadelphie.

PARAMETER ESTIMATION IN SIGNAL PROCESSING

J. Aguilar
LAAS-CNRS,
7, avenue du Colonel-Roche
31077-Toulouse   (France)

KEYWORDS:/ Non-Linear-Filtering / Poisson-process /
partitioned-estimation / radiactive-tracers / particle-counters

ABSTRACT: The filtering problem encountered in signal
processing uses implicitely a model of the signal generator.
The simultaneous estimation of the parameters of this model and
the signal can be solved by the partitioning approach if there
exist optimal, or at least sub-optimal filters for all possible
values of the unknown parameters. We develop here this approach
for the case of a Gauss-Markov signal observed through a
counting Poisson process, problem that commonly arises with
sensors using radiactive tracers.

I- INTRODUCTION

The so called "Filtering problem" consists on the construction
of a process X(t) that represents, in the best possible way,
another unobserved process x(t), using all the possible
observations {y(t), s≤t}. In a very general form we can
describe this situation as a search for the best attainable
model of process x(t). Of course in most of the classical
situations in signal processing something is known about the
real mathematical model that generates the signal y(t), but
this knowledge is either imprecise, or corrupted by other
processes called noise processes. Stochastic pocesses theory
gives an interesting and useful framework for mathematical
formulation of those concepts.

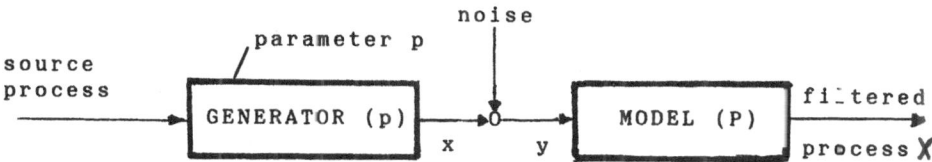

Figure 1

NATO ASI Series, Vol. F52
Sensor Devices and Systems for Robotics
Edited by A. Casals
© Springer-Verlag Berlin Heidelberg 1989

As represented in figure 1 the filtered process has to be generated by a MODEL of the GENERATOR, where p stands for the "true" parameters of the generator and P is an estimator of p that parametrizes the model.

It can be objected that most of the classical signal processing results (Wiener,1949), (Kalman-Bucy,1961), are linear filters and they do not explicitly modelize and estimate parameters of models, but they directly estimate process values; two remarks must be done to understand their particular apparent straightforward result: those linear filter results assume a very precise knowledge of the generator, and moreover this generator is linear and accepts all extrapolations in a very simple way. As soon as the linear models are variable or unknown, the well known "adaptative" signal processing becomes a nonlinear problem and as it was shown in (Lainiotis, 1971) an optimal solution is the so called "partitioning" that separates the parameter estimation problem from the filtering problem. This approach assumes that if the true values of parameters were known, the filtering would be perfect, or at least satisfactorily performing following the model shown in fig. 1, therefore this approach can give optimal results for all the situations when this happens. It is also necessary to assume that it is possible to build as many perfect filters as possible values of the parameter, or in practice at least to cover by means of interpolation all the possible points of the parametric space. Although the requisites of the optimal "partitioning method" are seldom fully satisfied, approximate versions give practical acceptable results and, in favorable situations, they accept theorical calculation of error bounds, or at least convergence properties.

II- PROBLEM STATEMENT

Our purpose is to fully developp a very important case where this parameter estimation technique gives useful results, its importance comes from the general measurement of the intensity, or mean rate, of point processes in sensors based in electron or photon counting, as is the case of the measurement of flows through radiactive tracers, of a classical use in medecine and biology.

As an example we shall briefly describe in Figure 2 a practical situation where this situation occurs.

```
---------------------------------------
                radiactive counter (Geyger) ------>
  (flow)        ========================================= Observed
-------->        .    .   .    ...          .     .      signal
 Source         ========================================, y(t)
 Signal
  x(t)          Liquid flow carrying radiactive particles
```

Figure 2

The mathematical formulation of the problem is the following: Let the signal accept a linear differential model with unknown parameters, it generates a gaussian-markovian process, the instantaneous value of this process drives the intensity of a point Poisson process observed by its counting for fixed sample intervals. The observed signal y(t) is called a doubly stochastic Poisson processes, as introduced in (Snyder,1975).

III- STOCHASTIC MODELISATION

Let $\{y(t),t>0\}$ be a counting Poisson process, and $\{x(t),t>0\}$ a non negative gauss-markov process, (strictly speaking such a process does not exist because gaussian distributions must accept negative values, nevertheless we call so a process such that the probability of negative values is negligible).

Definition 1: A "doubly stochastic Poisson Process" is a process y(t) such that:

-P1: $Pr[y(0)=>]=1$

-P2: Let us consider instants 0,a,b,c,d and such that $0 \leqslant a \leqslant b \leqslant c \leqslant d \leqslant t$, as well as an instant s such that $0 \leqslant s \leqslant t$, and let x(t) be a realisation of the signal process, then $(y(d)-y(c))$ and $(y(b)-y(a))$ conditioned by x(s) are almost surely integer and independent.

-P3: The conditional probability law of $(y(t)-y(0))$ with respect to x(t) is Poisson, with parameter

$$M(t) = \int_0^t L(x(s)ds$$

L(x) is a continous and measurable function of x, and, for simplicity, we shall only consider here the linear case where $L(x)=A.x+B$.

A doubly Stochastic Poisson Process shall be denoted $\{y(t), L(x(t)), t \geqslant 0\}$.

IV- FILTERING PROBLEM

Let $h(x,t)$ be a measurable function of for all t, the "filtering problem" consists on the estimation of $h(x,t)$ based on past observations of $y(s)$, $0 < s < t$. The well known theoretical solution says that the best estimator is the conditionnal expectation $E[h(x,t)/y(s); 0 < s < t]$.

Unfortunatelly this result assumes a complete knowledge of the conditionnal law, and generally this last depends on an infinite set of parameters.

A classical approach is to attempt to give an expression of the infinite series of moments of the conditional law. For the most general case this cannot be achieved, nevertheless we shall prove here that in the particular problem shown here, where the process $x(t)$ is gaussian-markovian with known parameters this is possible. Therefore, according to the "partitioning principle" we may developp a set of suboptimal filters for a given set of parameter values and so give a satisfactory approximate solution to the global signal processing.

Let us modelize the gauss-markovian process $x(t)$ by the first order Ornstein-Uhlenbeck stochastic differential equation of Ito type:

$$dx(t) = (a.x(t) + b.u(t)dt + r.dw(t)$$

where $W(t)$ is an unitary brownian motion and $u(t)$ a deterministic time function.

Theorem 1: The differential equations driving the sequence of centered moments $\{m[1], m[2], \ldots m[i], \ldots\}$ of the conditional probability $PROB[x(t)/y(s) 0 \leqslant s \leqslant t]$, are:

$$dm[1] = (a.m[1] + b.u)dt + R(dy - L(m[1].dt)$$

$$\ldots \ldots \ldots \ldots$$

$$dm[i] = (C_i^1 a.m[i] + C_i^2 r^2 m[i-2] - A(m[i+1] - C_i^1 m[2].m[i-1]\}).dt +$$

$$((R/m[2]).(m[i+1] + C_i^1 R.m[i] + C_1^2 R^2 m[i-1] + \ldots$$

$$\ldots + (-1)^{i} C_i^{i-2} R^{i-2} m[3] + (-1)^{i+1} (i-1).R^{i-1} m[1].dy+$$

$$(C_i^1 R.m[i-1] + C_i^2 R^2 m[i-2] - \ldots + (-1)^i C_i^{i-2} R^{i-2} m[2])).dy$$

$$\ldots\ldots\ldots\ldots$$

were $R = A.m[2]/L(m[1])$, and $C_i^j = (i!)/((i-j)!j!)$

The proof of this theorem can be found in [Jarachi-Aguilar,1986).

We notice that the estimated value of x is given by m[1], and for h(x,t) we have, in the linear case considered here, the optimal estimator h(m[1],t).

Although the filter shown in Theorem 1 is of infinite dimension it can be developed up to any order [n] and consequently this order can be chosen in such a way that it can take in account the dissymetry of the Poisson distribution, particulary apparent for small values of the mean of the distribution L(x).

Strictly speaking this value is unknown, but an estimator is calculated by L (m[1]). In practice truncature of the filter to an odd order is suitable and order [5] seems to give very good approximations. Moreover following the proposed corrections given in [Pugachev,1982],we propose to close the sequence making:

$$m[6] = 15.m[2].m[4] + 10.(m[3] - 3.m[3].m[3].m[3]$$

V- PARTITIONED ESTIMATION

The practical problem of signal processing that comes from the sensor configuration is not solved by the above result because in Theorem 1 we assume known the values of "A" and "B", as well as of "a", "b" and "r". Usully the gain and offset of the counting device enables some knowledge of "A" and "B", wereas the other parameters are intinsec to the source signal. Let us call the parameter vector p=[a,b,r], Theorem 1 gives in fact the optimal solution to the recurrent construction of PROB[x(t)/ {y(s);0≤s≤t}&p].Fro; an engineering point of view, we cannot attain this optimality but we can construct a good approximation obtained by a truncated version to an order [n], as high as needed.

We can therefore construct for any value pi an estimated probability law conditioned by the observations and the value of p,i.e. EST.PROB[x(t)/{y(s);0⩽s⩽t}&pi].

If we consider a finite set of parameter values {P}={p1,p2,...pi,..pN}, we dispose of a batterie of filters, or approximate filters in practice, that yields at each instant a finite set of estimated probability laws, such that the estimation of parameter p can be performed by a Bayesian procedure.

The following theorem states the sufficient conditions to derive an optimal estimator by this way:

THEOREM 2 : Let us consider the following hypotesis:

- H1: the true value of p is constant and belongs to the finite set {P}.
- H2: for any given value of p, {x(t/p),t⩾0} is Markov process and its realisations are p-measurables.
- H3: {y(t),L((x(t),t⩾0} is a doubly stochastic Poisson Process.
- H4: L(x)=A.x+B.
- H5: Conditional laws PROB[x(t)/{y(s);0⩽s⩽t&pi] are cal-culable for all i and all t.
- H6: Given an ordered sequence of instants {t1,t2,... ..,tj, tk}, for 0<i<j<k, the random variable (y(tj)-y(ti)) is an integer random variable.

Then the opttimal estimator, P(tk), of parameter p at instant tk is completely determined and moreover it is completely constructible.

The proof of this theorem is given in [Jarachi-Aguilar,1986] and consist in showing that with all those hypothesis the conditional expectation is determined and recursively constructed by theorem 1.

As it can be observed in theorem 2, we need a battery of filters that evolve in parallel and each one processes by itself the informative signal y(t), then a Bayesian estimator reconstructs all the conditional laws.

VI- ALGORITHM

Practically the optimal filters are not computable because of their infinite dimension, but they can be aproximated and therefore use an estimated value of the probability laws.

The following approximations must be intoduced in oder to give rise to a procedure of partitioned estimation suitable for this signal processing:

-A1: To relax H5 we use a truncated filter
-A2: We assume that for j=i-1, the time intervals Dt=(ti-tj) is small enough to justify the following equality: M(t+Dt)-M(t)=(A.m[1](t)+B)Dt
-A3: We assume that the true value of p belongs to the convex closure of {P}.

The partitionned filter and estimation problem is well posed if hypotesis H2, H3 and H4 are fulfilled, and if approximations A1, A2 and A3 are licit.

Then the algorithm consists on running a set of truncated filters, such that the conditional first n moments can be obtained, then to compose all the informations by a Bayesian procedure to finally calculate the conditional expectation of the parameter. By such an algorithm we obtain a good estimation of parameter p and the best filter for the signal can be chosen as the filter corresponding either to the expected value of the parameter, or to the most probable one, this last choice simplifies the composition of conditional probabilities but may produce undesirable jumps that are avoided by the smoothing propertie of conditionnal expectation.

VII- SIMULATION RESULTS

To illustrate the above considerations we chose the following parameter:

p= [-0.05,0.5,1.0]

The time interval for discretisation is chosen Dt=0.01 sec.

For simplicity the sensor parameters are set to A=1.0 and B=0 so that L is the identity.

In figure 3 we show the source signal corresponding to this simulated model (curve 1) as well as the estimated value (curve 2).

In figure 4 is shown the observed process Dy(t)=y(t)-y(t-Dt) from which estimation has to be performed.

The set of possible values for parameter p is given in the following table:

a	b	r
-.150	.1	.6
-.125	.2	.7
-.100	.3	.8
-.075	.4	.9
-.050	.5	1.0
-.025	.6	1.1
-.0	.7	1.2
-.025	.8	1.3
-.050	.9	1.4
-.075	1.0	1.5

The algorithm has been applied to an horizon of about 600 iterations and takes as estimator the most probable value.

In figure 5 it can be observed that the estimated probabilities jump between two parameter values, even for an advanced stage of the algorithm. To avoid this undesirable effect due to the approximations made, it is suitable to add a test of convergence that fixes the small probabilities to zero if they do not overcome a given threshold S during a given time T. In figure 6 we show the behaviour of the same algorithm for S=.001 and T=100 iterations, where the estimated value and the signal are undistinguishable.

VIII- CONCLUSION

Although parameter estimation is a problem that arises mainly for control problems, it appears that in signal processing the choice of the best model for the source signal is always suitable, therefore when the true values of the parameters of this models are insufficiently known their simultaneous estimation with the signal improves the quality of the filtered signal. The complete problem of filtering and estimation is fundamentally a non-linear problem that accepts seldom a fully constructible solution. The partitionning approach, by its well established structure enables to state the problem in terms of a set of simultaneous optimal or sub-optimal filters followed by a Bayesian estimator for the parameters. This approach is particularly adapted to the case when good conditional sub-optimal filters can be derived. In this paper we illustrate this assertion by applying the partitionning procedure to a Gauss-Markov signal observed through a Poisson Counting Process, and good results are obtained in situations where the classical Gaussian approximations are unacceptable because of the small intensity of the counting signal, and therefore the important dissymetry of the Poisson distributions involved.

ACKNOWLEDGEMENTS

The main results stated in this paper are part of the doctorate thesis made by F. Jarachi in L.A.A.S. with a grant from Ecole Mohhamedia d'Ingeniers de Rabat, Maroc. I am indebted to him for his excellent work, as well as in developping all the tedious filter equation as in simulation programming.

REFERENCES

(Wiener,1949) - Extrapolation, Interpolation and Smoothing of Stationnary Time Series with Engineering Applications. Wiley,NY.

(Kalman-Bucy,1961) - New Results in Linear Filtering and Prediction Theory. J. Basic Eng.ASME, Ser D, 83, 95-108.

(Lainiotis,1971) - Optimal Adaptative Estimation: Structure and Parameter Adaptation. I.E.E.E. trans AC April.

(Snyder,1975) - Random Point Processes. Wiley Interrscience, NY.

(Pugachev,1982) - Theorie des Probabilities et Statistique Mathematique. Editions MIR Moscou.

(Jarachi-Aguilar,1986) - Nonlinear Filtering of Counting Processes driven by Ornstein-Uhlenbeck Processes. International Journal of Control, VOL 44, No 5, 1193-1207.

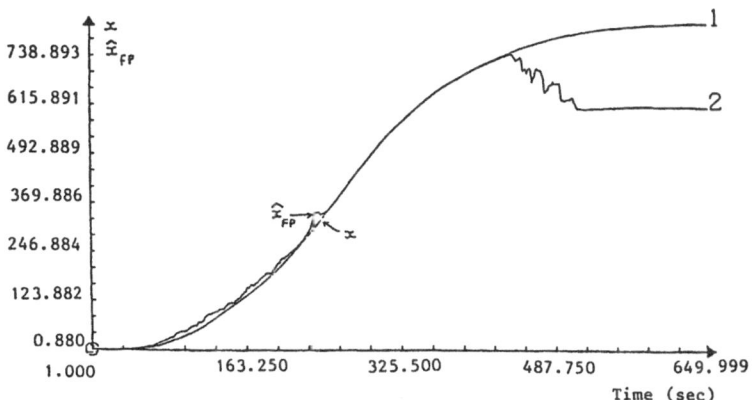

Figure 3

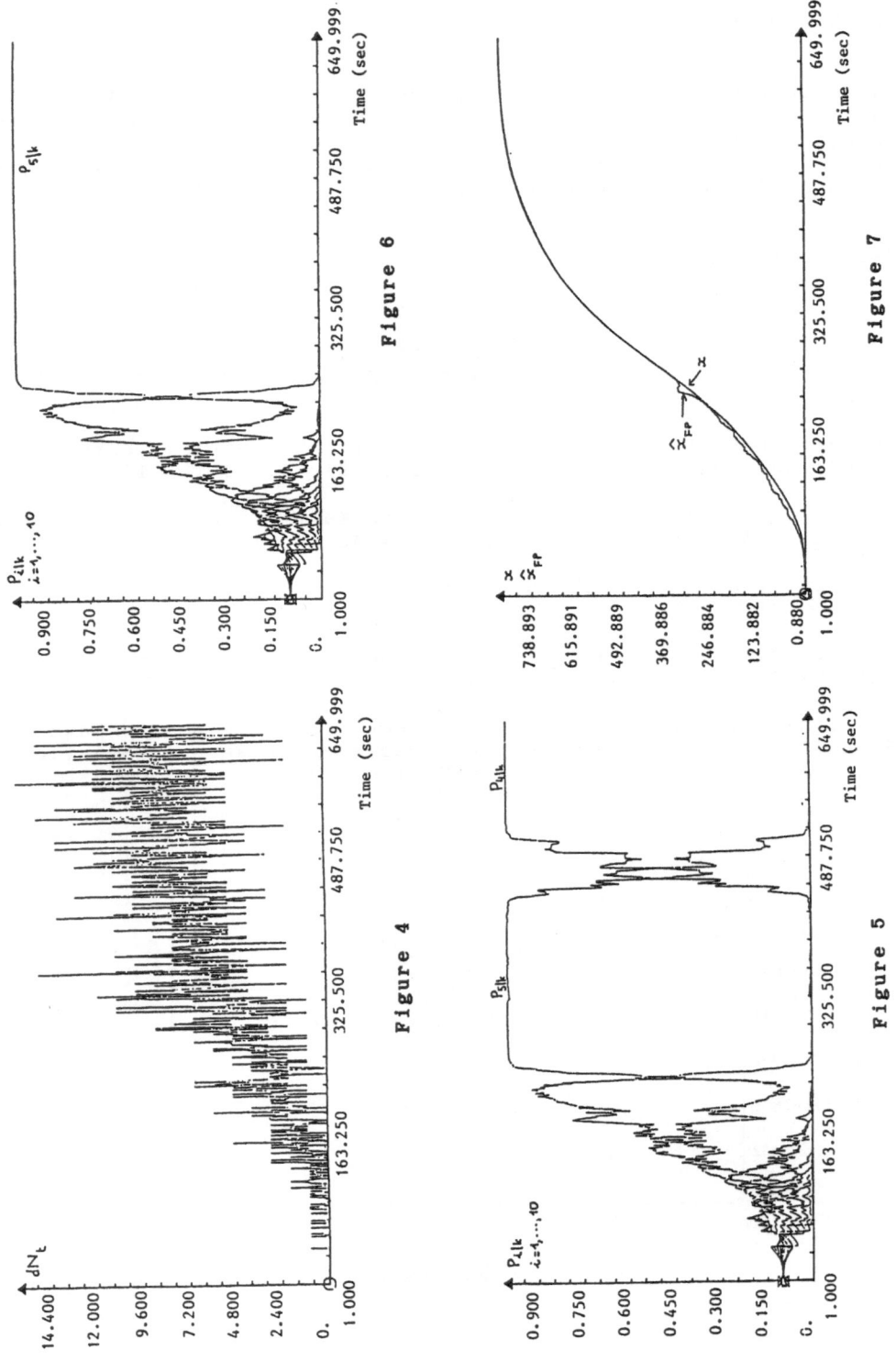

Figure 4

Figure 5

Figure 6

Figure 7

V. OTHER KIND OF SENSORS

DYNAMIC WEIGHING IN A PICK-AND-PLACE ENVIRONMENT

D.G. Whitehead, I.M. Bell, D.J. Mulvaney*, A. Pugh, P. Sweeting*
Department of Electronic Engineering
University of Hull
Hull, N.Humberside, HU6 7RX   (U.K.)

---

* D.J.Mulvaney and P.Sweeting are with Cambridge Consultants Ltd.

NATO ASI Series, Vol. F52
Sensor Devices and Systems for Robotics
Edited by A. Casals
© Springer-Verlag Berlin Heidelberg 198!

1.0 Introduction

The accurate weight measurement of food products during manufacture is of great importance as this significantly affects the amount of 'give−away' necessary to comply with weight legislation. Generally, in production this weight measurement is performed on a regular basis to ensure that the product is maintained to the target weight and within the target deviation. In the case considered here, the products dealt with are chocolate confectionary, but the problems faced are applicable to many production items. The manual method of on−line weight inspection uses an operator to remove the product from the production line, weigh it on an electronic balance, record the weight and return the product to the line. Some care is required in the removal and replacement of the units so that they remain ordered in complete rows.

The aim of the work currently being carried out at Hull University is to mechanise this process, both in the acquisition and weighing of the product. An important aspect of this work is the design of a weigh−head which can both acquire and provide weight information to the desired accuracy. In practice, it is required that a sensory tool be designed, to be mounted on an X−Y type gantry which will be capable of selecting products in a weight range of 50 − 200g from a conveyor belt, weighing, and then restoring them to their original belt positions. For development under laboratory conditions, a small conveyor belt is used and a PUMA 560 robot simulates the operation of the gantry, the weigh−head being mounted as the robot end−effector.

The accuracy required from the weighing process is ±0.3% or better. Operating constraints demand that the weighing process must be carried out while the gripper and product are moving, following the conveyor belt motion. It is this weighing aspect that will be described in detail and a technique is presented which allows the measurement of product weight to take place dynamically, that is, while the weigh−head is in motion and subject to considerable mechanical vibration and electrical noise.

2.0 Weigh−head Design

On past experience it has been found that a vacuum suction cup arrangement is the most appropriate tool for picking up chocolate products without damage. Such devices are also fairly tolerant to differing sizes and shapes. A robot tool based on the vacuum cup principle, and incorporating a tension−mode load cell has been designed and is shown schematically in figure 1. The load cell provides the only connection between two skeletonised plates through which a vertical force may be transmitted. The lower platform has four rigid mountings to which the elastomer vacuum cups are attached. To connect the rigid vertical tubes from the lower platform to the upper, inverted 'U' shapes of

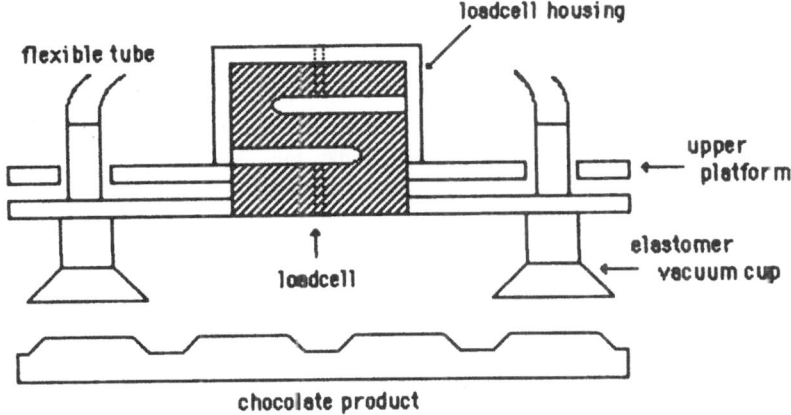

Figure 1 Schematic Diagram of the Weigh—Head

flexible tubing are employed. This provides the exit for the air, and directs any spring forces due to the tubing sideways, i.e. not in the axis of the load cell. The gap between the plates is approximately 3mm, though the transducer compresses by only 0.8mm under full load. Fine adjustment screws, (not shown), may be used to limit the compression to a safe value. Figure 2 shows a photograph of the weigh—head.

Figure 2 The Weigh—Head

The design of the tool is critically dependent upon available load cells. The unit chosen for this application is a Maywood Instruments U4000 strain gauge device operating in tension mode. This is capable of weighing up to 2kg with a linearity of 0.03%. Sensitivity is 5mV/V/kg and overload capacity is excellent. Signals from the load cell are amplified by a standard strain gauge amplifier (Radiospares 308−815).

3.0 The Weighing Process

The method of weighing the product is constrained by the process of acquisition and replacement with no more than 2−3 seconds being available for the measurement process. Because of this, the technique used for acquisition of the item to be weighed will be briefly described.

3.1 Product Acquisition

For reliable operation of the elastomer suckers, the position of the product must be known to an accuracy of 1mm immediately before acquisition. Coarse positioning is available since the conveyor belt utilises mechanical 'row−straighteners' to align the product which is carried along the conveyor belt in batches of 10 to 20 units. These straighteners force the units to form a common front. Since the conveyor belt speed is monitored and the position of the straighteners is known, coarse x−direction position information is available. In the y−direction, that is across the width of the belt, coarse information is obtained by knowing that the product will originate from moulds, and how many of these moulds are in use.

Fine position sensing is provided by proximity sensors attached to the weigh−head. Three such sensors are used and signals from these are sufficient to determine the accurate position of the product. Once the product has been located, it is lifted from the moving belt by means of the suction grippers and carried in the direction of belt travel while weighing takes place. It is then replaced in its original position relative to the conveyor belt. The speed of the conveyor belt and travel of the robot arm are the factors which limit the time available for the weight measurement to 2 − 3 seconds and during this time the product is in movement.

3.2 Product Weighing

The signals obtained from the load cell during the acquisition and replacement process are shown in figure 3. It can be seen that there is a considerable amount of noise present. The initial impulse and corresponding overshoot is due to the overload imposed upon the load cell when the vacuum suckers press onto the product. Whilst the the noise present

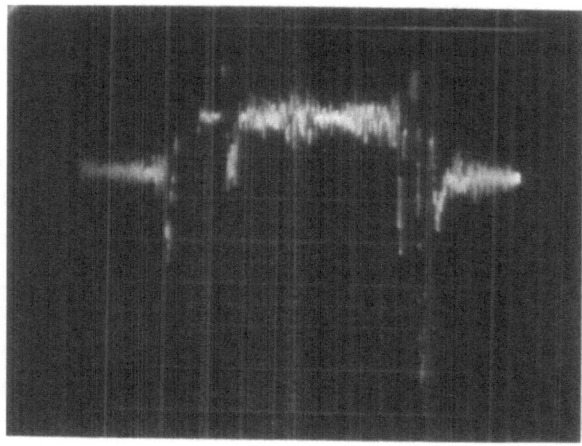

Figure 3 'Raw' Signals from the Loadcell   [50mV/div. 140g load]

during movement reflects the mechanical characteristics of the PUMA robot, any mechanical system used will impose similar disturbances.   Accordingly, a technique capable of resolving the true signal to within 0.3% from this simulated situation would be likely to succeed in any practical realisation of the process.

The technique chosen was one of filtering.   A straightforward single filter applied to the output of the load cell amplifier is not satisfactory due to the time constraints present. Figure 4 shows the output obtained from a 4th−order Sallen−Key low−pass filter (cut−off frequency 2.5Hz) when the signal of figure 3 is applied to its input.   Such a

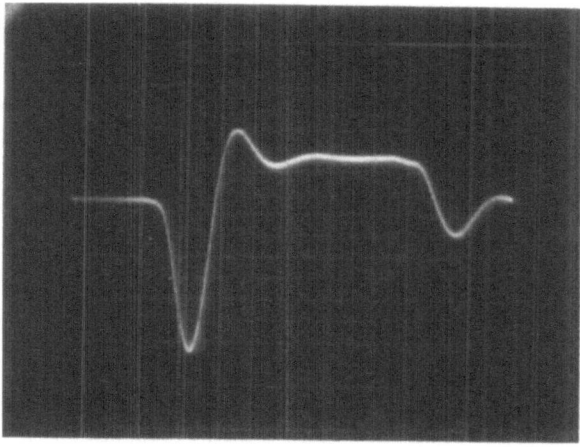

Figure 4 Output from the Low−Pass Filter

filter is capable of reducing the noise content to an adequate level, but the slew rate is too great in practice for the output to have reached a steady state before the product needs to be released.

Figure 5 shows the dual filter technique adopted. The weight value of the product is obtained by taking the difference between stored weight and zero weight signals. These signals are obtained by sampling a filtered version of the loadcell output at appropriate times. In order to prevent large step signals occurring at the inputs of the filters, with the attendant slew—rate problems, the sampled level is fed back to the filter input when the signal is not being sampled. Since the system will be repeatedly weighing nominally the same weight, the input level at each filter will be almost constant. A differential amplifier whose inputs are the zero and weight levels then generates an accurate representation of the true weight. Such a design has the additional advantage of compensating automatically for long term electrical or mechanical drift in the weighing system.

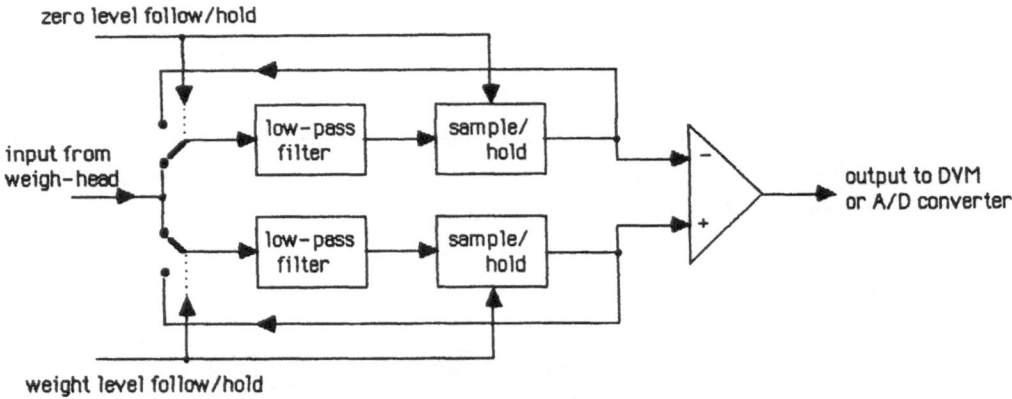

Figure 5 The Dual—Filter Technique

The weight (and zero) levels are sampled from just after the weight has been picked up (dropped) until just before it is dropped (picked up). There is a small delay imposed between dropping and in particular, picking up the weight to allow the output of the loadcell amplifier to settle, and to avoid the large negative overload transient which occurs when the robot presses the suckers on the object to be picked up. The sample/hold signals are generated by the robot control program and are taken from the controller's parallel output port. This ensures that these signals are synchronized with the actions of the robot.

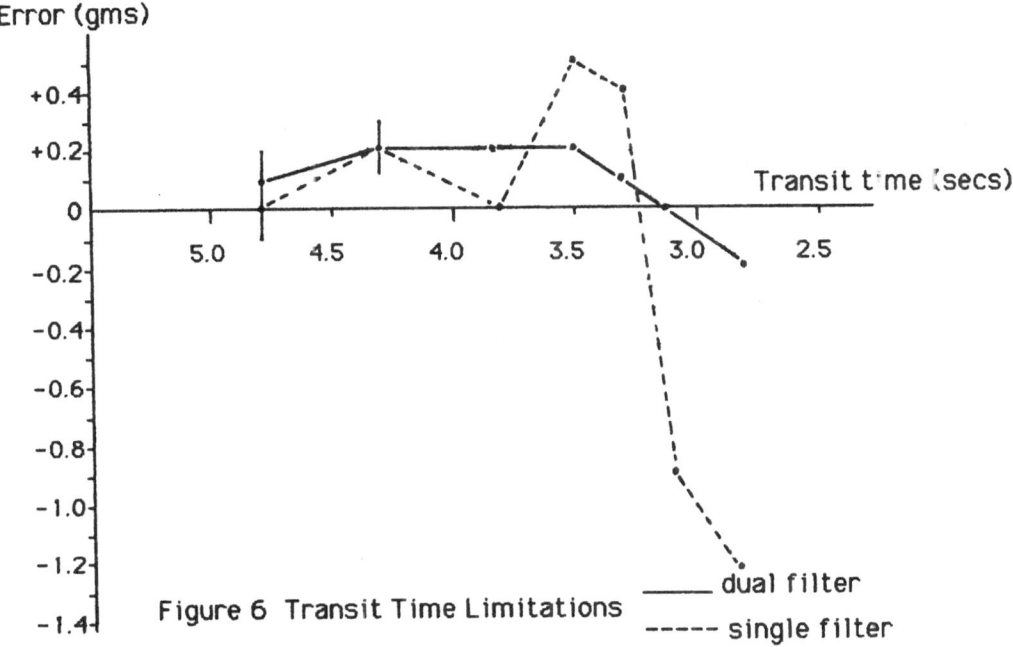

Figure 6 Transit Time Limitations

dual filter
----- single filter

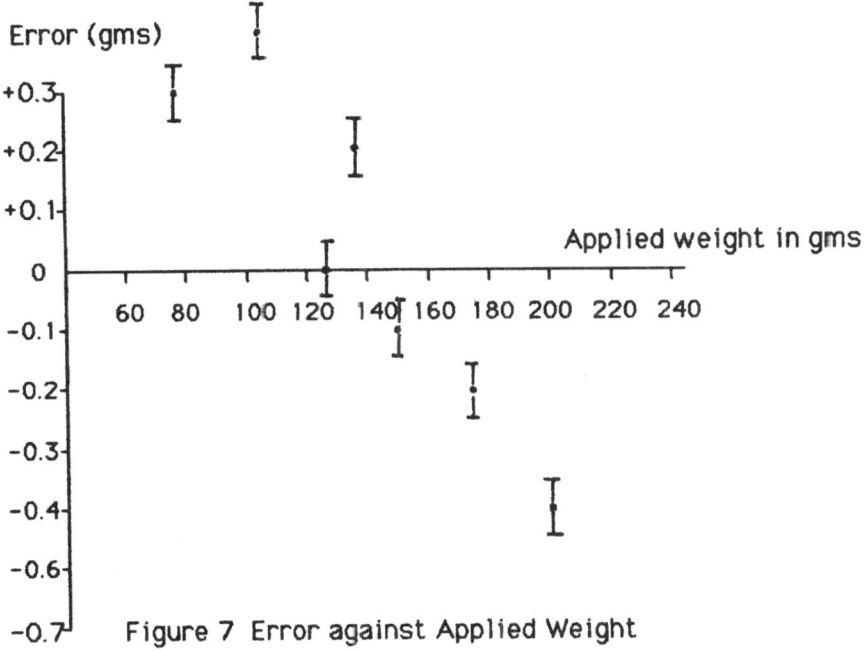

Figure 7 Error against Applied Weight

4.0 Experimental Results

Figure 6 shows the results obtained from a sequence of weight measurements using both a single—filter and the dual—filter approach. The speed of the pick—and—place operation was varied over the range of speeds likely to be encountered in practice. A PUMA robot with the weigh—head as an end—effector was programmed to pick up a known weight (in this case a simulated chocolate block of weight 134.7g). It can be seen that at the higher speeds, as to be expected, the single—filter became slew—rate limited with the output registering increasingly negative errors. The dual—filter system maintained an output consistant with the required accuracy. Each of the points on the graph was the average of five readings. The error bars indicate the accuracy to which readings could be taken (±0.05g).

Further tests were made, at a constant pick, weigh and place transit time of 2.8 seconds to measure the accuracy of the dual—filter over a range of weights. The results are shown in figure 7. It should be noted that the calibration of the check weights was within ±0.05g.

In practice, the system will be weighing the same approximate weight continuously with only small changes from unit to unit being expected. To simulate this the PUMA was programmed to repeatedly pick, weigh and place a known weight to which was added several smaller weights at intervals. The transit time was again 2.8 seconds. The results are shown in figure 8. For clarity, only the residuals are shown after the base weight

Residual (gms)

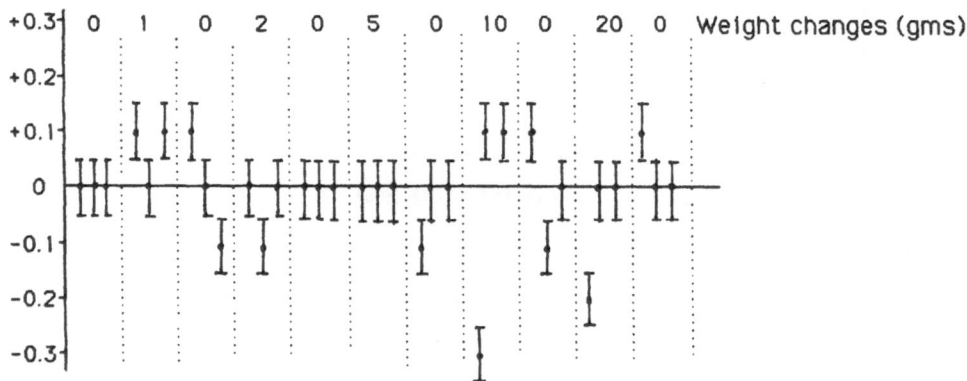

Figure 8 Response to Step Weight Changes on a Base Weight of 137.4g

and the added weights have been subtracted. It can be seen that the results show a high degree of consistency. The error bars indicate the accuracy to which the 'known' weights could be measured.

5.0 Conclusions

The work described, which is on−going has demonstrated that the design of a dynamic load sensor which can provide real−time weight information in an environment of high mechanical noise is a practicable proposition. The filtering system, presently implemented in hardware, will eventually be implemented using digital filters and storage. Such a system will not suffer from errors inherent in analogue systems (such as leakage in the 'hold' circuitry and amplifier gain drift) which could possibly introduce errors into the measurement process. A digital system will also be better able to recover from large changes in weight (say, due to an unsuccessful pick operation), by reloading the nominal weight level into the filter input from a backup location, or even swapping to a backup filter.

6.0 Acknowledgements

This work has been funded by the Science and Engineering Research Council ACME Initiative, contract number GR/C/91307. The authors would also like to thank Cadbury Schweppes plc for their help and support.

SENSOR FUSION IN CERTAINTY GRIDS FOR MOBILE ROBOTS

H.P. Moravec
Carnegie-Mellon University
Pittsburgh, PA 15213   (USA)

1 Abstract

A numerical representation of uncertain and incomplete sensor knowledge we call Certainty Grids has been used successfully in several of our past mobile robot control programs, and has proven itself to be a powerful and efficient unifying solution for sensor fusion, motion planning, landmark identification, and many other central problems. We had good early success with *ad-hoc* formulas for updating grid cells with new information. A new Bayesian statistical foundation for the operations promises further improvement. We propose to build a software framework running on processors onboard our new Uranus mobile robot that will maintain a probabilistic, geometric map of the robot's surroundings as it moves. The "certainty grid" representation will allow this map to be incrementally updated in a uniform way from various sources including sonar, stereo vision, proximity and contact sensors. The approach can correctly model the fuzziness of each reading, while at the same time combining multiple measurements to produce sharper map features, and it can deal correctly with uncertainties in the robot's motion. The map will be used by planning programs to choose clear paths, identify locations (by correlating maps), identify well known and insufficiently sensed terrain, and perhaps identify objects by shape. The certainty grid representation can be extended in the time dimension and used to detect and track moving objects. Even the simplest versions of the idea allow us fairly straightforwardly to program the robot for tasks that have hitherto been out of reach. We look forward to a program that can explore a region and return to its starting place, using map "snapshots" from its outbound journey to find its way back, even in the presence of disturbances of its motion and occasional changes in the terrain.

2 Introduction

Robot motion planning systems have used many space and object representations. Objects have been modelled by polygons and polyhedra, or bounded by curved surfaces. Free space has been partitioned into Voronoi regions or, more heuristically, free corridors. Traditionally the models have been hard edged - positional uncertainty, if considered at all, was used in just a few special places in the algorithms, expressed as a gaussian spread. Partly this is the result of analytical difficulty in manipulating interacting uncertainties, especially if

NATO ASI Series, Vol. F52
Sensor Devices and Systems for Robotics
Edited by A. Casals
© Springer-Verlag Berlin Heidelberg 1989

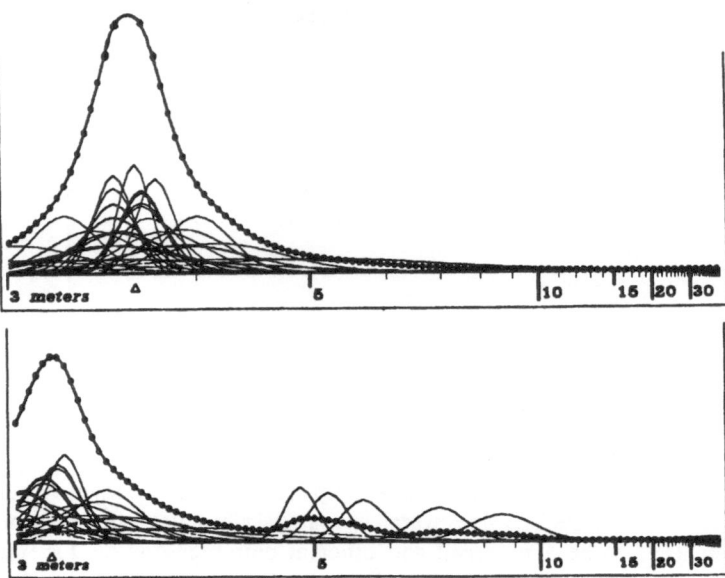

Figure 1: **Nine Eyed Stereo** - Identifications of a point on an object seen in nine different images taken as a camera traversed a track at right angles to its direction of view. Each pairing of images gives a stereo baseline, some short, some long. Long baselines have less uncertainty in the calculated distance. The distributions for all 36 possible pairings are added in a one dimensional "certainty grid", and the peak of the resultant sum is taken as the actual distance to the object. The top graph is for a case where all nine identifications of the point in the images are correct. The bottom is a case where one image is in error. The error produces eight small peaks at incorrect locations, but these are no match for the accumulation of correct values.

the distributions are not Gaussian. Incomplete error modelling reduces positional accuracy. More seriously, it can produce entirely faulty conclusions: a false determination of an edge in a certain location, for instance, may derail an entire train of inference about the location or existence of an object. Because they neglect uncertainties and alternative interpretations, such programs are brittle. When they jump to the right conclusions, they do well, but a small error early in the algorithm can be amplified to produce a ridiculous action. Most artificial intelligence based robot controllers have suffered from this weakness.

We've built our share of brittle controllers. Occasionally, however, we've stumbled across numerical (as opposed to analytic) representations that seem to escape this fate. One is deep inside the program that drove the Stanford Cart in 1979 [11]. Each of 36 pairings of nine images from a sliding camera produced a stereo depth measurement of a given feature, identified by a correlator, in the nine images. Some pairings were from short baselines, and had large distance uncertainty, others were from widely separated viewpoints, with small spread. The probability distributions from the 36 readings were combined numerically in a 1000 cell array, each cell representing a small range interval (Figure 1). Correlator matching errors often produced a multi-peaked resultant distribution, but the largest peak almost always gave the correct range. The procedure was the most error tolerant step in the Cart navigator, but it alone did not protect the whole program from brittleness.

A descendant of the Cart program by Thorpe and Matthies contained a path planner [10] that modelled floor space as a grid of cells containing numbers representing the suitability of each region to be on a path. Regions near obstacles had low suitability while empty space was high. A relaxation algorithm found locally optimum paths (Figure 2). The program represented uncertainty in the location, or even existence, of obstacles by having the suitability numbers for them vary according to extended, overlapping, probability distributions. The method dealt very reliably and completely with uncertainty, but also suffered from being embedded in an otherwise brittle program.

Our earliest thorough use of a numerical model of position uncertainty was in a sonar mapper, map matcher and path planner developed initially for navigating the Denning Sentry [8, 6, 7]. Space was represented as a grid of cells, each mapping an area 30 (in some versions 15) centimeters on a side and containing two numbers, one the estimated probability that the area was empty, the other that it was occupied. Cells whose state of occupancy was completely unknown had both probabilities zero, and inconsistent data was indicated if both numbers were high. Many of the algorithms worked with the differences of the numbers. Each wide angle sonar reading added a thirty degree swath of emptiness, and a thirty degree arc of occupancy, by itself a very fuzzy image of the world. Several hundred readings together produced an image with a resolution often better than 15 centimeters, despite many aberrations in individual readings (Figure 3). The resiliency of the method was been demonstrated in successful multi-hour long runs of Denning robots around and around long trajectories, using three second map building and three second map matching pauses at key intersections to repeatedly correct their position. These runs worked well in clutter, and survived disturbances such as people milling around the running robot.

Ken Stewart of MIT and Woods Hole has implemented a three dimensional version of the sonar mapper for use with small submersible craft. Initially tested in simulation in the presence of large simulated errors, Stewart's program provided extremely good reconstructions, in a $128 \times 128 \times 64$ array, of large scale terrain, working with about $60,000$ readings from a sonar transducer with a seven degree beam. Running on a Sun computer, his program can processed sonar data fast enough to keep up with the approximately one second pulse rate of the transducers on the two candidate submersibles at Woods Hole. The program was recently tested on real sonar data from a scanning transducer on a underwater robot that swam over the remains of the civil war battleship USS Monitor as part of a NOAA/Navy survey [1, 4]. The impressive results are shown in Figure 6.

Recently Serey and Matthies demonstrated the utility of the grid representation in a stereo vision based navigator running on our "Neptune" (Figure 7) mobile robot [5]. Edges crossing a particular scanline in the two stereo images are matched by a dynamic programming method, to produce a range profile. The wedge shaped space from the camera to the range profile is marked empty, cells along the profile itself are marked occupied. The resulting map is then used to plan obstacle avoiding paths as with the stereo and sonar programs mentioned above (Figure 4).

Matthies and Elfes combined improved versions of the sonar and stereo programs into a single one that builds maps integrating data from both sensors [2]. Their first results, also from a run of Neptune, are shown in figure 5.

In work in progress, In So Kweon of Carnegie-Mellon University has successfully demonstrated mapping of data from an ERIM scanning laser rangefinder into a three dimensional grid [3].

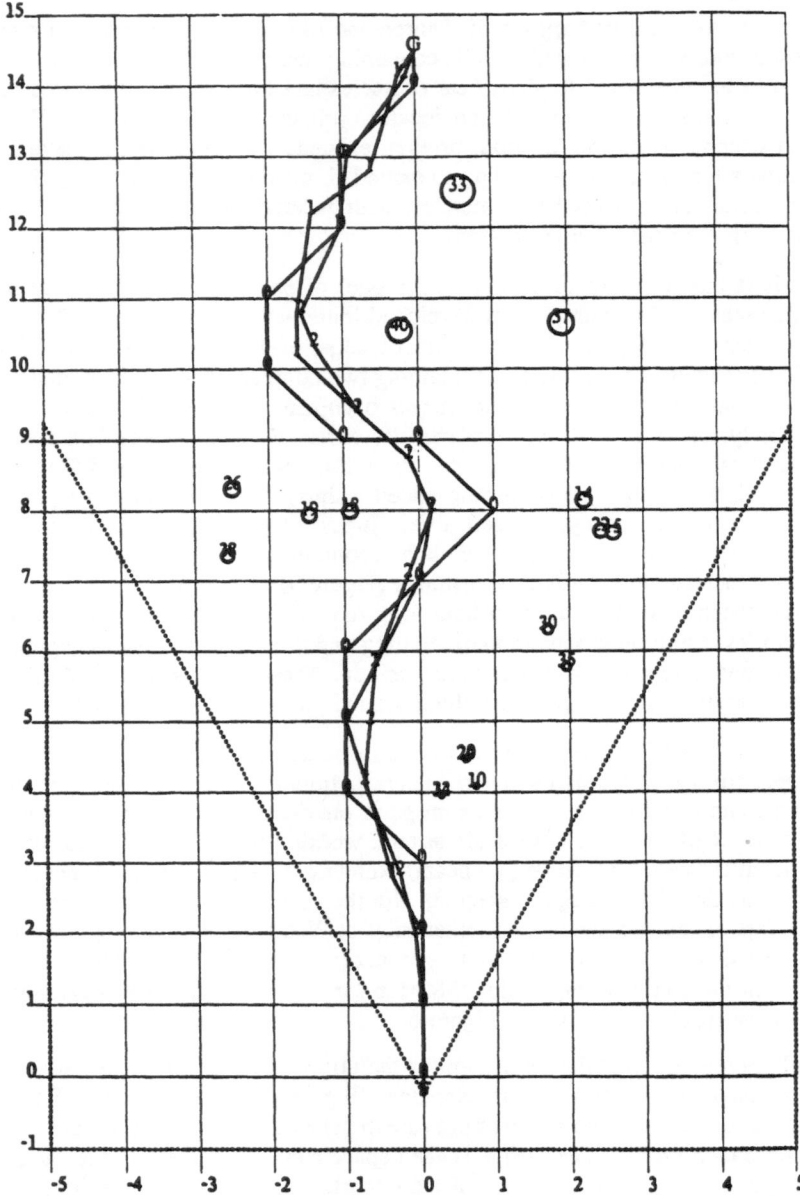

Figure 2: **Relaxation Path Planner** - A path is chosen that minimizes a given cost function in a Certainty Grid. Small perturbations are made in the vertices of the path in directions that reduce the cost.

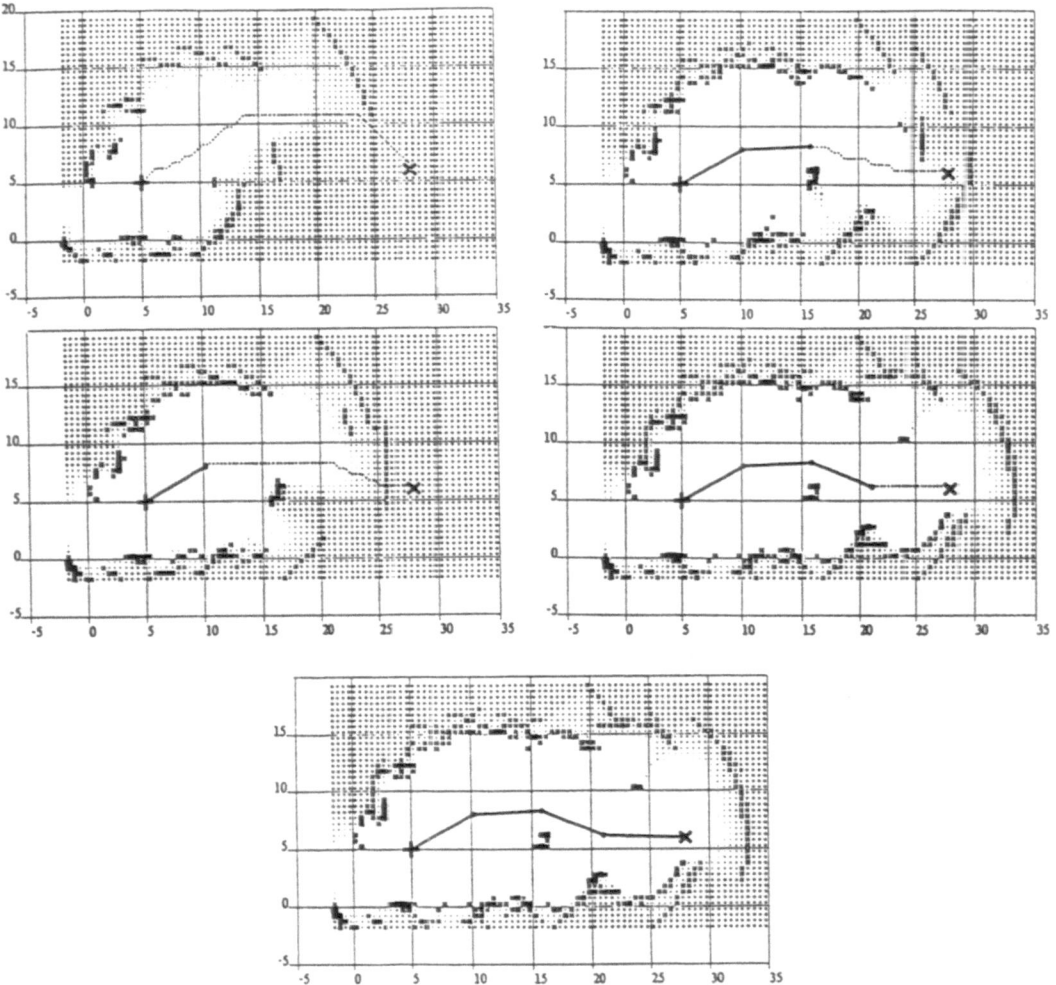

Figure 3: **Sonar Mapping and Navigation** - Plan view of the certainty grid built by a sonar guided robot traversing our laboratory. The scale marks are in feet. Each point on the dark trajectory is a stop that allowed the onboard sonar ring to collect twenty four new readings. The grid cells are white if the occupancy probability is low, dots if unknown, and × if high. The forward paths were planned by a relaxation path planner working in the grid as it was incrementally generated.

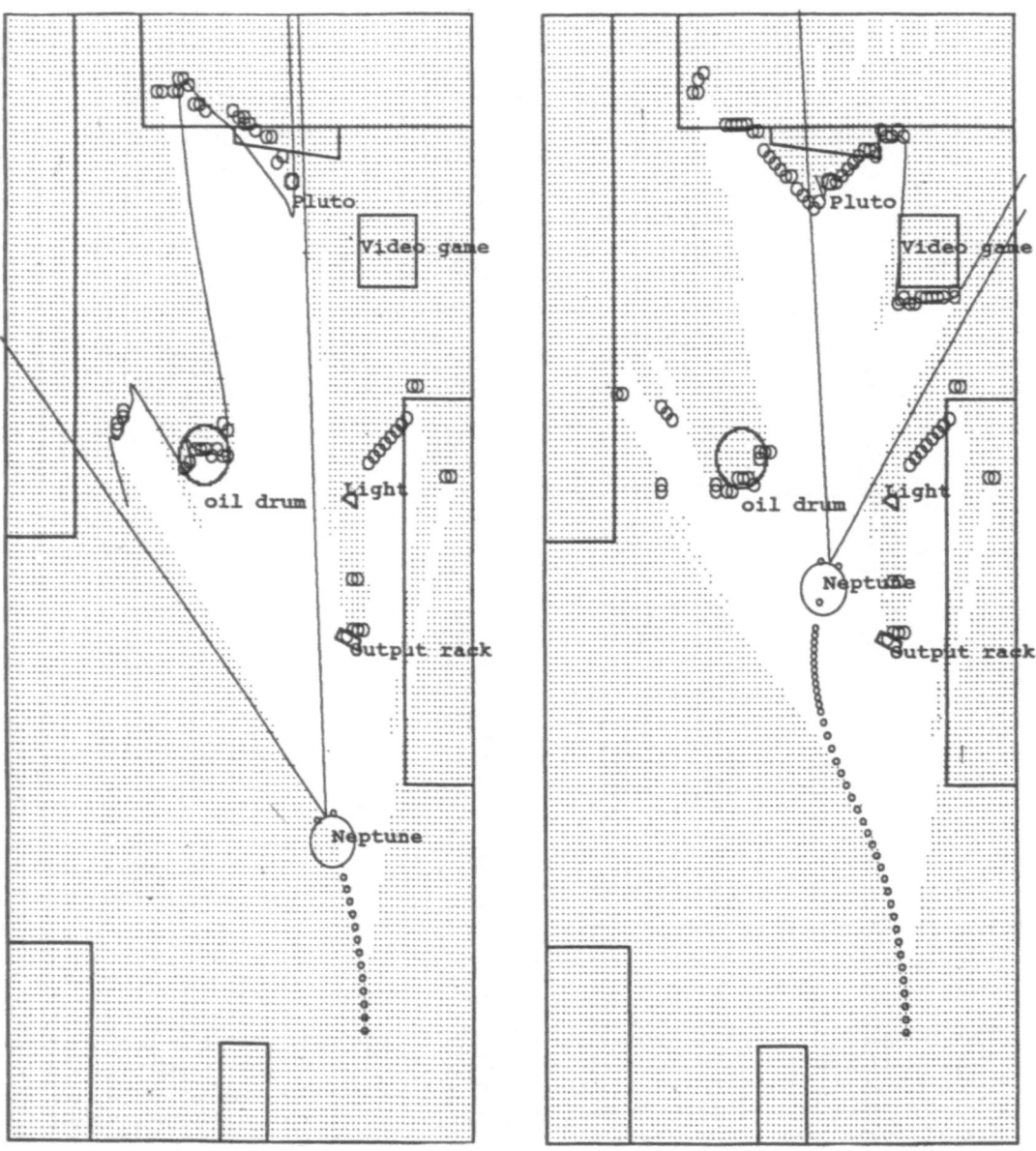

Figure 4: **Stereo Mapping and Navigation** - Plan view of the certainty grid built by a stereo guided robot traversing our laboratory. The situation is analogous to the sonar case of Figure 3, but the range profiles were gathered from a scanline stereo method using two TV cameras rather than a sonar ring.

Sonar   Stereo   Integration

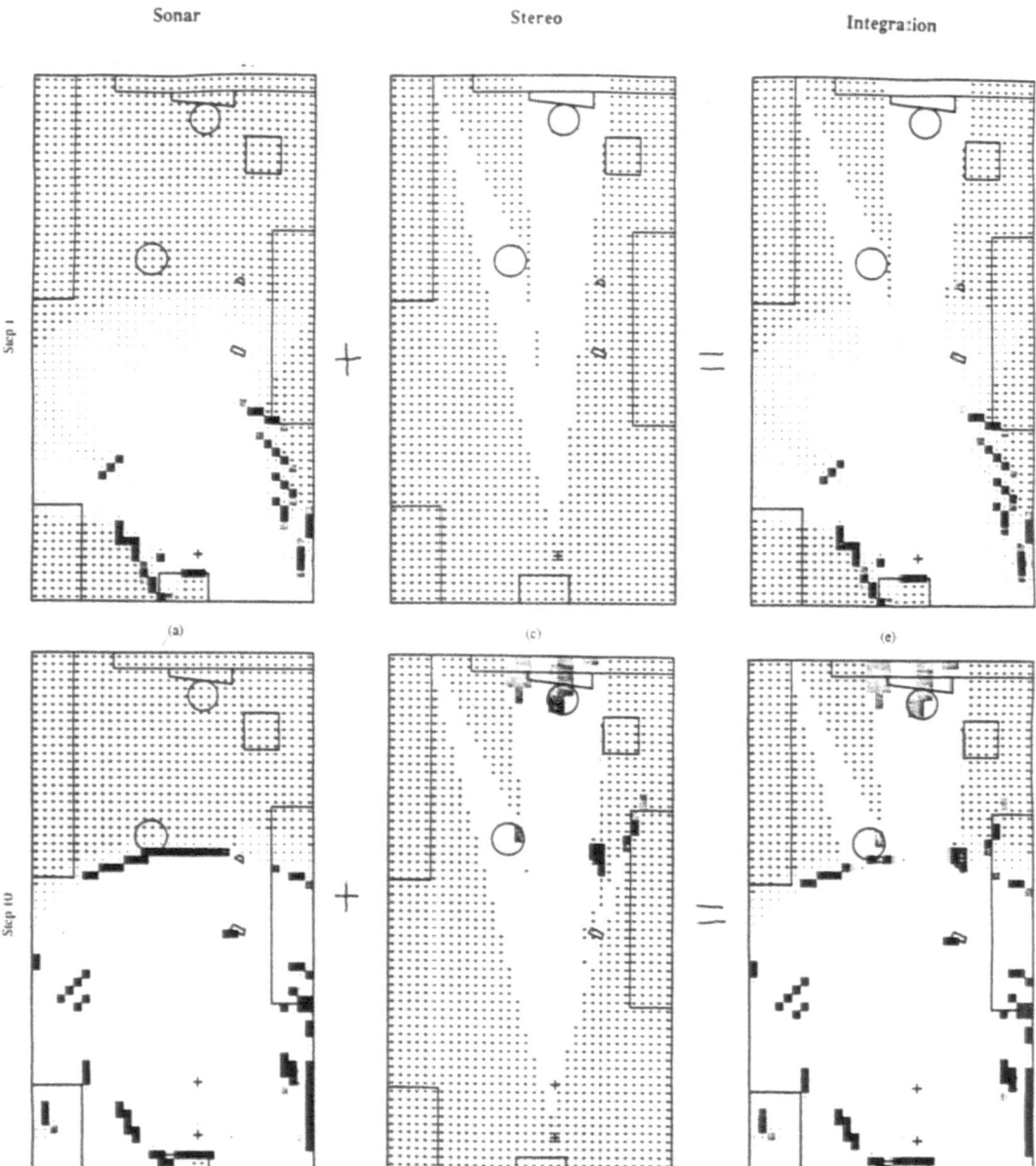

Figure 5: **Stereo and Sonar Sensor Integration** - Plan view of the certainty grids built on the first and tenth steps of a dual sensor run. The leftmost grids contain sonar data only, the center grid has stereo vision only, the rightmost is the combination of the two. Occupied regions are marked by shaded squares, empty areas by dots fading to white, and unknown territory by + signs.

Despite its effectiveness, in each instance we adopted the grid representation of space reluctantly. This may reflect habits from a recent time when analytic approaches were more feasible and seemed more elegant because computer memories were too small to easily handle numerical arrays of a few thousand to a million cells. I think the reluctance is no longer appropriate. The straightforwardness, generality and uniformity of the grid representation has proven itself in finite element approaches to problems in physics, in raster based approaches to computer graphics, and has the same promise in robotic spatial representations. At first glance a grid's finite resolution seems inherently to limit positioning accuracy. This impression is false. Cameras, sonar transducers, laser scanners and other long range sensors have intrinsic uncertainties and resolution limits that can be matched by grids no larger than a few hundred cells on a side, giving a few thousand cells in two dimensions, or a few million in three dimensions. Since the accuracy of most transducers drops with range, even greater economy is possible by using a heirarchy of scales, covering the near field at high resolution, and successively larger ranges with increasingly coarser grids. Besides this, the implicit accuracy of a certainty grid can be better than the size of its cell. The grid can be thought of as a discrete sampling of a continuous function. Extended features such as lines (perhaps representing walls) may be located to high precision by examining the parameters of surfaces of best fit. The Denning robot navigator mentioned above convolves two maps to find the displacement and rotation between them. In the final stages of the matching correlation values are obtained for a number of positions and angles in the vicinity of the best match. A quadratic least squares polynomial is fitted to the correlation values, and its peak is located analytically. Controlled tests of the procedure usually give positions accurate to better than one quarter of a cell width.

Our results to date suggest that many mobile robot tasks can be solved with this unified, sensor independent, approach to space modelling. The key ingredients are a robot centered, multi resolution, map of the robot's surroundings, procedures for efficiently inserting data from sonar, stereo vision, proximity and other sensors into the map, other procedures for updating the map to reflect the uncertainties introduced by imprecise robot motion, and yet others to extract conclusions from the maps. We've already demonstrated procedures that produce local and global navigational fixes and obstacle avoiding paths from such maps. Other tasks, such as tracking corridors, finding vantage points with good views of unseen regions, and identification of larger features such as doors and desks by general shape seem within reach.

3 The Representation

The sonar mappers mentioned above are our most thorough use to date of the certainty grid idea. Although our original implementations used two grids to represent occupancy knowledge (labelled $P_{occupied}$ and P_{empty}), Stewart's 3D system used only one. An analysis of the steps in our code revealed that one grid did indeed suffice, and this simplification maked clear several puzzling issues in the original formulation. Here I will present a sample of the methods we have used in most of our work so far. A new, Bayesian, approach follows after.

Before any measurements were made, the grid was initialized to a background occupancy certainty value, Cb. This number represented the average occupancy certainty we expected

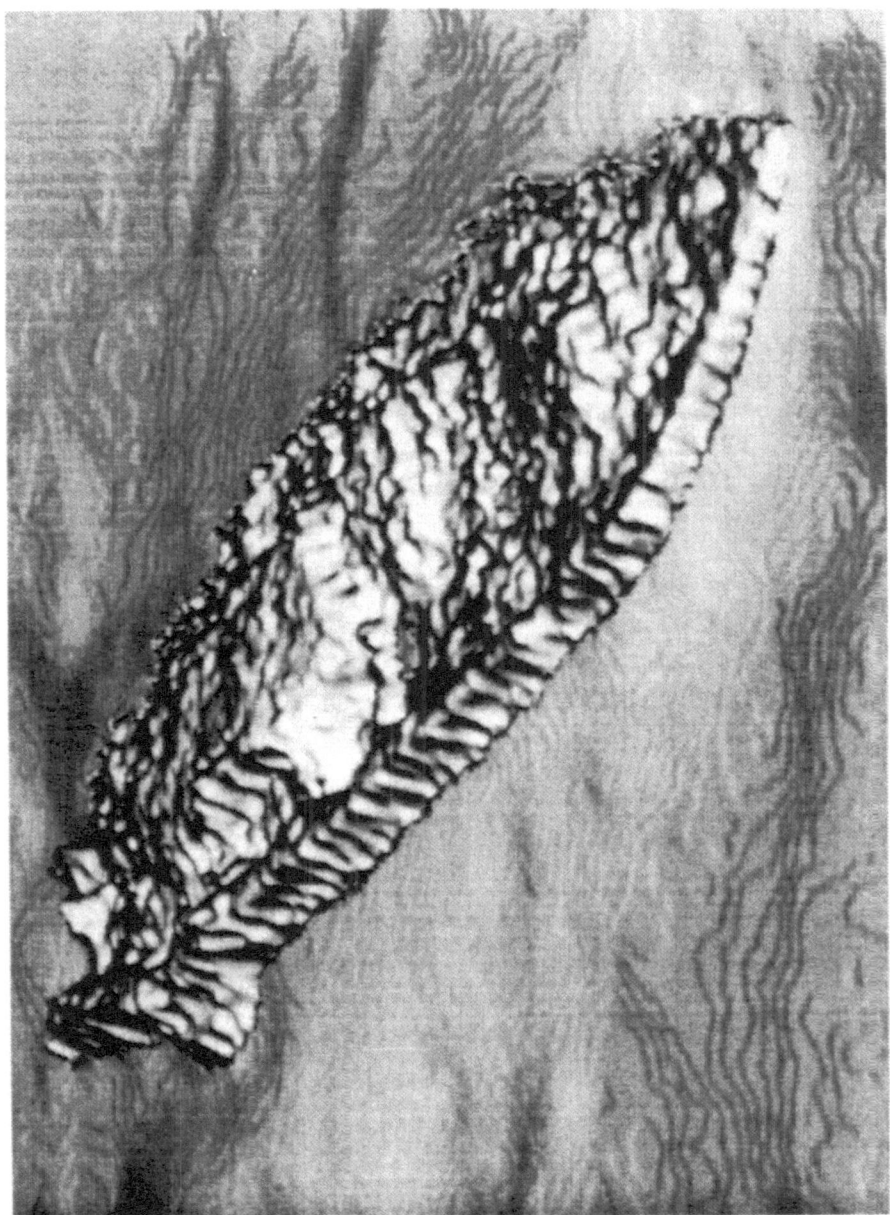

Figure 6: **Sonar Image of USS Monitor -** A surface has been extracted from a three dimensional certainty grid 128 cells on a side. The grid was built from approximately 100,000 readings from a 1.5 deg scanning sonar on a free swimming robot. The ship is lying upside down, with many parts of the hull collapsed.

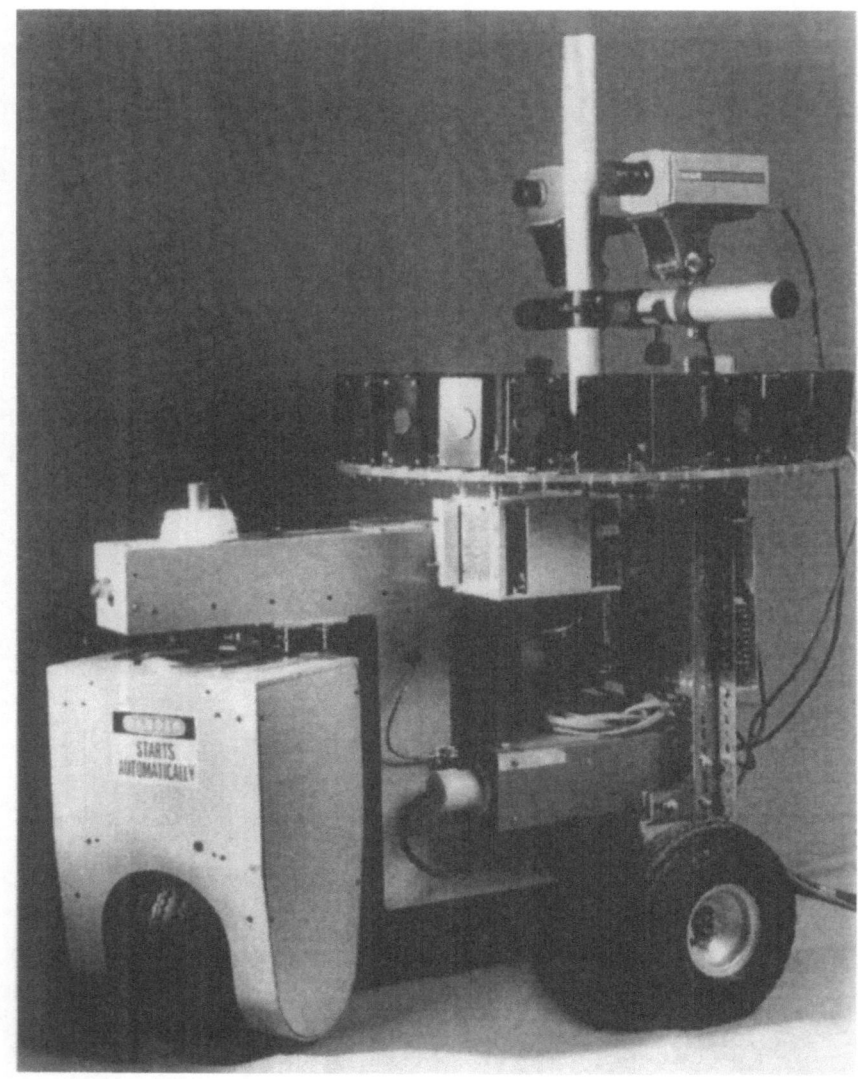

Figure 7: **The Neptune Mobile Robot** - Host for many early Certainty Grid experiments.

in a mature map, and encoded a (very) little bit of a-priori information we had about the world. In our lab a good Cb seemed to be about the number of cells in the perimeter of the grid divided by the total cells $(4 \times 32/(32 \times 32) = 1/8)$ in the case of the Denning code. If the space was very cluttered, Cb should be larger. As the map was used, values near Cb would indicate regions whose occupancy state was essentially unknown, while those much nearer zero would represent empty places, and those much nearer unity were likely to be occupied. Most of the planning algorithms that used the grid are better off if they do not make sharp distinctions, but instead numerically combine the certainty values from various cells to produce "goodness of fit" numbers for their various hypotheses. In this way the essential uncertainties in the measurements are not masked, and the algorithms do not jump to unnecessary, possibly false, conclusions.

4 Inserting Measurements

The readings of almost any kind of sensor can be incorporated into a certainty grid, if they can be expressed in geometric terms. The information from a reading can be as minimal as a proximity detector's report that there is probably something in a certain region of space, or as detailed as a stereo depth profiler's precise numbers on the countours of a surface.

The first step, in general, is to express the sensor's measurement as a numerical spatial certainty distribution commensurate with the grid's geometry. For an infrared proximity detector this may take the form of set of numbers P_x in an elliptical envelope with high certainty values in a central axis (meaning detection is likely there) tapering to zero at the edges of the illumination envelope. Let's suppose the sensor returns a binary indication that there is or is not something in its field of view. If the sensor reports a hit, cells in the certainty grid C_x falling under the sensor's envelope can be updated with the formula

$$C_x := C_x + P_x - C_x \times P_x$$

which will increase the C values. In this case the P values should be scaled so their sum is one, since the measurement describes a situation where there is something somewhere in the field of view, probably not everywhere. If the reliability of the sensor is less than perfect, the normalization may be to a sum less than unity. If, on the other hand, the detector registers no hit, the formula might be

$$C_x := C_x \times (1 - P_x)$$

and the Cs will be reduced. In this case the measurement states that there is nothing anywhere in the field of view, and the P values should reflect only the chance that an object has been overlooked at each particular position; i.e. they should not be normalized. If the sensor returns a continuous value rather than a binary one, perhaps expressing some kind of rough range estimate, a mixed strategy similar to the one described below for sonar is called for.

A Polaroid sonar measurement is a number giving the range of the nearest object within an approximately thirty degree cone in front of the sonar transducer. Because of the wide angle, the object position is known only to be somewhere on a certain surface. This range surface can be handled in the same manner as the sensitivity distribution of a proximity detector "hit" above. The sonar measurement has something else to say, however. The volume of the cone up to the range reading is probably empty, else a smaller range would have been returned. The empty volume is like the "no hit" proximity detector case, and can be handled in the same fashion. So a sonar reading is like a proximity detector hit at some locations, and increases the occupancy probability there, and like a miss at others, where it decreases the probability. If we have a large number of sonar readings taken from different vantage points (say as the robot moves), the gradual accumulation of such certainty numbers will build a respectable map. We can, in fact, do a little better than that. Imagine two sonar readings whose volumes intersect. And suppose the "empty" region of the second overlaps part of the range surface of the first. Now the range surface says "somewhere along here there is an object", while the empty volume says "there is no object here". The second reading can be used to reduce the uncertainty in the position of the object located by the first reading by decreasing the probability in the area of the overlap, and correspondingly increasing it in the rest of the range surface. This can be accomplished by reducing the range surface certainties R_x with the formula $R_x := R_x \times (1 - E_x)$ where E_x is the "empty" certainty at each point from the second reading, then normalizing the Rs. This method is used to good effect in the existing sonar navigation programs, with the elaboration that the Es of many readings are first accumulated, and then used to condense the Rs of the same readings. (It is this two stage process that led us to use two grids in our original programs. In fact, the grid in which the Es are accumulated need merely be temporary working space.)

The stereo method of Serey and Matthies provides a depth profile of visible surfaces. Although, like a sonar reading, it describes a volume of emptiness bounded by a surface whose distance has been measured, it differs by providing a high certainty that there is matter at each point along the range surface. The processing of the "empty" volume is the same, but the certainty reduction and normalization steps we apply to sonar range surfaces are thus not appropriate. The grid cells along a very tight distribution around the range surface should simply be increased in value according to the "hit" formula. The magnitude and spread of the distribution should vary according to the confidence of the stereo match at each point. The method used by Serey and Matthies matches edge crossing along corresponding scanlines of two images, and is likely to be accurate at those points. Elsewhere it interpolates, and the expected accuracy declines.

If the robot has proximity or contact sensors, its own motion can contribute to a certainty grid. Areas traversed by the robot are almost certainly empty, and their cells can be reduced by the "no hit" formula, applied over a confident sharp edged distribution in the shape of the robot. This approach becomes more interesting if the robot's motion has inherent uncertainties and inaccuracies. If the certainty grid is maintained so it is accurate with respect to the robot's present position (so called robot co-ordinates), then the past positions of the robot will be uncertain in this co-ordinate system. This can be expressed by *blurring* the certainty grid accumulated from previous readings in a certain way after each move, to reflect the uncertainty in that move. New readings are inserted without blur (essentially the robot is saying "I know *exactly* where I am now; I'm just not sure where I was before). The track in the certainty grid of a moving robot's path in this system will resemble the vapor trail of a high flying jet - tight and dense in the vicinity of the robot, diffusing eventually to nothing with time and distance.

5 Extracting Deductions

The purpose of maintaining a certainty grid in the robot is to plan and monitor actions. Thorpe and Elfes showed one way to plan obstacle avoiding paths. Conceptually the grid can be considered an array of topographic values - high occupancy certainties are hills while low certainties are valleys. A safe path follows valleys, like running water. A relaxation algorithm can perturb portions of a trial path to bring each part to a local minimum. In principle a decision need never be made as to which locations are actually empty and which are occupied, though perhaps the program should stop if the best path climbs beyond some threshold "altitude". If the robot's sensors continue to operate and update the grid as the path is executed, impasses will become obvious as proximity and contact sensors raise the occupancy certainty of locations where they make contact with solid matter.

As indicated in the introduction, we have already demonstrated effective navigation by convolving certainty grids of given locations built at different times, allowing the robot to determine its location with respect to previously constructed maps. This technique can be extended to subparts of maps, and may be suitable for recognizing particular landmarks and objects. For instance, we are presently developing a wall tracker that fits a least squares line to points that are weighted by the product of the occupancy certainty value and a gaussian of the distance of the grid points from an a-priori guess of the wall location. The parameters of the least squares line are the found wall location, and serve, after being transformed for robot motion, as the initial guess for the next iteration of the process.

For tasks that would benefit from an opportunistic exploration of unknown terrain, the certainty grid can be examined to find interesting places to go next. Unknown regions are those whose certainty values are near the background certainty Cb. By applying an operator that computes a function such as

$$\sum (C_x - Cb)^2$$

over a weighted window of suitable size, a program can find regions whose contents are relatively unknown, and head for them. Other operators similar in spirit can measure other properties of the space and the robot's state of knowledge about it. Hard edged characterizations of the stuff in the space can be left to the last possible moment by this approach, or avoided altogether.

6 A Plan: Awareness for a Robot

Uranus is the CMU Mobile Robot Lab's latest and best robot and the third and last one we intend to construct for the forseeable future. About 60 cm square, with an omnidirectional drive system intended primarily for indoor work, Uranus carries two racks wired for the industry standard VME computer bus, and can be upgraded with off the shelf processors, memory and input output boards. In the last few years the speed and memory available on single boards has begun to match that available in our mainframe computers. This removes the main arguments for operating the machine primarily by remote control. With most computing done on board by dedicated processors, enabling very high bandwidth and reliable connection of processors to sensors and effectors, real time control is much easier. Also favoring this change in approach is a realization by us, growing from our experience with robot control programs from the very complex to the relatively simple, that the most

Figure 8: **The Uranus Mobile Robot** - Great Expectations.

complicated programs are probably not the most effective way to learn about programming robots. Very complex programs are slow, limiting the number of experiments possible in any given time, and they involve too many simultaneous variables, whose effects can be hard to separate. A manageable intermediate complexity seems likely to get us to our long term goals fastest. The most exciting element in our current plans is a realization that certainty grids are a powerful and efficient unifying solution for sensor fusion, motion planning, landmark identification, and many other central problems.

As the core of the robot and the research we will prepare a kind of operating system based on the "certainty grid" idea. Software running continuously on processors onboard Uranus will maintain a probabilistic, geometric map of the robot's surroundings as it moves. The certainty grid representation will allow this map to be incrementally updated in a uniform way from various sources including sonar, stereo vision, proximity and contact sensors. The approach can correctly model the fuzziness of each reading, while at the same time combining multiple measurements to produce sharper map features, and it can deal correctly with uncertainties in the robot's motion. The map will be used by planning programs to choose clear paths, identify locations (by correlating maps), identify well known and insufficiently sensed terrain, and perhaps identify objects by shape. To obtain both adequate resolution of nearby areas and sufficient coverage for longer range planning, without excessive cost, a heirarchy of maps will be kept, the smallest covering a 2 meter area at 6.25 cm resolution, the largest 16 meters at 50 cm resolution (Figure 9). This map will be "scrolled" to keep the robot centered as it moves, but rotations of the robot will be handled by changing elements of a matrix the represents the robot's orientation in the grid. The map forms a kind of consciousness of the world surrounding the robot - reasoning about the world would actually be done by computations in the map. It might be interesting to take one more step in the

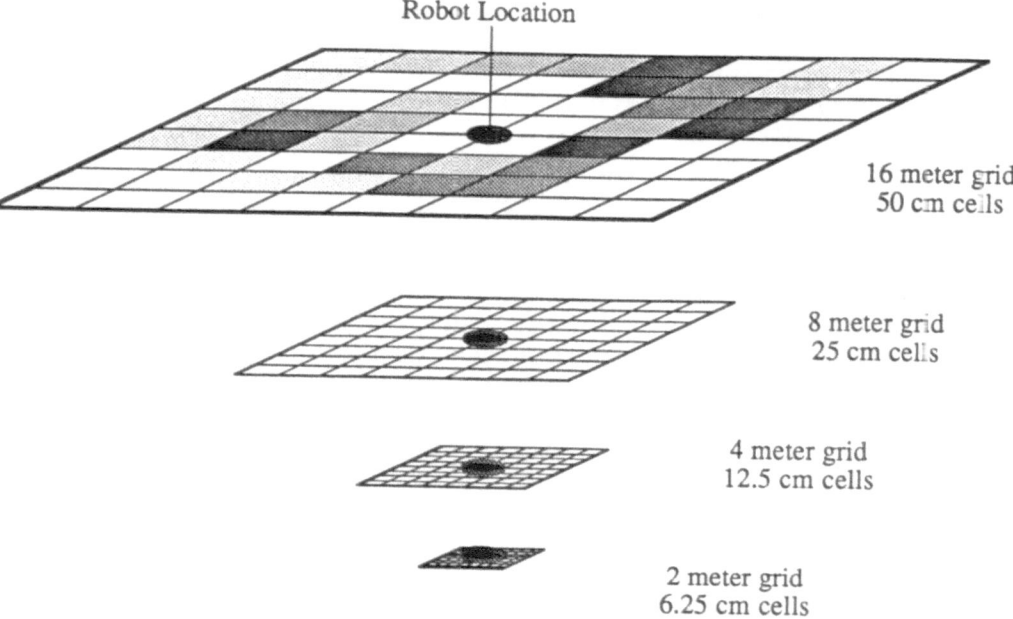

Robot Location

16 meter grid
50 cm cells

8 meter grid
25 cm cells

4 meter grid
12.5 cm cells

2 meter grid
6.25 cm cells

Figure 9: **Map Resolution Heirarchy** - Coarse maps for the big picture, fine ones for the fiddly details in the immediate environment. All the maps are scrolled to keep the robot in the center cells.

heirarchy, to a one meter grid that simply covers the robot's own extent. It would be natural to keep this final grid oriented with respect to robot chassis itself, rather than approximately to the compass as with the other grids. This change of co-ordinate system would provide a natural distinction between "world" awareness and "body" or "self" awareness. Such encoding of a sense of self might even be useful if the robot were covered with many sensors, or perhaps were equipped with manipulators. We have no immediate plans in that direction, and so will pass by this interesting idea for now.

Our initial version will contain a pair of two dimensional grid sets, one mapping the presence of objects at the robots operating height of a few feet above ground level. The other will map the less complex idea of presence of passable floor at various locations. The object map will be updated from all sensors, the floor map primarily from downward looking proximity detectors, though possibly also from long range data from vision and sonar. The robot will navigate by dead reckoning, integrating the motion of its wheels. This method accumulates error rapidly, and this uncertainty will be reflected in the maps by a repeated blurring operation. Old readings, whose location relative to the robot's present position and orientation are known with decreasing precision, will have their effect gradually diffused by this operation, until they eventually evaporate to the background certainty value.

It would be natural to extend the two-grid system to many grids, each mapping a particular vertical slice, until we have a true three dimensional grid. We will do this as our research results, and processing power permit. The availability of single board array processors that can be installed on the robot would help this, as the certainty grid operations are very amenable to vectorizing. The certainty grid representation can also be extended in the time dimension, with past certainty grids being saved at regular intervals, like frames in a movie film, and registered to the robot's current co-ordinates (and blurred for motion uncertainties). Line operators applied across the time dimension could detect and track moving objects, and give the robot a sense of time as well as space. This has some very thrilling conceptual (and perceptual) consequences, but we may not get to it for a while.

Even the simplest versions of the idea will allows us fairly straightforwardly to program the robot for tasks that have hitherto been out of reach. We look forward to a program that can explore a region and return to its starting place, using map "snapshots" from its outbound journey to find its way back, even in the presence of disturbances of its motion and occasional changes in the terrain. By funneling the sensor readings through a certainty grid, which collects and preserves all the essential data, and indications of uncertainties, and makes it available in a uniform way, we avoid the problem we've had, that for each combination of sensor and task a different program is required. Now the task execution is decoupled from the sensing, and thus becomes simpler.

7 Bayesian Reasoning

In most of our work to date, we have used *ad-hoc* formulas to update the certainty grid estimates. Recently a less arbitrary statistical approach derived from Bayes theorem [9] has captured our attention. Preliminary results using this approach are at least as good as those from the old formulas. Many puzzling aspects of the old scheme have been clarified in the process.

Let $p(A|B)$ represent our best estimate of the likelihood of situation A if we have received information B. By definition

$$p(A|B) = \frac{p(A \wedge B)}{p(B)} \tag{1}$$

Plain $p(A)$ represents our estimate of A given no special information. The alternative to event A will be referred to as $\overline{A}$ (not A). For any B

$$p(A|B) + p(\overline{A}|B) = 1 \tag{2}$$

A certainty grid is a regular finite element model of space. Each cell of the grid contains a liklehood estimate (our "certainty") of a given property of the corresponding region of space. Primarily we are concerned with simple occupancy of the region, represented by $p(o[x])$, the probability that region x is occupied. When a discussion contains only one particular x, we will drop the subscript, and refer simply to $p(o)$.

We will be considering data derived from wide angle sonar range measurements. A given measurement will be designated $M[i]$, with i being the sequential number of the reading. The intersection of a set of readings can be designated by a range in subscript, as in

$$M[< n] = \bigcap_{i=1}^{i<n} M[i] \tag{3}$$

or by a list as in

$$M[i, j, l] = M[i] \wedge M[j] \wedge M[l] \tag{4}$$

When only one reading $M[i]$ enters into a discussion, we will abbreviate its name to simply M. Each measurement has a value, a sonar range $R(M)$. The sonar sensor is quantized, and R will be an integer.

$P(o)$ is the probability that any particular cell is occupied, i.e. the average occupation density of our space. Our measurements don't give $P(o)$ directly, but it is approximately the overall average of the $p(o)$'s of all the cells of a typical map of the space. By definition, $P(\overline{o}) = \overline{P(o)} = 1 - P(o)$.

8   Fundamental Formulas

For two occupancy possibilities o and $\overline{o}$ of a cell, new information B and old information A, one form of Bayes theorem gives:

$$p(o|B \wedge A) = \frac{p(B|o \wedge A) \times p(o|A)}{p(B|o \wedge A) \times p(o|A) + p(B|\overline{o} \wedge A) \times p(\overline{o}|A)} \qquad (5)$$

$$p(\overline{o}|B \wedge A) = \frac{p(B|\overline{o} \wedge A) \times p(\overline{o}|A)}{p(B|o \wedge A) \times p(o|A) + p(B|\overline{o} \wedge A) \times p(\overline{o}|A)} \qquad (6)$$

the "odds" formulation of this is compact and convenient for computation, and will be important later:

$$\frac{p(o|B \wedge A)}{p(\overline{o}|B \wedge A)} = \frac{p(B|o \wedge A)}{p(B|\overline{o} \wedge A)} \times \frac{p(o|A)}{p(\overline{o}|A)} \qquad (7)$$

Formulas 5 and 6 are somewhat complicated for repeated use. Formula 7 is better, it is formulated in terms of a product of odds. When the odds ratio involves a probability and its complement, odds and probabilities can be interconverted by the relationship:

$$Odds(A) = \frac{p(A)}{p(\overline{A})} = \frac{p(A)}{1 - p(A)} \qquad (8)$$

$$p(A) = \frac{Odds(A)}{1 + Odds(A)} \qquad (9)$$

The $p(B|\ldots)$ ratio in 7 is not of this form. To compute its value, both numerator and denominator must be evaluated separately. To make apparent this difference, the ratio $p(B|o \wedge A)/p(B|\overline{o} \wedge A)$ will be referred to as $Odds(B||o \wedge A)$. Once the ratio is obtained, however, it can be treated as any other odds number.

If we deal exclusively with odds, all the combining operations become multiplications. Formula 7 is expressed:

$$Odds(o|A \wedge B) = Odds(B||o \wedge A) \times Odds(o|A) \qquad (10)$$

An additional transformation streamlines the computation more. Let $L(A)$ represent the logarithm, to some suitable base, of $Odds(A)$. The formula then becomes a simple addition.

$$L(o|A \wedge B) = L(B||o \wedge A) + L(o|A) \qquad (11)$$

The terms can be integers if the base of the logarithms is chosen well. Perfect certainty ($p(o) = 0$ and $p(o) = 1$) can no longer be represented, but such values are probably a mistake in any representation, since they are unalterable by any input. Only God should be absolutely certain of anything!

9   Combining Formula

Bayes' theorem is a formula that combines independent sources of information A and B into an estimate of a single quantity $p(o|A \wedge B)$. The new information B, occurs in terms of the probability of B in the situation that (in our case of interest) a particular cell is or is not occupied, $p(B|o)$ and $p(B|\overline{o})$. This inversion is central to the usefulness of Bayes' theorem. But consider the problem of generating a map from information A and B when each has already been individually processed into a map, i.e. find $p(o|A \wedge B)$ given $p(o|A)$ and $p(o|B)$.

Bayes' formula (7) applied to information B and null A (i.e. only global information from the A side) makes the relationship between $p(o|B)$ and $p(B|o)$ clear:

$$\frac{p(B|o)}{p(B|\overline{o})} = \frac{p(o|B)}{p(\overline{o}|B)} \times \frac{P(o)}{P(\overline{o})} \tag{12}$$

thus

$$\frac{p(o|B)}{p(\overline{o}|B)} = \frac{p(B|o)}{p(B|\overline{o})} \times \frac{P(\overline{o})}{P(o)} \tag{13}$$

It would be nice to substitute this ratio back into the original version of formula (7) to produce a formula giving $p(o|A \wedge B)$ in terms of $p(o|A)$ and $p(o|B)$, thus allowing maesurements A and B to be incorporated into maps independently, and combined afterward. This is not possible in general because the two measurements may interact in some way - for instance, either A or B alone may indicate a high probability of o, but taken together, they may confirm some other hypothesis, and reduce the probability of o. If, however, we make a strong assumption of independence of B from A:

$$\frac{p(B|o \wedge A)}{p(B|\overline{o} \wedge A)} = \frac{p(B|o)}{p(B|\overline{o})} \tag{14}$$

we can use equations (7,13,14) to produce a *map combining formula*:

$$\frac{p(o|A \wedge B)}{p(\overline{o}|A \wedge B)} = \frac{p(o|B)}{p(\overline{o}|B)} \times \frac{p(o|A)}{p(\overline{o}|A)} \times \frac{P(\overline{o})}{P(o)} \tag{15}$$

While the non-informative reading in equation 7, that which leaves the original $p(o)$ estimate unchanged, is found when $p(B|o)/p(B|\overline{o}) = 1$, the non-informative case in equation 15 happens when $p(o|B)/p(\overline{o}|B) = P(o)/P(\overline{o})$, i.e. when the cell density estimated from the reading is the same as the average cell density of the whole map.

Formula 15 is most important because it provides a means for combining maps of the same area obtained by different means, for instance by independent scans of different sensors. It can also be used to incorporate individual sensor readings as we build a map, but is in general inferior to formula 7 for this purpose because it precludes the use of any knowledge we may have of sensor interactions. In odds and log odds form it becomes:

$$Odds(o|A \wedge B) = \frac{Odds(o|A) \times Odds(o|B)}{Odds(o)} \qquad (16)$$

$$L(o|A \wedge B) = L(o|A) + L(o|B) - L(o) \qquad (17)$$

10   Sonar Wedges

Simple reasoning from first principles can produce estimates for $\mathbf{p}(M|o)$ and $\mathbf{p}(M|\overline{o})$ for a sonar reading. These values can be used to update maps by direct application of the Bayes formula 7. In a later section we will show how to systematically incorporate in the possibility of errors in the readings. For now assume that the sensor always perfectly returns the range to the nearest occupied cell within its angle of sensitivity. Define the sonar regions according to the diagram:

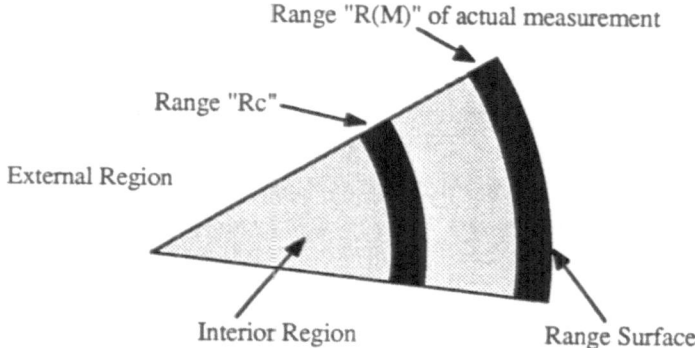

Range "R(M)" of actual measurement

Range "Rc"

External Region

Interior Region

Range Surface

When incorporating a new reading B, the map built from prior readings A can be used to help provide an estimate for the $\mathbf{p}(B|o \wedge A)$ and $\mathbf{p}(B|\overline{o} \wedge A)$ of formula 7.

Let $\mathbf{P}(R)$ be the probability, prior to a the reading, that the next sonar measurement will result in a given range R. $\mathbf{P}(R)$ can be approximated by stepping through the possible ranges $R_1, R_2, \ldots R_n$, starting with the shortest, R_1, and multiplying the occupancy probability $\mathbf{p}(o|A)$ of each cell on the range surface for that range by the probability that the sonar would detect an occupied cell at that location. The sum of these products at R_1, call it $S(R_1)$, is $\mathbf{P}(R_1)$. Now $\mathbf{P}(R_2) = S(R_2) \times (1 - \mathbf{P}(R_1))$ since an echo from R_2 can happen only if the sonar pulse has not already been intercepted at the shorter range R_1. In general

$$\mathbf{P}(R_i) = S(R_i) \times (1 - \sum_{j<i} \mathbf{P}(R_j)) \qquad (18)$$

By definition $\sum \mathbf{P}(R)$ over all R is unity. We will suggest later exactly how to compute S, and how the detection probabilities required in the calculation above can be determined empirically by collecting statistics from many maps of the correlation of individual sensor readings with the composite maps they helped create.

10.1 External Region

Consider a cell in the external region. $p(M|o)$ is our estimate that a measurement range of $R(M)$, as opposed to some other range, will occur if we happen to know only that the cell is occupied. Since the cell is outside the sonar cone, its state of occupancy has no effect on the reading, and we can refer only to the uniform range distribution to conclude that

[external region probabilities]

$$p(M|o) = p(M|\overline{o}) = P(R(M)) \qquad (19)$$

and

$$\frac{p(M|o)}{p(M|\overline{o})} = 1 \qquad (20)$$

inserting these into formula 7 gives

$$\frac{p(o|M[1:i])}{p(o|M[1:i])} = \frac{P(R(M))}{P(R(M))} \times \frac{p(o|M[<i])}{p(\overline{o}|M[<i])} = \frac{p(o|M[<i])}{p(\overline{o}|M[<i])} \qquad (21)$$

so, the occupancy certainty is unchanged, as it should be, since the sonar reading contains no information about the external cell.

10.2 Range Surface

Now consider a cell on the range surface, and suppose this surface covers n cells in all. By definition the range at this surface is $R(M)$.

If the cell is occupied, then a perfect sensor would detect it, and so could not return a greater range than $R(M)$. All ranges beyond the occupied cell, thus, would be "short circuited" by the cell. It would, however, be possible for it to return shorter ranges, if closer cells within the sonar beam angle happened to be occupied. The probabilities of the shorter ranges should not be changed by the presence of the farther occupied cell. The probability of getting just the range reading would be unity minus the a-priori probability of getting a lesser reading, or equivalently, the a-priori probability of getting a reading greater than or equal to $R(M)$.

[range surface probabilities]

$$p(M|o) = 1 - \sum_{R<R(M)} P(R) = \sum_{R\geq R(M)} P(R) \qquad (22)$$

If, on the other hand, the cell were unoccupied, the original distribution is hardly altered, except that the particular $R = R(M)$ is slightly less likely to occur, since one possible way to achieve it is eliminated. The cell in question is one of "n" cells we assumed on the range surface, so we can say the chance of $R(M)$ happening is reduced by the factor $(n-1)/n$. The probabilities of other ranges are not affected directly, so the result is a slight "notching" at $R = R(M)$ of the a-priori $P(R)$ distribution. Simply notching it, however, reduces its total area, which must then be renormalized by the proper factor to restore the area of the distribution to unity, increasing all the other probabilities slightly:

$$p(M|\bar{o}) = \frac{P(R(M)) \times (n-1)/n}{1 - P(R(M))/n} = P(R(M)) \times \frac{n-1}{n - P(R(M))} \qquad (23)$$

10.3 Interior Region

If we assume a perfect sensor, an occupied cell in the wedge at a range less than $R(M)$ would always be detected and thus prevent a the reading M. So in this case

[interior region probabilities]

$$p(M|o) = 0 \qquad (24)$$

if the cell in the interior is unoccupied, then the range "Rc" at which it occurs is less likely than other ranges. Suppose there would be k cells occupied by a range arc at distance Rc. By reasoning like that of formula 23, the probability of Rc would be reduced by a factor of $(k-1)/k$, but the overall distribution would have to be normalized by dividing by $(1 - P(Rc)/k)$. This normalization raises the probability of $R(M)$ from its a-priori value:

$$p(M|\bar{o}) = \frac{P(R(M))}{1 - P(Rc)/k} \qquad (25)$$

11 Measurement Errors

Of course our sensors are not perfect in the sense of the last section. Sometimes they fail to respond to an occupied location, at other times they give a spurious indication of occupancy. It is easy to modify the perfect case formulas for this. Suppose there is a small chance e_{hit} that an empty cell will act as if it were occupied. Suppose also that there is a small possibility e_{miss} that an occupied cell will fail to be detected. We can then construct the formulas with error from the error-free cases using the general formulas:

$$p(M|o)_{error} = p(M|o) \times (1 - e_{miss}) + p(M|\bar{o}) \times e_{miss} \qquad (26)$$

$$p(M|\bar{o})_{error} = p(M|\bar{o}) \times (1 - e_{hit}) + p(M|o) \times e_{hit} \qquad (27)$$

If $e_{hit} = e_{miss} = 1/2$, then the sensors are returning random readings. This is captured in (15) as $p(M|o)_{error} = p(M|\bar{o})_{error}$, the non informative case, similar to formula (12). As e_{hit}

and e_{miss} grow closer to $1/2$, the reading becomes less and less informative. If $e_{hit} = e_{miss} = 1$ the sensor is perfect, but the hit and miss indications are mistakenly swapped. In that case (15) restores the correct pairing.

A real sensor can be modelled, somewhat redundantly, by varying e_{hit} and e_{miss} over the sensed area. Towards the edges of the beam they will creep towards $1/2$, achieving this value perfectly outside the beam, and perhaps at the boundary of the interior and range surface regions.

If $0 \leq e_{hit}, e_{miss} \leq 1$, the error formulas can only reduce or leave unchanged the amount of information provided by the "perfect" estimates of $\mathbf{p}(M|\mathbf{o})$ and $\mathbf{p}(M|\overline{\mathbf{o}})$.

The quantities e_{hit} and e_{miss} must be initialized to some reasonable values when a system begins operation. They can then be adjusted by an iterative "learning" process. Each time a map is constructed from many sensor readings, the correlation of each cell in each individual sensor profile with the area of the total map it overlays is recorded. This correlation is averaged over many maps. If the map often reports a high occupancy probability in a location where the sensor profile indicates occupancy, e_{hit} will be low for that position in the sensor profile, otherwise it will be high. Similarly if the map usually has low probabilities at a site indicated empty in the sensor profile, e_{miss} will be low, otherwise it will be high.

The detection probability required to compute the function S in equation 18 is given by the expression $1 - e_{miss}$. A more accurate computation of S would also take into account the chance that an empty cell would falsely register as occupied. So a good estimate for S would be:

$$S(R) = \sum_R \mathbf{p}(\mathbf{o}|A) \times (1 - e_{miss}) + \mathbf{p}(\overline{\mathbf{o}}|A) \times e_{hit} \qquad (28)$$

12 Stereo Range Measurements

Another approach to incoprorating individual sensor readings uses the independent combining formula 15. In general this will result in inferior maps, but it requires less computation. Here's an example from a consideration of a stereo vision measurement. One of our stereo approaches matches edges crossing a given scanline in the left and right images. A range is deduced from the relative horizontal shift of each edge between the two images. The edges cannot be located with infinite precision, so there is some uncertainty in depth, given by a distribution $P(D|m)$, the probability the object ranged by stereo measurement m is actually at distance D. Note that this is not the probability that location d is occupied, because d may contain an object other than the edge being ranged. Let $\mathbf{p}(d|m)$ represent the probability, given measurement m, that d is occupied by *any* object. We can estimate $\mathbf{p}(d|m)$ directly by the following reasoning.

Consider the possible case that the ranged edge is at a particular distance D (this possibility has probability $P(D|m)$). In this case, and assuming a perfect sensor as we did in the sonar example, we can conclude that

$$\mathbf{p}(d|D) = 0 \qquad when \qquad d < D \qquad (29)$$

since a perfect sensor would detect an intervening object

$$\mathbf{p}(d|D) = 1 \qquad when \qquad d = D \qquad (30)$$

since the object is at D

$$p(d|D) = p(o) \quad when \quad d > D \tag{31}$$

since the edge blocks the view behind it, the measurement gives us no special information beyond d=D.

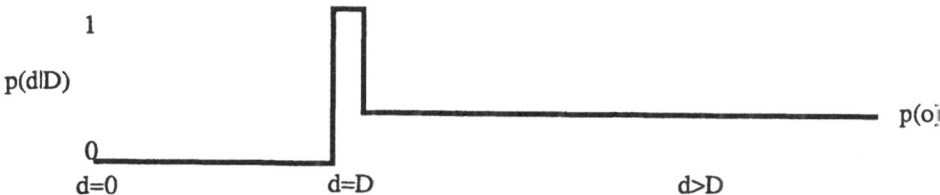

Graphically:

Relationsip (16) holds only if the edge is exactly at distance D. Measurement m tells us only that there is a probability $P(D|m)$ that this is the case. To get the overall probability, we must sum the $p(d|D)$s over all possible D's, each weighted by its probability of actually being the case. Thus

$$p(d|m) = \sum_D (p(d|D) \times p(D|m)) \tag{32}$$

This will generally have the form

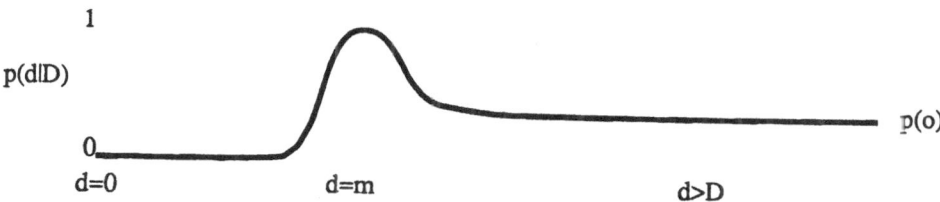

Though if the spread of $p(D|m)$ is very large the hump at $d = m$ will be attenuated, and the curve will have the approximate shape:

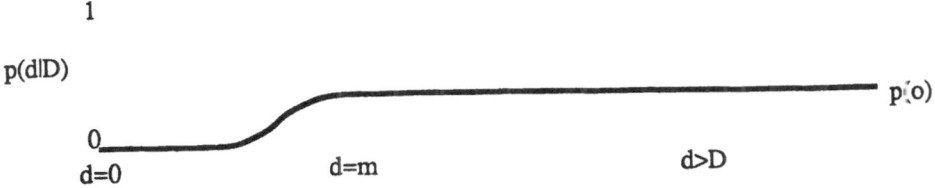

The values of $p(d|m)$ can be used directly to update maps by use of the combining formula 15.

13 Acknowledgement

This work has been supported since 1981 by the Office of Naval Research under contract N00014-81-K-503. I wish to thank Peter Cheeseman for setting my nose firmly in the Bayesian groove.

14 References

1. Stewart, W. K., **A Model-Based Approach to 3-D Imaging and Mapping Underwater**, to be presented at the 7th International Conference and Exhibit on Offshore Mechanics and Arctic Engineering (OMAE), Houston, Texas, February 7-12, 1988.

2. Matthies, L. H. and A. E. Elfes, **Sensor Integration for Robot Navigation: Combining Sonar and Stereo Range Data in a Grid-Based Representation**, to be presented at the IEEE Decision and Control Conference, Los Angeles, CA, December 1987.

3. Kweon, I. S., personal communication, Robotics Institute, Carnegie-Mellon University, Pittsburgh, PA, 15213, October 1987.

4. Stewart, W. K., **A Non-Deterministic Approach to 3-D Modeling Underwater**, 5th Symposium on Unmanned Untethered Submersible Technology, University of New Hampshire, June, 1987.

5. Serey, B. and L. H. Matthies, **Obstacle Avoidance using 1-D Stereo Vision**, CMU Robotics Institute Report, November, 1986.

6. Elfes, A. E, **A Sonar-Based Mapping and Navigation System**, Workshop on Robotics, Oak Ridge National Lab, Oak Ridge, TN, August, 1985 (invited presentation), in the proceedings of the 1986 IEEE International Conference on Robotics and Automation, San Francisco, April 7-10 1986 also to appear as an invited paper in IEEE Transactions on Robotics and Automation.

7. Kadonoff, M., F. Benayad-Cherif, A. Franklin, J. Maddox, L. Muller, B. Sert and H. Moravec, **Arbitration of Multiple Control Strategies for Mobile Robots**, SPIE conference on Advances in Intelligent Robotics Systems, Cambridge, Massachusetts, October 26-31, 1986. In SPIE Proceedings Vol 727, paper 727-10.

8. Moravec, H. P. and A. E. Elfes, **High Resolution Maps from Wide Angle Sonar**, proceeding of the 1985 IEEE International Conference on Robotics and Automation, St. Louis, March, 1985, pp 116-121, and proceedings of the 1985 ASME conference on Computers in Engineering, Boston, August, 1985.

9. Berger, J. O., **Statistical Decision Theory and Bayesian Analysis**, 1985.

10. Thorpe, C. E., **Path Relaxation: Path Planning for a Mobile Robot**, CMU-RI-TR-84-5, Robotics Institute, Carnegie-Mellon University, April, 1984, also in proceedings of IEEE Oceans 84, Washington, D.C., August, 1984 and Proceedings of AAAI-84, Austin, Texas, August 6-10, 1984, pp. 318-321.

11. Moravec, H. P., **Robot Rover Visual Navigation**, UMI Research Press, Ann Arbor, Michigan, 1981. also available as **Obstacle Avoidance and Navigation in the Real World by a Seeing Robot Rover**, Stanford AIM-340, CS-80-813 and CMU-RI-TR-3.

A SIX DEGREES OF FREEDOM POSITIONAL DEVIATION SENSOR FOR THE TEACHING OF ROBOTS

F. Dessen, J.G. Balchen
Norwegian Institute of Technology
Division of Engineering Cybernetics
7034 Trondheim   (Norway)

Abstract: A device for physically guiding a robot manipulator through its task is described.   It consists of inductive, contact-free positional deviation sensors. The sensor will be used in high performance sensory control systems.   The paper describes problems concerning multi-dimensional, non-linear measurement functions and the design of the servo control system.

1. INTRODUCTION

Over the years, several principles have been introduced for programming industrial robots. The methods most commonly used, involve showing the robot a sequence of positions by physically leading the robot through its task. This has been done either by remote control through a teach pendant or by moving the arm manually. The advantage of these methods compared to off-line programming is the continuous feedback given to the programmer when watching the manipulator moving through space.   This is most clear in the case of manual programming where the operator not only receives visual feedback close up, but also has direct control of the manipulator through his hands. The method is very popular in applications such as paint spraying, where it allows trajectory recording in real time. However, the success of this approach relies on careful design of the manipulator. The arm should be lightweight, preferably gravitional effects should be compensated, and joint friction during the teach process must be low. These considerations may increase the price of the robot in question.

NATO ASI Series, Vol. F52
Sensor Devices and Systems for Robotics
Edited by A. Casals
© Springer-Verlag Berlin Heidelberg 1989

Lately, several approaches have been made to include the
possibility of manual control in arbitrary industrial robot
systems. This is usually done by mounting a force sensing
handle on the manipulator end effector. The force applied is
then used to indicate the desired motion of the manipulator.
Unfortunately, this approach leads to a stability problem due
to interaction between the operator and the manipulator
(Hirzinger, 1982). The bandwidth of such a servo assisted
manual control system may often be low, and it is still
necessary to apply considerable force in order to make the
manipulator move.

The ideal system for manual lead-through programming would be a
system where the operator can hold the tool freely in his own
hands and complete the task by himself. By this method, no
special care needs to be taken when designing the manipulator.
The programmer will not be hampered at all by the manipulator,
only by the weight and mass of the tool. This approach relies
on the existence of a special positional measurement system.
One possibility is using a lightweight, low friction dummy
arm with servo displacement sensors. A second possibility is
using remote sensing devices based on optical, magnetic,
sound- or RF-wave measurements.

This paper presents a new method for contact-free positional
sensing of a tool or teach-handle moved freely by the operator
(Balchen, 1986). The system consists of a small-range
positional sensor mounted at the tip of the manipulator. The
approach is outlined in the next section of this paper. Later
sections describe the positional sensor and the tracking system
in more detail. These sections also describe some of the
problems that were met during the design of the system.

2. OUTLINE OF THE SYSTEM

Consider a robot manipulator with a sensor mounted at its tip.
Let the sensor be able to measure the position of a tool, which

may be held by an operator. The situation is illustrated in Fig. 2.1. Here the position of the manipulator wrist relative to its base is represented by the homogeneous transformation matrix T. In the same way the tool position relative to the wrist and the base are denoted by E and C respectively. The situation is described by the transformation equation

$$C = T E \qquad (2.1)$$

In Fig. 2.2, matrix E is split into two fixed transformations, M and F, and one of three alternative displacement matrices, $^T D$, $^M D$ or $^E D$. Here, M defines the sensor frame, F is a tool description matrix, and $^M D$ expresses tool displacement as seen from the sensor frame. $^T D$ and $^E D$ are related to $^M D$ by

$$^T D = M \ ^M D \ M^{-1} \qquad (2.2)$$

$$^E D = F^{-1} \ ^M D \ F \qquad (2.3)$$

Using the displacement matrix $^M D$, (2.1) can be written

$$C = T M \ ^M D \ F \qquad (2.4)$$

Since T may be computed from the manipulator joint displacements, and $^M D$ from the sensor output, tool position relative to base may at any time be found using (2.4). The computed tool motion may be recorded in order to create a robot program.

Imagine that the positional sensor has a limited range, or that the quality of the measurements is best when $^M D \approx I$. In this case we want the manipulator to follow the tool as closely as possible. This tool tracking is done by a positional control system.

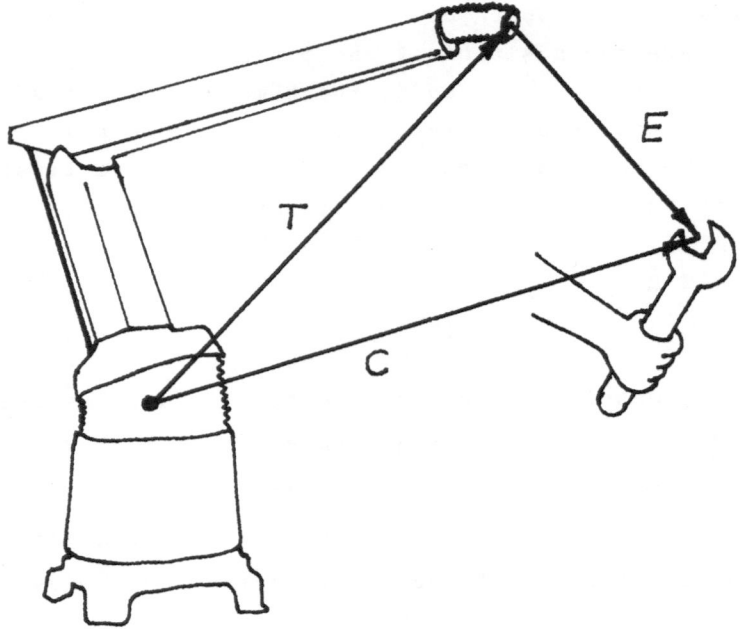

Figure 2.1.

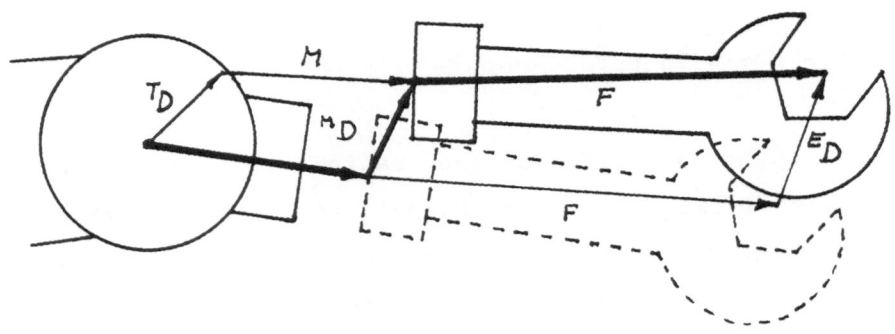

Figure 2.2.

Motion recording

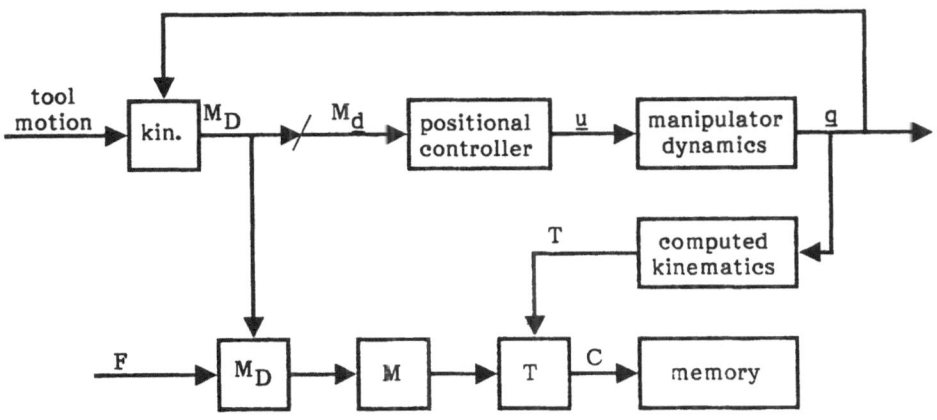

Motion playback

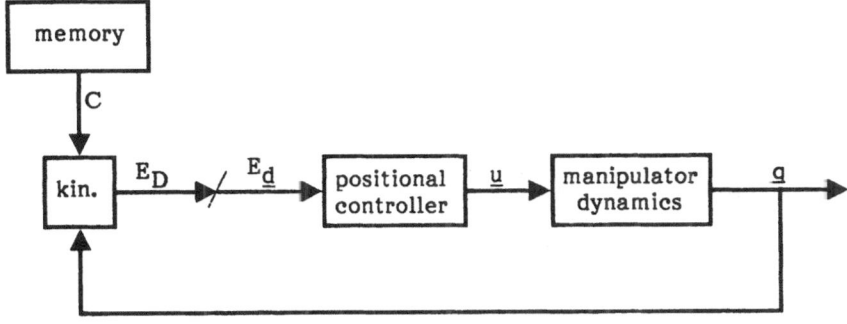

Figure 2.3.

A complete system for robot trajectory recording is shown in Fig. 2.3. The input to the positional controller is represented by the vector

$$^M\underline{d} = [d_1, d_2, d_3, \delta_1, \delta_2, \delta_3]^T \qquad (2.5)$$

which may be extracted from the tool displacement matrix, $^M D \approx I$ (Paul, 1981). The generalized coordinates, $\underline{q}$, are used to compute the wrist position matrix, T.

3. DESCRIPTION OF THE SENSOR

So far, very little has been said about the displacement sensor itself; only that it is mounted at the tip of the manipulator, and that it possibly has a limited range.

The device being described in this section is shown in Fig. 3.1. It consists of two cylindrical parts on which magnetic sensors are mounted. The larger part, called the sensor housing, is mounted on the manipulator. A magnetically conductive pin, which moves inside the housing, is mounted on the tool.

The way sensors are mounted, plane motion in two cross sections of the housing may be monitored. This is illustrated in Figure 3.2. Additionally, translation along- and rotation about the z-axis of the pin is monitored. From this it is possible to determine pin motion relative to the housing. As in Sec. 2, this motion may be represented by the transformation matrix $^M D$. Expressing $^M D$ in terms of the 5 translational displacement

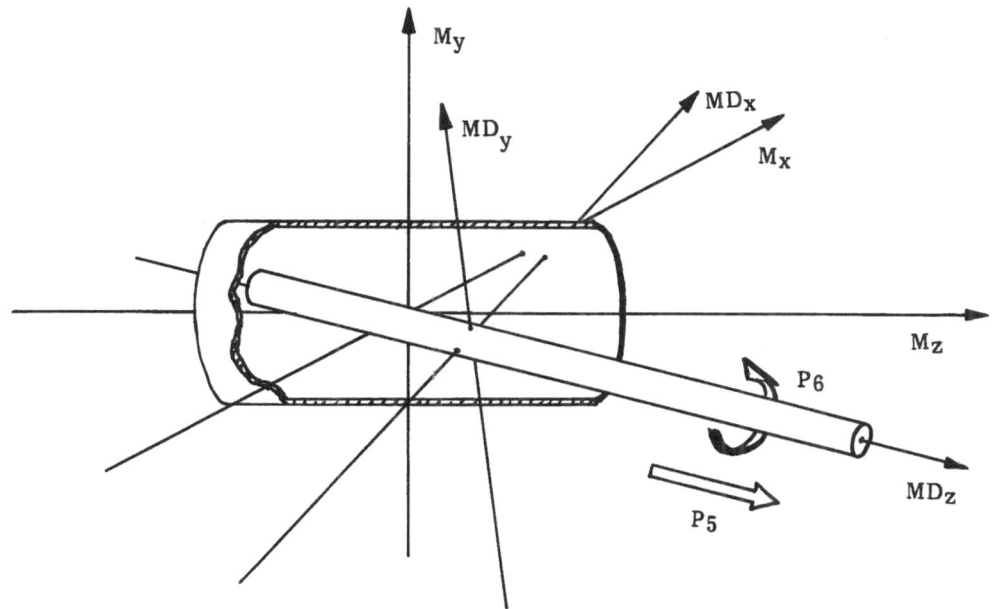

Figure 3.1.

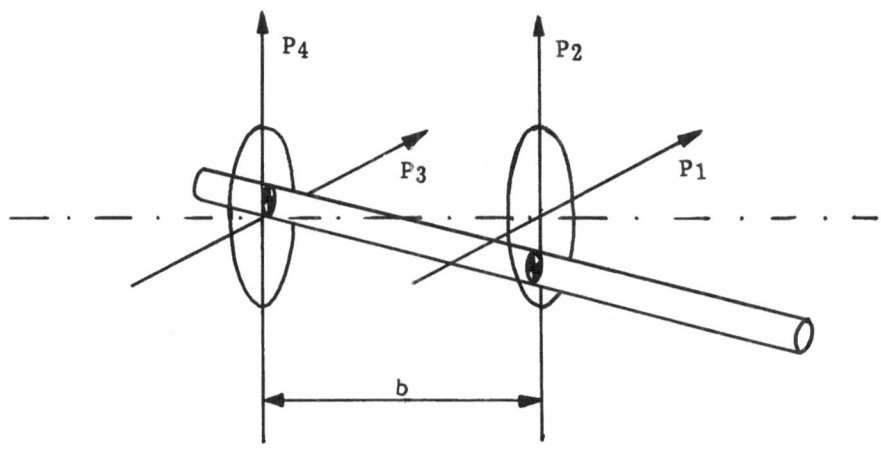

Figure 3.2.

quantities, $p_1 \dots p_5$, and the rotational displacement, p_6, is straight forward. However, the complete expression is quite bulky, and will not be quoted here.

For tracking purposes, i.e. to make the sensor housing follow the pin, only an approximation of $^M D$ will be needed. From Figs. 3.1 and 3.2 it is seen that the first order expansion of $^M D$ near $^M D = I$ is

$$^M D \approx I + {}^M \Delta = \begin{bmatrix} 1 & -\delta_3 & \delta_2 & d_1 \\ \delta_3 & 1 & -\delta_1 & d_2 \\ -\delta_2 & \delta_1 & 1 & d_3 \\ 0 & 0 & 0 & 1 \end{bmatrix} \qquad (3.1)$$

where

$$d_1 = \frac{p_1 + p_3}{2} \quad ; \quad d_2 = \frac{p_2 + p_4}{2} \quad ; \quad d_3 = p_5$$

$$\qquad (3.2)$$

$$\delta_1 = \frac{p_2 + p_4}{b} \quad ; \quad \delta_2 = \frac{p_1 - p_3}{b} \quad ; \quad \delta_3 = p_6$$

These matrix elements also correspond to the elements of vector $\underline{d}$, in (2.5).

For recording purposes, the exact value of $^M D$ must be computed. These computations again rely on the exactness of the sensor output model, i.e. how sensor output, $\underline{y}$, is related to the intermediate displacement vector, $\underline{p}$. This relationship is represented by mapping $Q: R^6 \rightarrow R^6$ so that

$$\underline{y} = Q\underline{p} \qquad (3.3)$$

During the design of the displacement sensor, much effort was made to make the mapping as linear as possible, and with little coupling between vector elements. The attempt was fairly successful, but in order to compute $^M D$ exactly, both non-linearity and coupling must be taken into account.

The discussion of sensor equations will be restricted to the first four elements of p. This is possible since, because of the symmetry, variations in p_5 or p_6 has no effect on the first four elements of the measurement vector, y. The subsystem may be represented by a mapping $\tilde{Q} : R^4 \rightarrow R^4$ so that

$$\tilde{y} = \tilde{Q}\ \tilde{p} \tag{3.4}$$

where $\tilde{y} = [y_1,\ y_2,\ y_3,\ y_4]^T$ and $\tilde{p} = [p_1,\ p_2,\ p_3,\ p_4]^T$

Fig. 3.2 defines the elements of $\tilde{p}$. These displacements are sensed by devices resembling differential transformers, with outputs denoted by $\tilde{y}$.

The mapping, $\tilde{Q}$, may in some applications be approximated by a scalar constant:

$$\tilde{y} = c_0\ \tilde{p} \tag{3.5}$$

A more precise diagonal matrix approximation includes elements depending on $\tilde{p}$.

$$y_1 = c_0[1+(c_1+c_2\sin2\psi_{12})(p_1^2+p_2^2)][1+c_3(p_1-p_3)^2]p_1 \tag{3.6a}$$

$$y_2 = c_0[1+(c_1+c_2\sin2\psi_{12})(p_1^2+p_2^2)][1+c_3(p_2-p_4)^2]p_2$$

$$y_3 = c_0[1+(c_1+c_2\sin2\psi_{34})(p_3^2+p_4^2)][1+c_3(p_1-p_3)^2]p_3$$

$$y_4 = c_0[1+(c_1+c_2\sin2\psi_{34})(p_3^2+p_4^2)][1+c_3(p_2-p_4)^2]p_4$$

(3.6b)

Here $\psi_{12} = $ atan2(p_1, p_2) and $\psi_{23} = $ atan2(p_3, p_4).

In (3.6), the first two factors in each equation represent amplification as a function of translational deviation, and are computed for each of the two cross-sections in Fig. 3.2 separately. The third factor represents the influence from rotational deviation.

Sensor calibrations show that the first two factors are the most significant ones, so that $\tilde{Q}$ may be split into two loosely coupled subsystems. This fact may be used to simplify the inversion of (3.4). Two possible approaches are now outlined:

1. Iterative solution: (3.6) may be inverted using a Newton-Raphson scheme. In this case, the inverse Jacobian matrix involved may be approximated by neglecting the last factor of (3.6).

2. Approximate, symbolic solution: The two subsystems in (3.6) may be inverted separately. Afterwards, new, approximate coupling terms may be inserted and adjusted using an off-line curve fitting scheme.

The final shape of (3.6), or its inverse, will depend on the need for accuracy, and the kinematical structure, in each application. For instance, in the tracking system described in sections 2 and 4, only (3.5) is used. On the other hand, situations may exist where even more accurate expressions than (3.6) are necessary.

4.   CONTROL SYSTEM PROTOTYPE

The sensor described in Sec. 3 allows a radial pin displacement of ± 1 cm, and an axial displacement of ± 2 cm approx.   The success of the complete system relies on the presence of a control system which makes the manipulator follow the tool closely at all times. This section deals with the design of the control system.

Before going into details, it is of interest to have a rough idea of the performance that may be expected. For this purpose we look at a simple 1 d.o.f. positional control system, as shown in Fig. 4.1. The controller may be of the proportional (P) or proportional + integral (PI) type. Asymptotical plots of the transfer function from positional reference to control error, $N(s) = e(s)/x_0(s)$, are shown in Fig. 4.2. It is assumed that the bandwidth ω_0 is feasible.   By studying the asymptotical behaviour of $N(s)$ for $\omega < \omega_0$, it is seen that, with a proportional controller,

$$\lim_{s \to 0} \frac{1}{s} N(s) = \frac{1}{\omega_0} \qquad (4.1)$$

and with a PI-controller,

$$\lim_{s \to 0} \frac{1}{s^2} N(s) = \frac{1}{\omega_0^2} \qquad (4.2)$$

Since the range of the deviation sensor is limited to ± 1 cm approx., it is found that maximum velocity obtained when using a P-controller will be $v_p = 1$ cm ω_0.   Under   PI-control, a maximum acceleration of $a_{p1} = 1$ cm ω_0^2 is obtained. With a band-

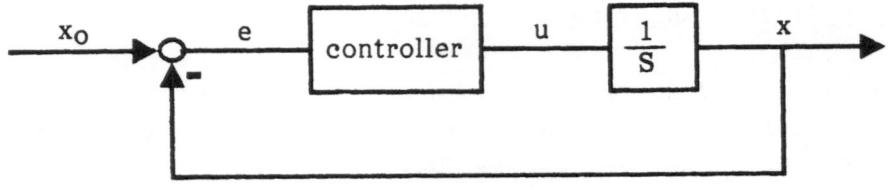

Figure 4.1.

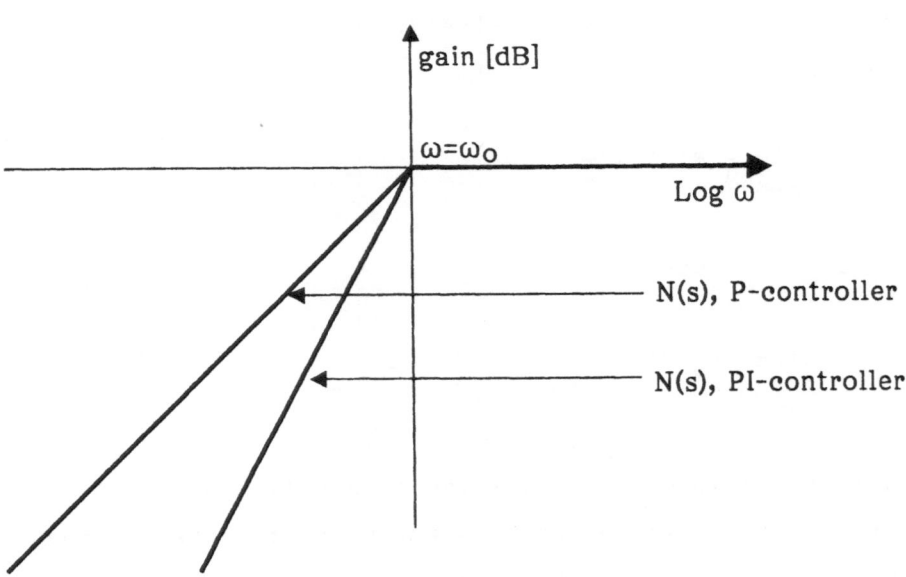

Figure 4.2.

width of ω_0 = 10 rad/s, which is typical for a hydraulically driven paint spraying robot, we obtain the values

$$\hat{v}_p = 10 \text{ cm/s} \qquad \hat{a}_{p1} = 100 \text{ cm/s}^2 \qquad (4.3)$$

It is seen that the mere use of proportional control gives low performance. It is also worth noting that maximum acceleration under PI-control is proportional to ω_0^2, which makes it worthwhile to increase ω_0 as much as possible. This will be considered in the next section.

The tool tracking system has so far been implemented on a Trallfa TR-400S paint spraying robot. The first version of the control scheme is designed to resemble the original, joint coordinate, positional control scheme as much as possible. This implies that the positional deviation measurements are converted into an equivalent vector, $\underline{e}$, of servo control errors. This vector can be used by the original controllers.

The conversion is done by using the transformations listed below:

$$\underline{p} \leftarrow \underline{y} \qquad (3.5)\text{extended}$$

$$^M\Delta \leftarrow \underline{p} \qquad (3.1)(3.2)$$

$$^T\Delta \leftarrow {}^M\Delta \qquad (2.2)(3.1)$$

$$\Delta\underline{q} \leftarrow {}^T\Delta \qquad \text{manipulator Jacobian}$$

$$\underline{e} \leftarrow \Delta\underline{q} \qquad \text{servo to joint transformation}$$

The complete scheme has been implemented in integer assembly code on a TMS 99105 microprocessor. A sampling rate of 100 Hz is used, and the program is finished in less than 1 ms. Site tests indicate that the control scheme works, and that the reasoning which leads to (4.3) is valid.

This first implementation also shows that it may not be wise to include the original joint controllers in the tool tracking scheme. This is due to the following:

1: Design criteria for the tool tracking system and the original play-back system differ. The original controllers are designed to follow a high speed reference trajectory without overshooting. This implies that the control error integrator is restricted, which in turn reduces the steady state acceleration (4.3). In the case of tool tracking, it is not so important whether the manipulator overshoots or not.

2: Optimal controllers would, in the two cases, differ because of differences in the underlying control performance indices. An optimal playback controller would be designed considering the characteristics of the tool used and of the task itself, whereas the tool tracking system must conform to the limitations of the displacement sensor used.

3: During playback, joint displacement is compared to a reference in order to compute the servo control error. During tool tracking, error signals are produced at the tip of the manipulator. Whenever manipulator link elasticities are significant, this may cause an additional stability problem (Cannon, 1984).

The first point is easily solved by creating separate tool tracking controllers with higher error integration rates. Points 2 and 3 may be solved by considering more complex system models. This will in turn lead to more complex control schemes.

5. CONTROL SYSTEM IMPROVEMENTS

Thus far the complete system for robot programming has been outlined, and the displacement sensor has been described in some detail. In Sec. 4 the tool tracking system was described, and some notes were made on tracking performance. To conclude this paper, some suggestions will be made on how to improve the performance.

From the discussion leading to (4.3) it is evident that the steady state tracking speed or acceleration depends on sensor range and servo bandwidth. Hence control system improvements may be divided into two classes:

1. Optimizing the use of the sensor's "workspace"
2. Maximizing the control system bandwidth

Besides actually increasing the sensor range, point 1 leads to the problem of designing a controller which minimizes some performance index

$$B = f(^{M}\underline{d}) + g(\underline{x}, \underline{u}) \qquad\qquad (5.1)$$

Here the functional f expresses the sensor's work space in some way, and g the effort made by the robot manipulator. It may be difficult to obtain an exact solution to the problem, however, sub-optimal controllers may be designed by modifying well known control schemes, such as the motion rate (Whitney, 1969) or the acceleration control (Luh, Walker and Paul, 1980a) schemes.

Increasing system bandwidth, as in point 2, involves consider-ing more complex dynamic models for the system. The model may be expanded in three directions:

a) Servo dynamics
b) Manipulator kinetics

c) Link and joint elasticities

Point b) involves designing variable controllers. This may be done either by finding a means of making the existing controllers vary according to the manipulator configuration, or by applying complete schemes such as the "computed torque method" (Luh, Walker and Paul, 1980 b). According to (Cannon, 1984), positional sensing at the manipulator wrist may lead to a special stability problem due to link elasticities. Calculations show that this reduces the bandwidth of the Trallfa TR400 manipulator by a factor of 2 approx. The bandwidth may be increased by taking the effect of elasticity into account.

It is seen from the previous discussion that several approaches can be made in order to increase the performance of the robot programming system described. It is the authors' opinion that the system may soon be used for the real time teaching of high speed paint spraying robots. In this case, some, or all of the points made in this concluding section must be considered.

REFERENCES

Balchen, J.G. (1986). Norwegian Patent Application 870419.
Cannon, R.H., Schmitz, E. (1984). Precise Control of Flexible Manipulators. Robotics Research: The First International Symposium, New Hampshire, August, pp 841-862.
Hirzinger, G. (1982). Robot-teaching via Force-torque-sensors. 6th Eur. Meeting on Cybernetics and Systems Research EMCSR 82, Vienna, April 13-16.
Luh, J.Y.S., Walker, M.W., Paul, R.P.C. (1980 a). Resolved Acceleration Control of Mechanical Manipulators. IEEE Trans. Autom. Control, 25, pp 468-474.
Luh, J.Y.S., Walker, M.W., Paul, R.P.C. (1980 b). On-line Computational Scheme for Mechanical Manipulators. J. Dynamic Systems, Measurement and Control, 102, pp 69-76.
Paul, R.P.C. (1981). Robot Manipulators. MIT-press.
Whitney, D.E. (1969). Resolved Motion Rate Control of Manipulators and Human Prostheses. IEEE Trans. Man-Machine Syst., 10, pp 47-53.

VI. APPLICATIONS

FORCE FEEDBACK STRATEGIES AND THEIR APPLICATION TO ASSEMBLY

H. Van Brussel
Katholieke Universiteit Leuven
Department of Mechanical Engineering
Clestijnenlaan 300B
B-3030 Leuven   (Belgium)

KEYWORDS/ABSTRACT : force feedback / robotic assembly / accommodation / compliance / force and tactile sensors / compliant motion.

Force sensors have great potential for improving the autonomy and flexibility of robots in assembly systems. Although force sensors are sufficiently developed in order to be reliably used in industrial environments, the control aspects involving force feedback are not yet fully understood. The aim of the paper is to assess and to compare some potentially useful schemes involving force control.

INTRODUCTION

Vision systems are generally recognized as being potentially helpful in robotic applications by providing state feedback from a partly unstructured or time varying robot environment. Of equal importance for increasing robot flexibility is force and tactile feedback. Force controlled robot systems can be based upon either passive or active accommodation. Accommodation is defined as the process in which the contact forces between the parts held by the manipulator and the environment modify their relative position or motion. In passive accommodation systems the position correction is generated by the contact forces themselves, whereas in active accommodation the forces are a source of information from which the position corrections are calculated.

Passive systems are simple, easy to use and fast in operation. However they have limited flexibility. Active force accommodation systems are much more flexible since

NATO ASI Series, Vol. F52
Sensor Devices and Systems for Robotics
Edited by A. Casals
© Springer-Verlag Berlin Heidelberg 1989

their behaviour is determined by the way the force sensing loop is closed in the controller.

In this paper, several workable force control concepts will be discussed and illustrated with practically proven case studies.

FORCE CONTROL CONCEPTS

Force feedback applications can be divided into two types.

Type I : applications whereby the contact forces are not essential to the process, but can be used as a source of information on the actual position of the robot endpoint (gripper) relative to the environment. Most assembly operations belong to this category.

Type II : applications whereby the contact force is an essential part of the process itself, like in grinding, polishing, deburring, contour tracking, etc. In these cases, both the position of the end effector and the force exerted by the end effector on the environment have to be controlled simultaneously.

In both types of applications, the robot is in physical contact with the environment, in contrast to many traditional handling operations where the robot is moving in free space. It can be readily understood that in both cases the robot will behave differently and that different control approaches are required. The tasks where the robot is in contact with the environment are called compliant motion tasks, compared to free motion tasks where there is no contact.

Type I - force control

The dexterity of a human assembly worker is not only due to his tactile perception capabilities but to an equally large extent due to the inherently present compliance in the arm and in the workpiece/hand-interface. Also in robots this compliance is required. An infinitely stiff robot would, due to its finite position resolution, never be able to perform close-tolerance assembly operations.

For applications when the force is not a goal to realise but an aid in positioning parts, the most successful technique is therefore position control with controlled end-point impedance. End-point impedance can be defined as the ratio of the contact force variation to the programmed position variation of the end point. The total impedance is a cascade of three types of impedance (Fig.1):

* Impedance of the mechanical structure.
  $$F = Z_R (X_R - X_C)$$
  where F is the force exerted at the end point, X_R is the real position of the end point and X_C is the position of the end point calculated from the measured joint encoders.

* Impedance of the servo controllers.
  $$F = Z_S (X_C - X_S)$$
  where X_S is the input to the servo controller.

* Impedance as a result of active force feedback.
  $$F = Z_A (X_S - X_P)$$
  where X_P is the programmed end-point position.

The total impedance Z_E at the end point is given by:
$$Z_E^{-1} = Z_A^{-1} + Z_S^{-1} + Z_R^{-1}$$
An impedance at the end position with known characteristics facilitates the prediction of the contact forces and, if the impedance is optimised for the problem at hand, the contact forces themselves will provoke a desired position correction.

This technique, whereby the contact forces between the parts held by the manipulator and the environment modify their relative position, is called accommodation. Depending on the type of impedance that is relied upon, three types of accommodation can thus be distinguished:

* Purely <u>passive accommodation</u> (based on Z_R).
* <u>Adaptive</u> or programmable <u>passive accommodation</u> (based on Z_S).
* <u>Active accommodation</u> (based on Z_A).

If restricted to real impedances (quasi-static behaviour), the relation between the real position X_R of the end point and the programmed position X_P is given by:

$$X_R = X_P + C\ F$$

where C is the flexibility matrix of the end point of the robot. An accommodation system has to be designed such that for the application in mind the term C F adds the proper correction to the programmed path X_P.

Purely passive accommodation

Passive accommodation is a frequently used and often unconsiously applied technique for tackling the problems caused by the uncertainties of the manipulator in relation to its working environment. In nearly every robot application the proper functioning of the system depends to a more or less high degree on a compliance in the setup. Two kinds of passive accommodation can be distinguished:

* Passive accommodation due to inherently present compliance. This can only be applied under restrictive conditions. The main objection to this type of accommodation is that the end-point impedance is not exactly known. A second drawback is that the allowable error window is very small. Errors larger than a few tenths of a millimetre will normally result in unacceptably high contact forces.

* Specially designed passive compliant structures. A lot of ingenuity has been put into the design of fixtures, grippers and tools with a compliance matrix optimised for one particular problem. One of the most interesting examples in this class of special purpose tools is the Remote Center Compliance (RCC) developed by the Charles Stark Draper Laboratories [1]. This gripper has a built-in compliance with a flexibility matrix optimised for the insertion of cylindrical beveled shafts into chamfered holes. The most salient features of the RCC or comparable structures are their ease of application and speed of operation. On the other hand they remain special purpose tools with inherent lack of adaptability.

Active accommodation

Active accommodation systems have a high degree of flexibility since their behaviour can be influenced by choosing the way the force sensing loop is closed in the controller. A second benefit of active systems is that the allowed errors or the regions of uncertainty can be much larger since the original path of the manipulator can be changed on-line. These two elements, an easy changeover to another task and a higher adaptability, have to be weighed against the increase in complexity of a robot with active force feedback.
Some feasible implementation schemes of active accommodation will be discussed hereafter under Type II - force control.

Adaptable passive accommodation

This method, which is based on a programmable control of the dc-gain of the servo controllers, tries to combine the universality of active accommodation and the bandwidth of purely passive accommodation [2]. The basic principle is illustrated in fig.2(a) for a controller operating in cartesian space; the normal position controller (RP) can be switched off and replaced by a linear matrix relation between the position errors and the cartesian force command to the robot.

Here, KE represents the desired end-point stiffness. FZ is a feedforward term compensating the gravity forces working on the robot. Fig.2(b) gives the same scheme for the more practical configuration of an open-loop cartesian controller and a closed-loop joint controller. The stiffness of the joint controller KJ can be calculated from the desired cartesian stiffness KE:

$$KJ = J^T KE J$$

KJ depends on the instantaneous position of the robot and is not diagonal.

There is a basic difference in the causality of the spring-like behaviour of active accommodation versus adaptable passive accommodation. In the active controller, the offset in position of the end point is caused by an additional input to the controller based on the measured contact force. In the adaptive passive controller shown in fig.2, the contact forces themselves directly create the offset. In principle there is no force measurement, hence the name passive accommodation. Due to this change in causality the adaptive accommodation controller is not subjected to the stability problems of the active controller in making contact with the environment and to the resolution problem of the position encoders (see further). There is no need for an extra hardware compliance while the end-point impedance remains universally programmable.

The scheme shown in fig.2 can only be implemented if the mechanical structure conforms with the principles of force/position causality, as stated above. This means that the mechanical transmissions have to be reversible and a force exerted at the end point of the robot has to result in a detectable displacement of the robot. The friction forces, reduced to the end point have to be an order of magnitude smaller than the desired range of contact forces.

For the existing commercially available robots this is not the case. The principle of adaptive passive accommodation is perfectly suitable for wrists, multi-finger hands and directly driven robots. If a direct drive system is not feasible due to torque limitations, a quasi-reversibility can be obtained by means of an additional feedback in every axis controller from the torque measured beyond the transmission.

Application - Precise assembly with adaptive passive accommodation.

How the principle of end-point impedance can facilitate a task formulation is now illustrated [3]. Fig.3 gives the flowchart of the insert operation of a cylindrical peg into a hole with very close tolerances. This experiment has been carried out with a five degrees of freedom cartesian robot with reversible drives and programmable dc gains. The approach phase is rather simple in the case of beveled edges. Care has to be taken not to convert the lateral positioning error into an angular error. This can be done by programming the stiffness as indicated in the flowchart. The same sliding behaviour can be obtained by means of a simple negative feedback matrix (active accommodation). Contact forces can be kept much lower in this case. For non-beveled parts only active approach algorithms are possible [3].

The actual insertion is a much more severe problem especially in the case of jamming, caused, for example, by surface roughness. In the second part of the flowchart a directional search algorithm has been used to find the angle of misalignment between the peg and hole. If the disturbing influence of the torques in the rotational degrees of freedom is eliminated by programming the stiffnesses as indicated in fig.3, the ratio of the Y to X force component gives an indication of the angle of misalignment. The higher the level of the lateral force, the more reliable this indication will be. On the other hand, it is not necessary to correct exactly in the direction of misalignment in order to realign the peg.

For these reasons the force space has been divided into three zones: the zero force zone, the inner zone or the zone of uncertainty where there is an ambiguous relation between the lateral force vector and the angle of misalignment, and the outer zone. Fig.4 gives the evolution of the force during a typical insertion both in the time domain and in the force space.

Type II - force control

The basic control law for applications of Type II has been formulated by Mason [4]. In a task-related cartesian frame two orthogonal sets of directions have to be defined; a set of directions for which a desired force can be specified and a set of directions for which a desired position can be specified. Both sets of directions have to be controlled with different criteria.

This task specification has been extended and adapted to non-ideal situations in [5], in an attempt to rigorously separate the task specification level from the control level.

Three basic control schemes can be distinguished:
* Position and force as parallel loops.
* Position inner loop, force outer loop.
* Force inner loop, position outer loop.

Position and force as parallel loops

There are very few examples of control schemes that are based explicitly on the division of the cartesian space in two orthogonal subsets. The basic problem is the transformation of the task space to the joint space: the subsets after transformation are no longer orthogonal, such that every axis of the robot influences the force vector and the position of the endpoint simultaneously. The free-joint method [6] ignores this non-orthogonality at first and during every sampling instant divides the robot axes into axes controlled in position mode and axes controlled in force mode. The errors in force and position resulting from this approximation are taken into account at the next sampling instant. In the hybrid force/position controller [7] every

axis contributes both to the control of the force and the
position. The controller (Fig.5) has a completely parallel
structure with a force control law (RF) for the force
directions and a position control law (RP) for the position
directions.
The selection matrix, defined in the cartesian task space,
defines the position controlled and the force controlled
directions. The exact formulation of the problem in terms of
the selection matrix S is of tantamount importance for the
proper functioning of the scheme shown in fig.5. For real
environments with friction in the position directions and
surfaces with finite stiffness in the force directions, the
parallel controller has to be completed with feedforward
terms based on a priori knowledge of the environment (Z_C),
(Fig.6).

Position inner loop, force outer loop [8]
By choosing this configuration the complex control problem is
split up into two much more manageable subproblems. The
internal position control system takes care of the
manipulator dynamics, whereas the external loops control the
interaction with the environment. Figure 7 shows the block
diagram for a one-dimensional manipulator in contact with its
environment. The contact force, F, is measured and is fed
back in an external loop closed around the position control
system.

k_0 represents the contact stiffness, i.e. the total stiffness
of robot and environment measured at the contact point. In
figure 7 it is assumed that: 1) the contact force disturbance
which acts on the position control loops is negligible, or is
compensated for by direct feedback of the measured force to
the actuator torque; 2) the sensor dynamics can be neglected.

Since in most applications the position of the environment,
x_0, is not exactly known, the force control law has to
contain integration in order to avoid large force errors.
Several control laws can be applied (e.g. PI-control, PI-

state feedback, etc.). Their design can be optimized using conventional control techniques provided the position loops dynamics, $h(s)$, are properly identified. Theoretically, the force loop can be made arbitrarily fast using state feedback. However, according to our practical experience, it is very difficult if not impossible to obtain a force loop which is much faster than the inner position loop.

The control strategy used offers a very interesting solution for the approach phase, i.e. the transition phase between motion in free space and motion in contact with the environment. This approach phase is executed under force control. A desired force is applied to the force controller. In the no contact situation this results in a constant approach velocity, due to the integration in the force control law. After collision with the environment a force response with acceptable overshoot results.

Practical systems have limited position measurement accuracy, x. This causes force limit cycles in a static situation (constant desired contact force, no motion). An upperbound for the peak-to-peak amplitude of these limit cycles is given by $k_O.x$. The requirement of a passive compliance k_O^{-1} in order to obtain sufficient force accuracy, is generally pointed out as a drawback of the external control method. However, this compliance is also required for other reasons.

It is evident that the force control gain is proportional to k_O^{-1}. Hence :
- the speed of the force loop is proportional to k_O^{-1};
- the attenuation of disturbances (both position disturbances x_O, and force disturbances in the actuation system) is proportional to k_O^{-1};
- the force accuracy in the presence of finite position resolution is proportional to k_O^{-1}.

As a conclusion, high structural flexibility is advantageous in active force control, since it offers a high speed of

operation, a high force accuracy, and a high disturbance
rejection. On the other hand, too much flexibility causes a
reduction of bandwidth due to actuator limitations. In
addition it is clear that high structural flexibility is
disadvantageous with respect to motions in free space.

In order to extend the control strategy to multiple
dimensions, the following assumption has to be made: the
robot positioning system is supposed to be dynamically
decoupled in the task space. This assumption offers an
acceptable working model for many industrial manipulators
provided with decentralized controllers, as long as the
bandwidths and damping ratios of all independent joints
control systems are almost identical. In this case the
positioning system can be modeled as :

$$\underline{x}^t(s) = h(s).I.\underline{x}_{dt}(s).$$

The remaining coupling between the task frame directions then
stems from the contact stiffness matrix, K_{ot}. By introducing
its inverse in the multidimensional force control law,
decoupled and identical systems result in all task frame
directions. These systems can be controlled using a set of
independent one-dimensional force controllers as depicted in
figure 8.

Force inner loop, position outer loop
The problems caused by the limited resolution of the position
encoders can be circumvented by closing the position loops
around the force loops. Position errors are then converted
into force commands. The robot axes are now essentially
force controlled, by which position limit cycles are avoided
and consequently a passive compliance is not required
anymore. Nevertheless, this compliance is still required
upon making contact between robot and environment, due to the
limited bandwidth of the servo loops.

Applications of Type II - force control
As described in [9], successful deburring, contour tracking,
assembly and palletizing applications have been performed

based upon the force-around-position loop approach, showing its general validity even in realistic disturbance-prone environments.

REFERENCES

[1] Drake, S., Watson, P. and Simunovic, S. 1977. High speed robot assembly of precision parts using compliance in lieu of sensory feedback. In, Proc. 7th Int. Symp. on Industrial Robots, pp.87-97. JIRA, Tokyo.

[2] Simons, J. 1980. Force Feedback in Robot Assembly using and Active Wrist with Adaptable Compliance. PhD Thesis, Katholieke Universiteit Leuven, Belgium.

[3] Van Brussel, H. and Simons, J. 1979. The adaptable compliance concept and its use for automatic assembly by active force feedback accommodation. In, Proc. 9th. Int. Symp. on Industrial Robots, pp.167-181. SME, Dearborn, MI, USA.

[4] Mason, M. 1979. Compliance and Force Control for Computer Controlled Manipulators. MIT Artificial Intelligence Lab., Memo 515.

[5] De Schutter, J.and Van Brussel, H. 1986. A Methodology for Specifying and Controlling Compliant Robot Motion. Proc. 25th IEEE Conf. on Decision and Control, Athens, pp.1871-1876.

[6] Paul, R. and Shimano, B. 1976. Compliance and control. In, IEEE Joint Automatic Control Conference, pp.694-699. IEEE, New York.

[7] Raibert, M. and Graig, J. 1982. Hybrid position force control of manipulators. Trans. ASME, Dynamic Systems, Measurement and Control, 102 (June): 126-132.

[8] De Schutter, J. and Van Brussel, H. 1987. Compliant Robot Motion. II. A Control Approach Based on External Control Loops, Int. Journal of Robotics Research.

[9] De Schutter, J. 1986. Compliant robot motion: task formulation and control. PhD Thesis, Katholieke Universiteit Leuven, Belgium.

307

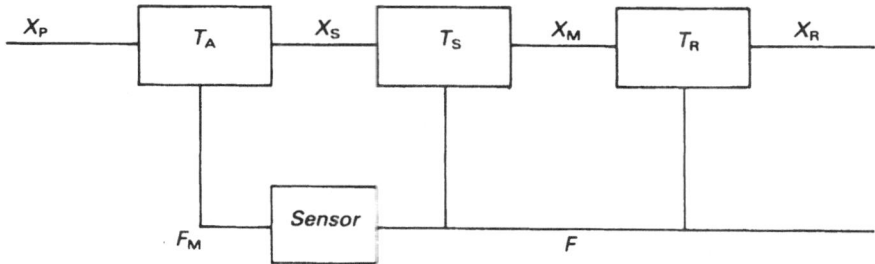

Fig.1. The end-point impedance can be considered as a series connection of three impedances. T_A is the transmittance between the programmed position and the position modified through active force feedback; T_S is the transmittance between the modified position presented to the servo controllers and the position measured by the joint encoders; and T_R is the transmittance between the position determined via the joint encoders and the real end-point position.

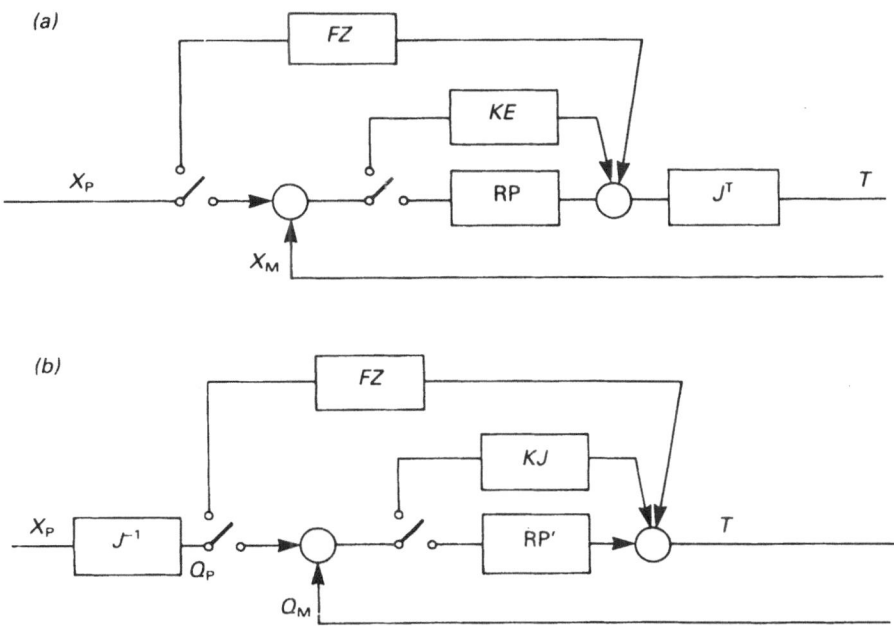

Fig.2. Adaptive passive accommodation by (a) adjustment of the servo controllers represented in the cartesian space, and (b) adjustment of the controllers in the joint space.

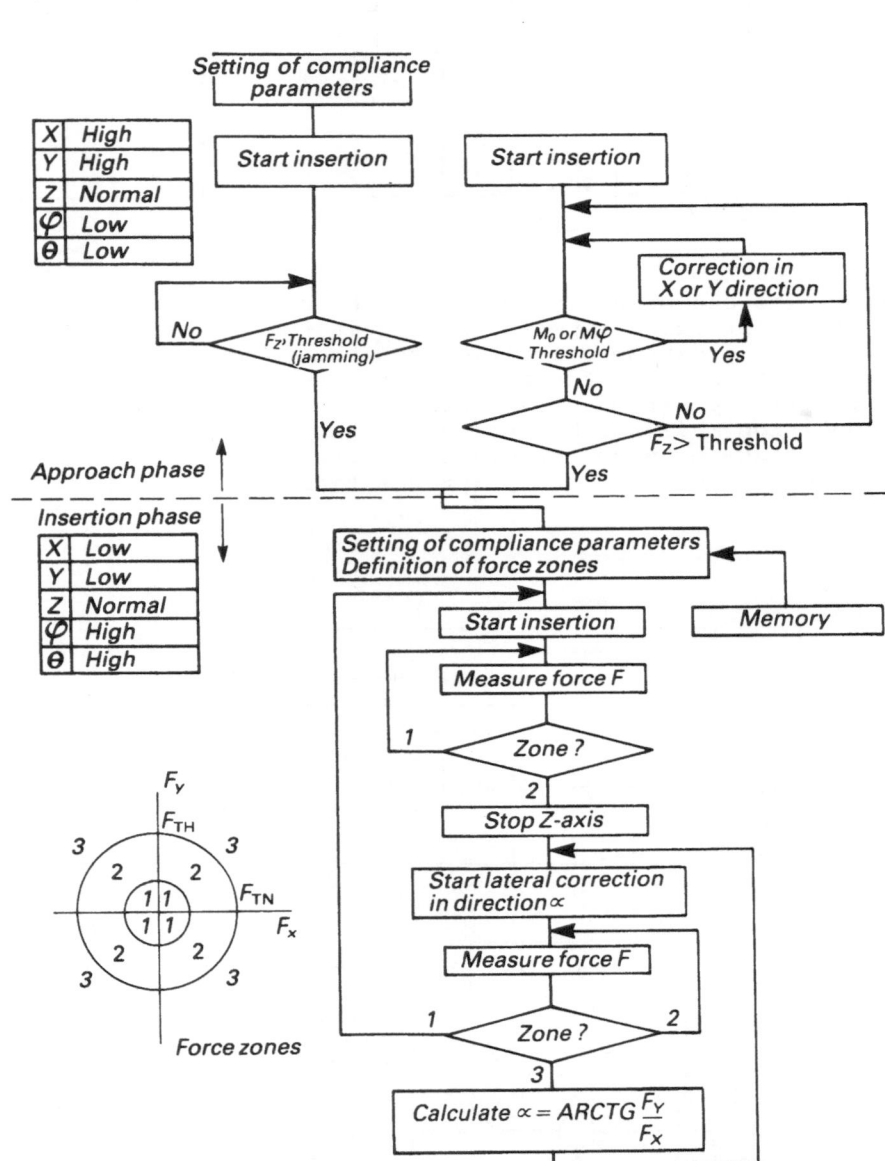

Fig.3. Flow chart of peg/hole assembly strategy (approach and insertion) based on adaptive passive accommodation.

309

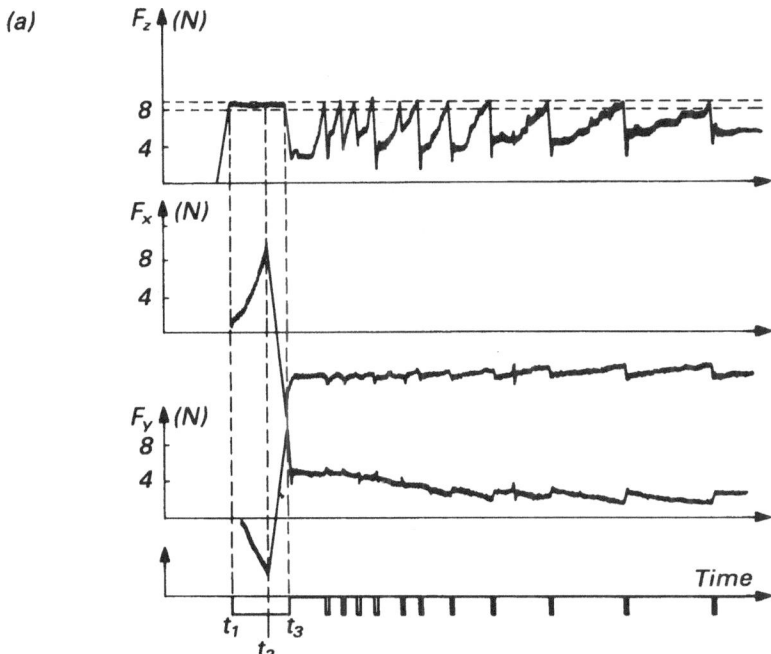

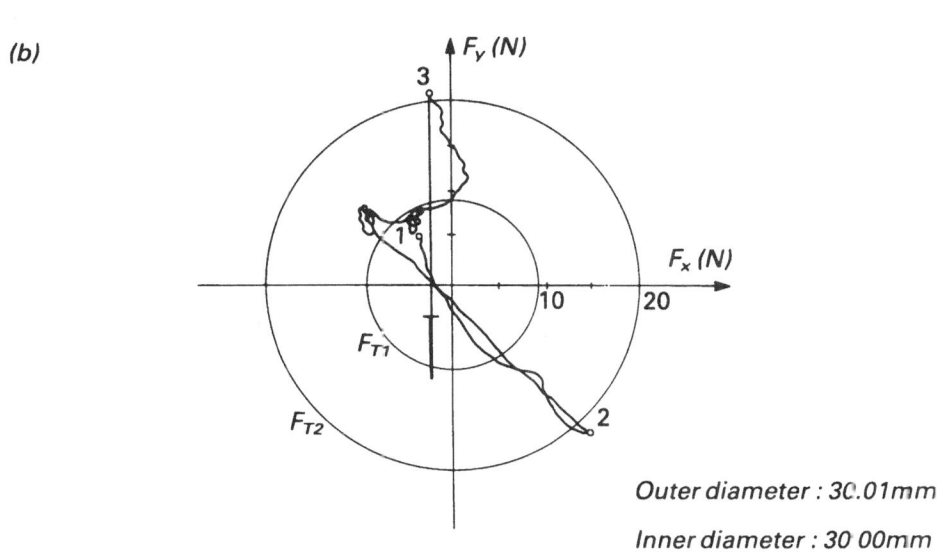

Outer diameter : 30.01mm

Inner diameter : 30 00mm

Fig.4. Evolution of the measured force vector during assembly, in the time domain (a) and in the force space (b).

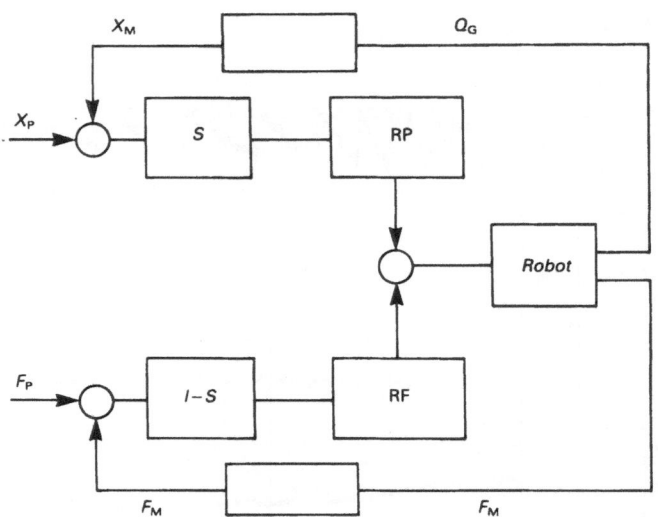

Fig.5. Hybrid force/position controller with two parallel
loops, described by Craig [7].

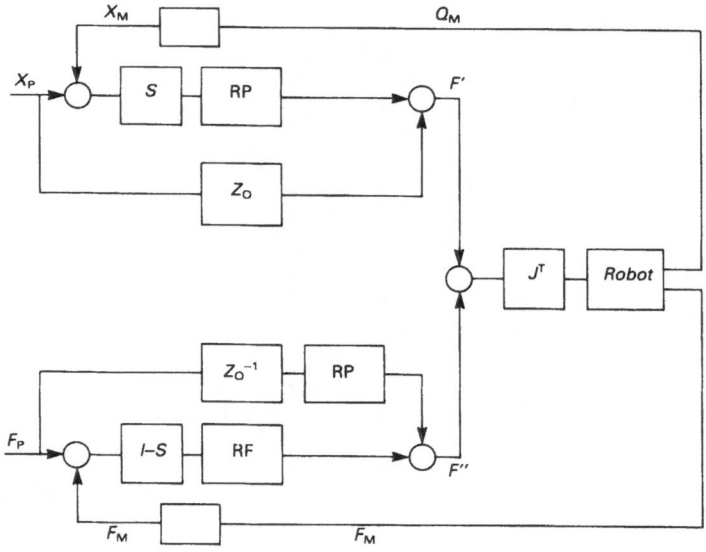

Fig.6. Improved hybrid force/position controller. A priori
information concerning the displacements in the force
direction and concerning the forces to be exerted in the
position direction are taken into account by a feedforward
loop.

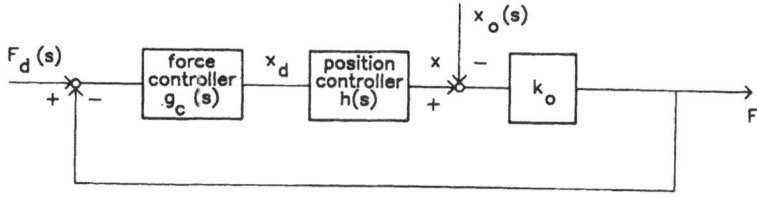

Fig.7.   One-dimensional force-around-position control.
         Simplified scheme.

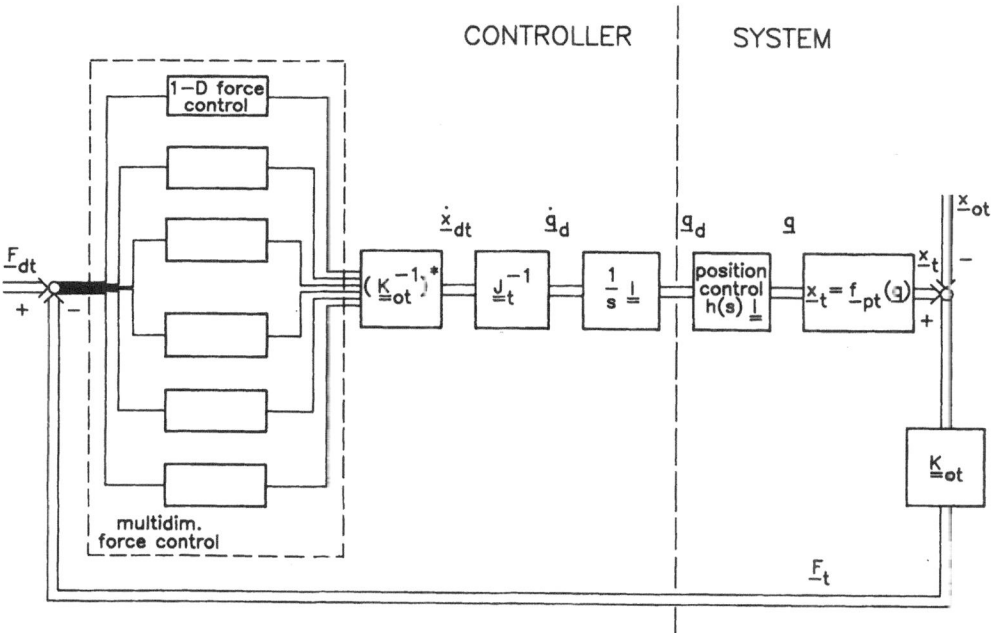

Fig.8.   Multidimensional decoupled force control.

APPLICATION OF LASER RANGE FINDER
TO ROBOT VISION

Y. Shirai
Electrotechnical Laboratory
Umezono Sakuramura
Niiharigun Ibaraki   (Japan)

1. INTRODUCTION

Sensing technology plays an important role in order to realize flexible robots that can perform tasks in a variety of situations. Sensors are divided into two categories: sensors for local information such as tactile sensors, and sensors for global information such as vision sensors. The former is effectively used by actuators while they are interacting with the environment. While, the latter is useful for higher level planning by providing information of the global environment. The main role of robot vision is, therefore, understanding the environment of the robot.

The study of robot vision includes the development of the following themes: input devices, feature extraction, object recognition, and so forth. Recently, range data input and processing have been intensively studied because of the following reasons:
   Range data is reliable for three-dimensional vision.
   More range finders have become available now.
   Range data can be processed more quickly now.

We have long been working on robot vision for recognition of complex scenes with multiple objects occluding one another. This paper describes some of recent research works in this framework.

2. RANGE FINDER

The direct measurement of range data is very useful for robot vision. One of the ranging methods is based on the active

triangulation, i.e. a plane of light is projected onto the
objects from one direction, and a camera detects its image
from another direction [1]. This type of range finder is now
practically used for classification of objects which are too
difficult to be discriminated with a two-dimensional image
[2].

Instead of scanning a plane of light in a scene, multiple
planes of light can be projected. While the planes of light
are turned on and off, the images of reflected light are
taken. This method is called time coded pattern projection.

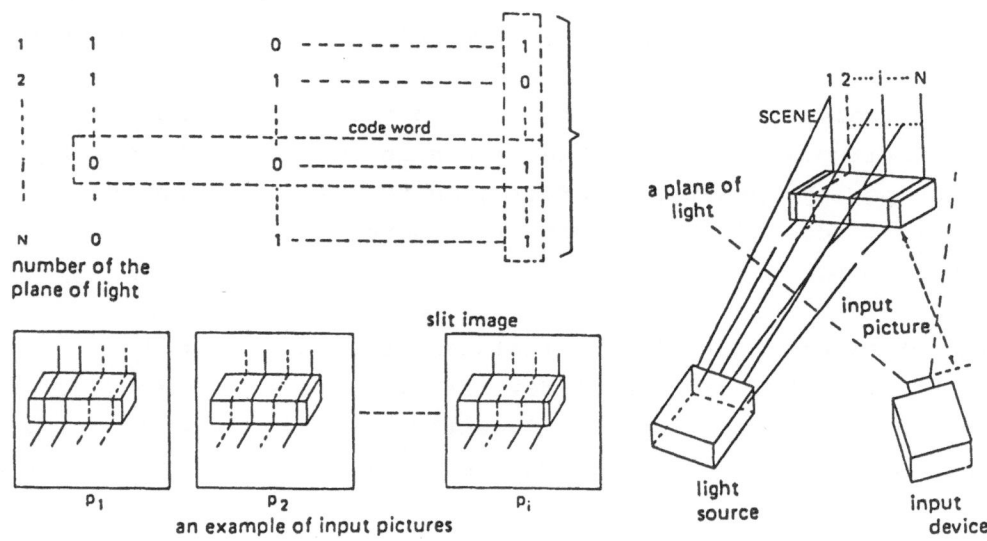

Fig. 1. Principle of time coded light pattern method

The principle of the time coded pattern projection method is
illustrated in Fig. 1 [3]. Suppose that vertical planes of
light is projected and that a camera is placed in the right of
the light source. Let N denote the resolution of the vertical
pattern, and n denote the number of images taken by the
camera. Each vertical plane is coded by n on-off patterns of
light planes. The coding may be binary or error correcting
coding depending on the situation. The intersections of all N
vertical planes with object surfaces are sequentially observed

as n binary images. Each vertical plane is located by decoding n observed images. Then, the range of the corresponding surfaces are obtained by triangulation. Algorithms were proposed to extract center positions of stripe images [3].

One of the most difficult problems with this method is how to project time coded patterns in a short time. Conventional range finders which make patterns with slide projector suffer from this problem. An improvement was proposed which employs a liquid crystal mask to change patterns electronically [4]. The horizontal resolution of this range finder is limited by the number of the pattern elements (currently 128).

In order to increase the resolution, we have developed a new range finder. Fig. 2 shows the principle. It consists of many light sources which are turned on and off according to a specified sequence. A light plane is easily generated with a laser diode and a cylindrical lens. In order to get a high resolution data (e.g., 512 lines), the same number of laser diodes may be used. It is, however, difficult to place many diodes in a small area. High resolution range data is obtained by rotating the mirror and observing the for each mirror direction.

Instead, the system employs 15 laser diodes and a rotating mirror [5]. For a mirror orientation, time-coded patterns are taken while controlling on-off of laser diodes, and then another set of images are similarly

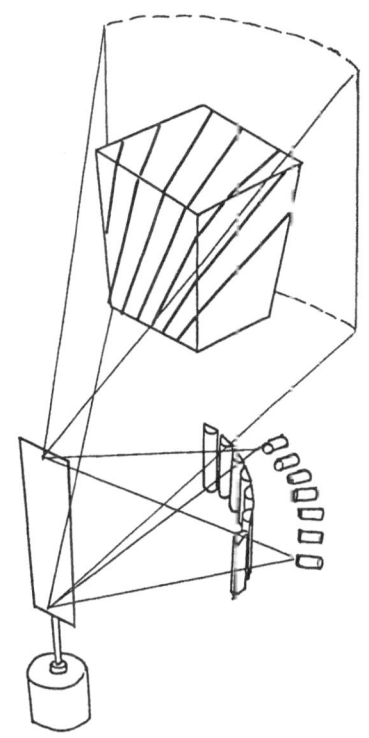

Fig.2. Principle of
new arange finder

taken after rotating the mirror
by a certain small angle. By
adjusting a step angle of the
mirror, range data with
arbitrary resolution can be
obtained. Fig. 3. shows a
pattern projector which
consists of laser diodes and a
rotating mirror.

If smaller light sources (or
optical fibers) are available,
more light sources can be used.
Then, input time may be greatly
reduced.

Fig.3. Pattern projector of
range finder

We have also developed special hardware for computing the
center position of stripes from video signal. It detects
candidates of stripes from each horizontal scanning line.
During a horizontal return period it determines a right number
of stripes based on the area of stripe patterns, and compute
the center position for each stripe as that of the center of
gravity. Since all operations are executed online, stripe
positions of an image can be computed in a frame scanning
period. Note that images of stripe patterns need not be
stored.

3. OBJECT RECOGNITION

Generally objects are recognized by matching a scene
description to models of objects. The object recognition
method is divided into two categories: scenes are predicted
and not predicted. Most industrial applications fall into the
former case. In the latter case, object models should include
enough information to be matched to scene descriptions viewed
from any directions. The scene description should also
contain enough information to be matched to unambiguous
models.

Methods of model representation and matching are closely related to scenes and the the task of the vision system. We are taking a general apprcach; a basic model of an object is a three-dimensional geometric model which can reconstruct the shape of the object completely and objects in a scene are recognized by matching scene description to the models.

3.1 Models
The geometric model is suitable for representing manufactured objects. It has long been studied in a field of CAD/CAM. Geometric modelers have also been developed for CAD/CAM systems [7].

In our laboratory a modeler GEOMAP [8] has been used for vision and manipulation research. We have developed a new modeler called SOLVER [9] by augmenting GEOMAP with useful functions for vision research such as generation of convex-hulls, description of visible part of faces, and identifying parts of models. Fig. 4 shows an example of visible parts of a face.

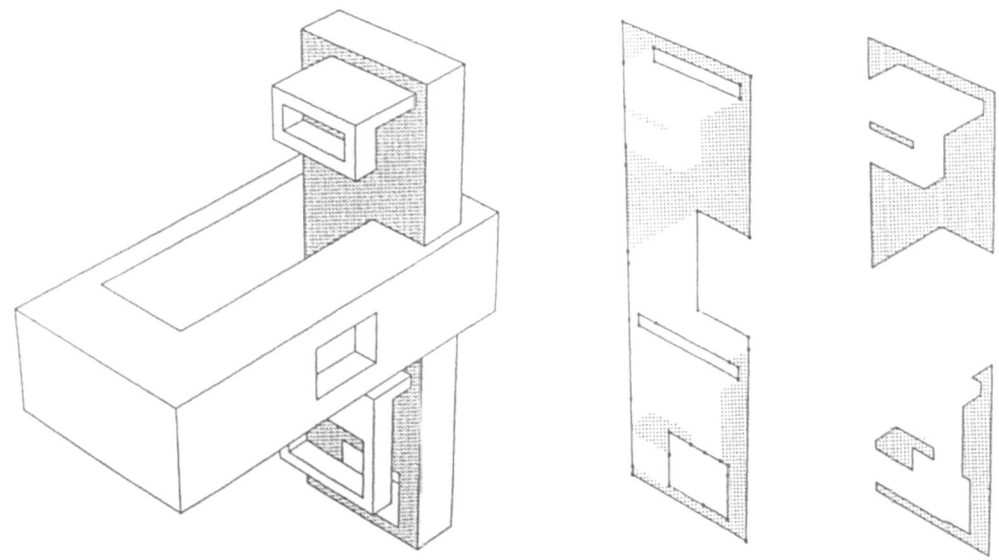

Fig.4. Visible parts obtained by SOLVER

Once a three-dimensional shape is represented in computer, many useful properties for matching to a scene description can easily be derived from the representation. These properties may be calculated beforehand or when necessary.

3.2 Scene Description

A vision system first extracts smooth surface regions (corresponding to visible surfaces) using a region merging method [10]. The border lines of those regions can be easily classified into jump edges, discontinuous edges, and corner edges [10]

A description of the scene is generated from the regions and edges in terms of the properties of regions and the relations between regions. In order to recognize objects easily, the system adds some extra properties which are directly derived from the original description.

The properties of a region consist of the followings:
1. Surface type fitted to the region. The type is classified as planner, curved or undefined. The curved surface is further divided into ellipsoid, cone, or other surface.
2. Surface parameters. The shape of a surface is described by surface parameters. Those of a plane, for example, are coefficients of the plane equation, and those of a cone are the position of the top, the axis orientation and the angle of the tangent plane to the axis.
3. Maximum radius and maximum area.

The relations between two regions consist of:
1. Type of intersection. The type is classified as convex, concave, or mixed.
2. Angle between regions. The angle is defined only if they have a common corner edge. If a region if not a plane, the angle is defined for a plane which is fitted to points near the common edge.

We propose additional properties of regions, called

meta-properties as follows:

1. Accuracy index. The accuracy index is defined for regions
   which are represented by parameters (planes and parametric
   curves). The index represents how accurately a region is
   approximated by the parameters. The index is determined
   based on the following facts. The more data points are
   included in the region, the more accurately the parameters
   are obtained. A region facing to the camera is more
   accurately approximated. A region with less parameters is
   more accurately approximated.

2. Interest index. The interest index represents how much
   the region contributes to identifying an object. If the
   region has many neighbor regions of the same object, it
   provides many constraints in matching to object models.
   Thus the index is the number of neighbor regions sharing
   common corner edges,

In order to recognize objects from those properties,
corresponding model properties of regions are computed
beforehand from the geometric models. The computation is easy
because the model contain enough information.

Matching process is in principle a tree search process; a
node of a search tree corresponds to a pair of matching
regions [11]. The search process, however, is not efficient
if a search tree is arbitrarily generated.

A proposed search method generates an efficient search tree by
making use of meta-properties described above. The system
first sorts regions obtained from a scene (observed regions)
according to the interest index. Regions with the same
interest index are further sorted according to the accuracy
index. A region with the highest index (called kernel region)
is picked up first and matched to models. The tolerance of
the matching is determined by the accuracy index. The child
nodes are generated which correspond to matched model regions
as shown in Fig. 5.

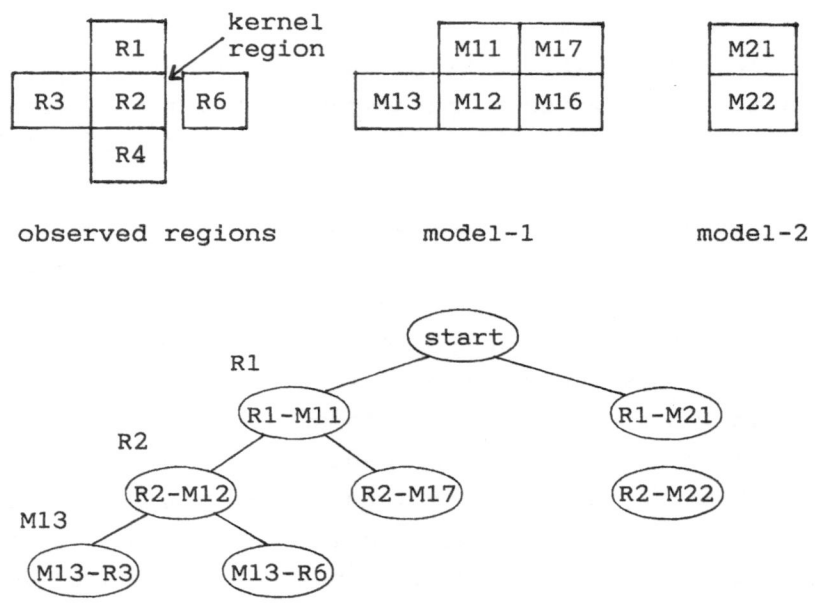

Fig. 5. Example of search tree

The next step depends on the interest index of the kernel region. If the region has neighbor regions connected to it (interest index > 0), the best neighbor region is picked up, and matched to model regions. In the Fig. 5, three nodes are generated by matching region R2. This data-driven matching process continues until no other neighbor regions of the kernel are found. Then a model-driven matching process starts; for each corresponding model region, a neighbor model region is selected and matched to observed regions (nodes R3-M13 and R5-M13 in the figure). Note that this process may match observed regions which are not connected to known regions (such as R6 in the figure). If multiple regions are matched successfully before this process, the affin transformation from the model object to the observed object is obtained. Then the rest of the search process is very efficient because the position and orientation of matching regions are predicted.

If a kernel region has no connected neighbor regions, model-driven matching starts as illustrated in Fig. 6. This process is generally less efficient when the affin transformation is not known. In most cases, however, many observed regions are already successfully matched to model regions before this process is activated. That is, less interesting regions cannot be kernel regions earlier than more interesting regions.

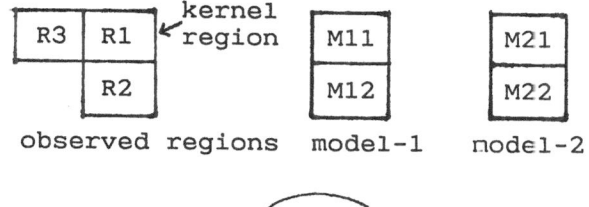

observed regions   model-1   model-2

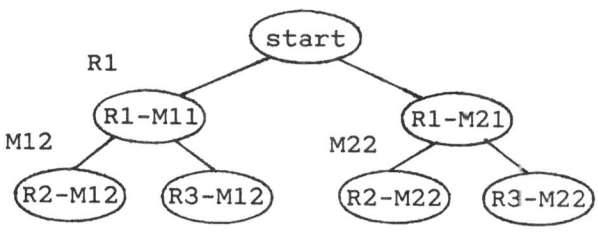

Fig. 6. Example of model-driven search

The search tree, therefore, does not grow large because three are usually a small number of observed regions left.

When a set of observed regions is uniquely identified, the above process is repeated for the rest of observed region. If a kernel is matched to multiple regions, the process is repeated for the same observed regions. Sometimes the multiple matching may be disambiguated in the later process. If not, the set of regions has multiple interpretations.

4. CONCLUSION

A new range finder and a method of range data processing have been described for a flexible robot vision. By using thresholds of matching derived from a accuracy index, objects are reliably recognized. A tree search for matching is efficiently performed guided by an interest index. Although the proposed method is still in the stage of research, it may be soon applied to practical works.

REFERENCES

1. Shirai, Y. and Suwa, M.: Recognition of Polyhedrons with a Range Finder. Proc. 2nd Int. Joint Conf. on Artificial Intelligence, pp. 80-37 1971

2. Ozeki, O., Nakano, T. and Yamamoto, S.: Discrimination
   Method for Three-Dimensional Object Using Range Data.
   Trans. IECEJ, Vol. J69-D, pp. 111-112 (in Japanese) 1986

3. Minou, M., Kanade, T. and Sakai, T.: A method of
   Time-Coded Parallel Planes of Light for Depth Measurement.
   Trans. IECE of Japan, Vol. E64, pp. 521-528 1981

4. Sato, K. and Inokuchi, S.: Range-Image System Utilizing
   Nematic Liquid Crystal Mask. Proc. 1st Int. Conf. on
   Computer Vision, pp.657-661 1987

5. Yoshimi, T. and Ohshima, M.: A Pilot Model of a Range
   Finder with Multiple Light Sources. Proc. 4th Annual Conf.
   of Robotics Society of Japan, pp. 291-292 (in Japanese)
   1985

6. Yoshimi, T. and Ohshima, M.: Real-Time Pre-Processing
   circuit for Multi-Laser Range Finder. Proc. 34th Annual
   Conf. of Information Society of Japan, pp.1725-1726 1987

7. Requicha A.A.G. and Voelcker H.B.: Solid Modeling: A
   Historical Summary and Contemporary Assessment. Computer
   Graphics and Application, Vol. 2, pp. 9-24 1982

8. Kimura, F. and Hosaka, M.: Program Package GEOMAP. Proc.
   Geometric Model Project Meeting 1978

9. Koshikawa, K. and Shirai, Y.: A 3-D Modeler for Vision
   Research. Proc. '85 ICAR, pp. 185-190 1985

10. Shirai, Y. Three-Dimensional Computer Vision. Springer
    1985

11. Grimson, W.E.L.: The Combinatorics of Local Constraints
    in Model-Based Recognition and Localization from Sparse
    Data. J.ACM, Vol.1, pp.89-94 1986

VISUAL INSPECTION SYSTEM WITH QUALITATIVE ANALYSIS CAPABILITIES

J. Amat, A. Casals
Dep. Enginyeria de Sistemes, Automàtica i Informàtica Industrial
Facultat d'Informàtica (UPC), c/ Pau Gargallo, 5
08028-Barcelona   (Spain)

1. INTRODUCTION

In every production system, inspection constitutes a very relevant task. Even now, many visual inspection tasks are done in a non automatic way. Human visual inspection is characterized by its ability to detect easily qualitative faults even if they are very small, for instance a selection of wood based on its knots and waters or by the faults on the finish of its surface. On the other hand, a person has great difficulties in appreciating, by means of his vision, quantitative parameters such as precise measures.

During the last years, computer vision has begun to be used in automatic inspection systems. Vision systems have the advantages derived from being non contact sensors. Moreover, recent progress in the field of microelectronics tend to decrease both cost and operation time of vision systems.

However, the limited resolution of TV cameras restricts the use of computer vision systems for inspection tasks, based on accurate measures, to the cases in which only a reduced zone of the scene has to be analysed, because this zone can be seen in great detail. Vision systems can be used also efficiently in applicactions in which only a qualitative test of a part or object is needed. For instance, to determine the existence or not of some defined components or features of the image.

On the other hand, it is not easy to develop general purpose vision systems for images having a certain degree of complexity. In these cases, it is more convenient to develop special purpose vision systems focused to solve specific applications. These systems require in many cases special light conditions.

Nowadays, most vision systems are based on the detection and measure of some features in objects such as : perimeters, vertices, areas, holes, distances between holes, etc..

In the same field but following another direction, the work realized has been focused on the development of a measure system, basically oriented to perform a qualitative analysis of the scene but with high speed capabilities, in order to be

NATO ASI Series, Vol. F 52
Sensor Devices and Systems for Robotics
Edited by A. Casals
© Springer-Verlag Berlin Heidelberg 1989

used in processes that needs a 100 percent inspection.To attain this goal it has been necessary to develop a specialized processor that realize by hardware a part of the processing.

2. REQUERIMENTS OF AN INSPECTION SYSTEM

The problems to be faced in an inspection system based on visual perception rely on two different principles : qualitative estimation or quantitative measurements.

In the inspection systems based on a qualitative estimation of the image , an assessment of the possible faults in the parts or in the inspected material is carried out. As a result , the parts or materials inspected are accepted or refused. The assessment of these faults can be done following different techniques : a first one can consist on an analysis of the signal, this is the case of parts classification based on their colour. Another technique is based on the comparison, pixel to pixel, of every element of a model image with those corresponding to the obtained image. With these comparisons, for instance, the missing of a component or the wrong positioning of an element in a printed board can be detected. A third technique consists on analysing the image in order to detect the missing of a part, as for instance a hole. Furthermore, the analysis of the image allows to detect faults such as the wrong positioning of a set of elements of a part. Finally, another technique is based on the analysis of fulfiling of some defined rules, this is the case of the quality control of printed boards in which the minimum distances between paths, minimum thickness,etc. are analysed.

In the systems based on a quantitative inspection, a measurement of certain parameters of the image are carried out. The acceptance, rejection or classification of parts or materials is done by comparison of these parameters measured with some predefined boundary values. One of the techniques of quantitative inspection used is based on the counting of the number of elements or some defined characteristics of an element allowing to detect abnormalities in a process as for instance the number of parts packed on a box. Another quantitative technique used is based on the measurement of some given distances or boundaries, as for instance the thickness of a wire, the dimensions of a part, the level of filling up of a bottle or even the correct positioning of a label on a bottle. The measurement of such magnitudes and their trends and the deviations of some characteristics allows to detect if the part inspected is inside tolerances.

The degree of complexity of an inspection system based on visual perception can vary enormously depending on the kind of parts inspected, the kind of faults to be detected, as well as, on the way in wich they appear on the image. The degree of difficulty that carry with it the analysis of a scene in wich only a part appears is less than the one in wich more than a part appear simultaneously. The degree of difficulty is even greater when the parts are different. On the other hand, the aspect of the background can help the extraction of the needed information from the image. For this reason, in the implantation of a visual perception inspection system, it is important to try to get the maximum contrast between the background and the number of elements to analyse in order to attain the maximum system efficiency and reliability.

Commercial systems available nowadays, are generally oriented to carry out quantitative measures over the scene, using windows with dimensions and positioning previously fixed, in order to adapt themselves to every different kind of scene.

3. SYSTEM STRUCTURE

The system assayed is based on a specialized processor oriented to perform a qualitative analysis of a scene, able to detect faults that would be hardly described by parameters, and then not suitable to be detected by the systems available at present.

The information elaborated and compacted by the special purpose processor is treated by a microcomputer that performs the calculus of the boundary values and the average ones with which it is possible to detect the eventual local deviations appearing during the analysis of the scene.

In figure 1 the functional structure of the system is shown. In the learning phase stage, a map of some determined and very elemental patterns localized in the image is elaborated.In the process of analysing the image, a texture function consisting on the number of 1ns contained in a determined window is generated.

From this information, the microcomputer evaluates the possible errors appearing on the analized image, corresponding to a scene having a repetitive texture. Errors, failures in the material, lack of elements etc., are detected by the deviation of any of the texture function parameters.

3.1 Preprocessor description

Using a 16 lines memory as well as a shift register, with which it is possible to do the serial to parallel conversion of the pixels of a line, the data corresponding to the neighbours of a pixel are obtained.

In this way a cross-shaped window is intended to detect patterns having vertical and horizontal symmetry, as would be the case for rounded holes.

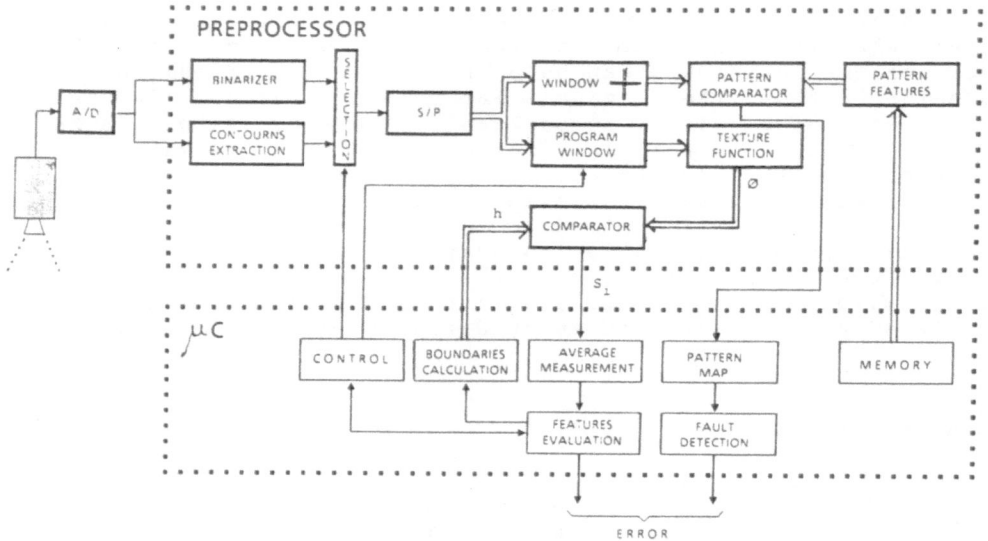

Fig.1   System   structure

On the other hand a   selector allows to configurate a 32 x 4
pixels window, which   lengh can be reduced as in a bit slice
structure. Fig.2

To attain a greater sensitivity of the texture function in
spite of being so simply obtained, the binarized image or the
contour one can be chosen. The binarized image is obtained by
using a comparator having a dynamic threshold automaticaly
calculated. The contour image is obtained by means of a
gradient operator

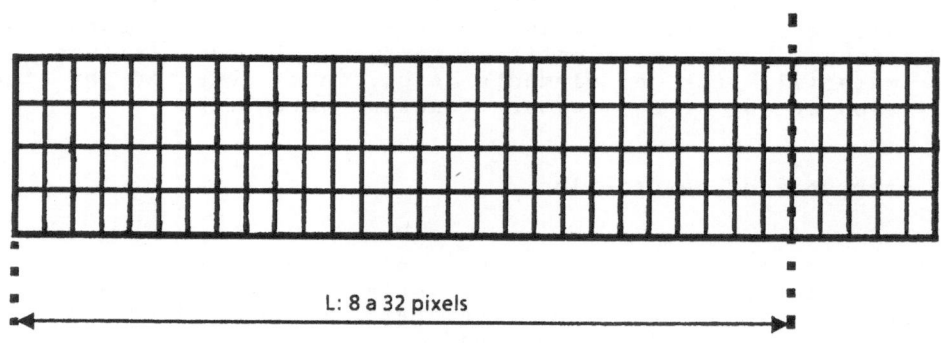

L: 8 a 32 pixels

Fig.2   Programmable window

3.1.1 Patterns detector unit

This operator performs in real time the detection of some patterns in the image and their localization. A very simplified operator based on a 16 x 16 cross-shaped window has been chosen. One can consider that the window operator is placed over the centre of a pattern when its central pixel differs from the texture of the surface of the objet, and the first pixels of the object found in the four directions are considered equidistant to this central pixel.

The measure of these distances are achieved by means of priority encoders that compact the information corresponding to the 16 + 16 pixels, to 4 distances coded with 3 bits. Fig.3. With this codification, the processor with which the equidistances are compared, admitting 1 bit of tolerance, can be implemented by means of a ROM. In the same way, an order corresponding to the value of some determined boundaries of these 4 measures, to be considered, can be used. These values are dependent on the kind of pattern analized.

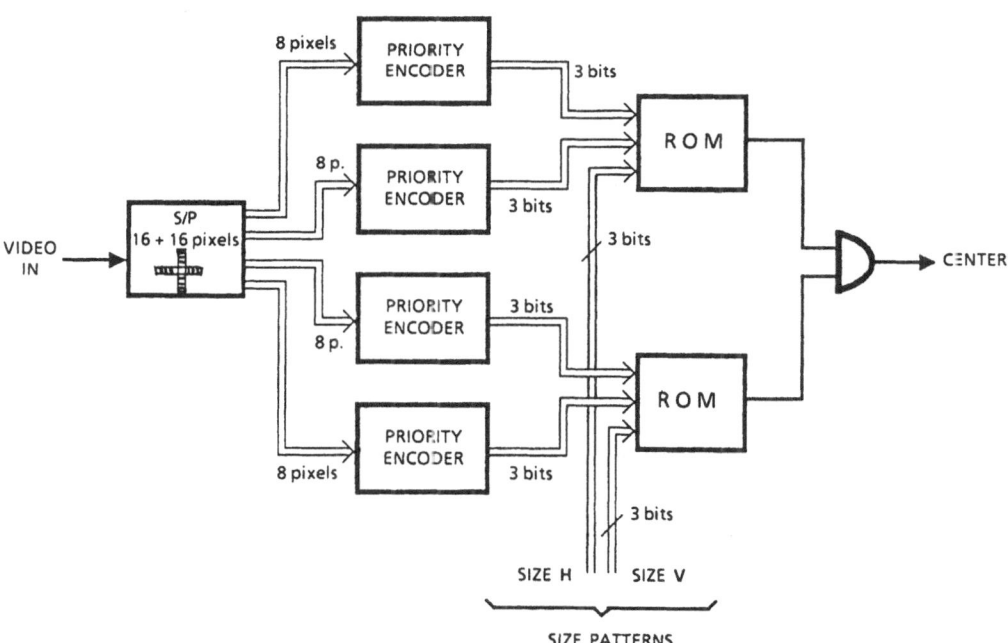

Fig.3 Coding of the radies by means of priority encoders

Using different values of the boundaries of the measures accepted, corresponding to the vertical and horizontal arms of the cross-shaped window, for example, holes and lenghthen patterns can be detected.

Using this pattern detector system, all the visible holes of an observed part can be located and their presence verified. If the sizes of the patterns are larger than the size of the used window, they can not be detected.

The low resolution attainable by the system prevents it to carry out a test based on quantitative measures, corresponding to their sizes , but on the other hand the number of holes and their approximate localisation can be detected.

3.1.2. Texture Analysis

In some inspection tasks one has to deal with scenes having repetitive geometric textures. This is the case of analysing a mesh, a reticule, a coog or a sequence of identical elements as for instance the pins of a connector.

In such applications, it is usually of some interest not only to count the existence of all the elements but also their equidistance. The developed processor analizes the repetitive textures of a scene, extracting as its features, the ratio of black/white pixels inside a determined window and analysing its variation all over the image,fig.4. Most of the possible defects on this kind of surfaces analysed produce a local variation of this rate,fig.5, conspicuous enough to be detectable.This variation can be detected either by the analysis of the variable corresponding to the mentioned ratio along a line, or by the period of its fluctuation along the image.

In order to perform this detection operation, an operator that analyses the whole image by means of a 32 pixels window of programmable format is used.(Fig.2). When the position of the window operator moves along a line of the image, the value of the variable $\emptyset$ obtained will vary relatively few if the lengh of the window is great enough in relation to the periodicity of the texture.On the other hand during the line to line advancement of the window, the $\emptyset$ function will vary appreciably due to the fact that the window shape is predominantly horizontal. This variation will be periodic and of relatively low resolution since the increases and decreases of the function will be produced basicaly when the window changes from one line to the next, that is every 64 μs.

Location of failures can be perfomed by detection of the periodicity variations of the function $\emptyset$ (y) obtained by shifting the window all over the image. With the use of several counters it is possible to obtain either the period T of this function as well as the time intervals $t_1, t_2 \ldots t_i$

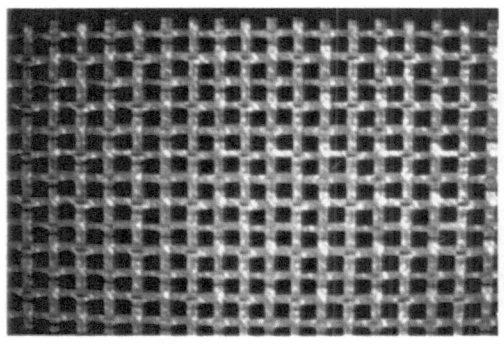 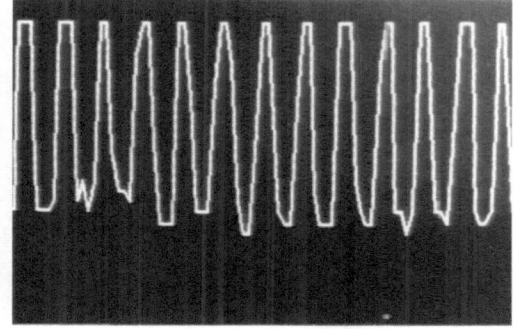

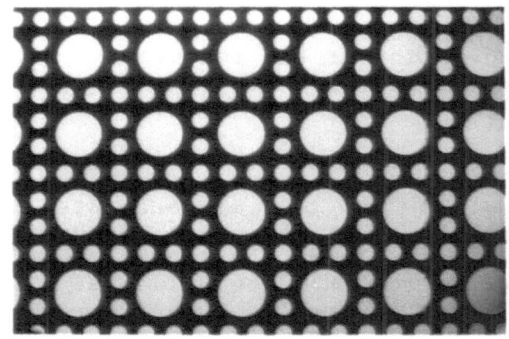 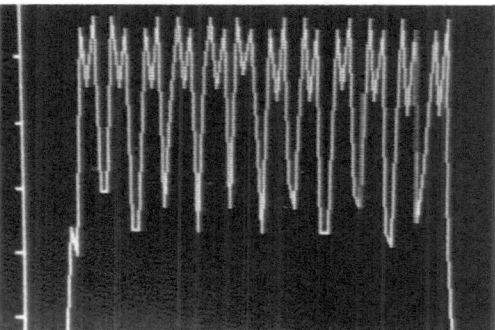

Fig.4   Texture functions obtained by application of the
        window operator along different patterns

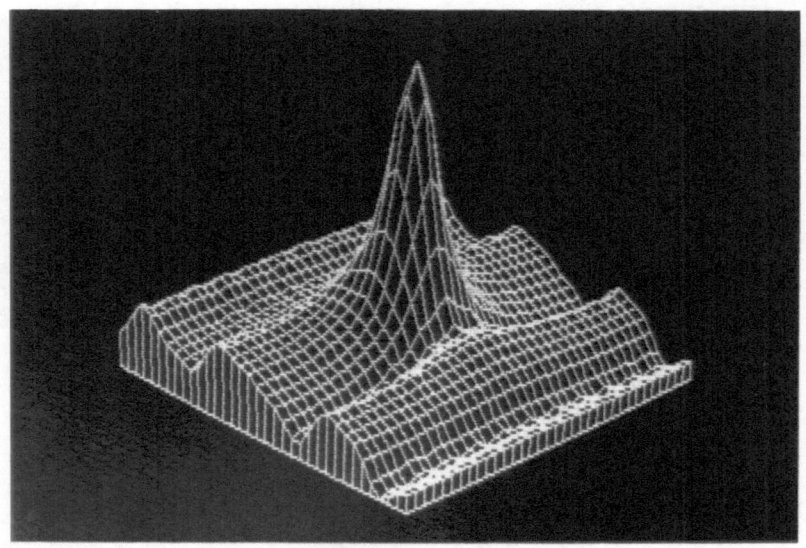

Fig.5 Detection of a local variation of the texture
        function in a faulty material

corresponding to the times in which $\emptyset$ surpasses or not some
defined thresholds $\emptyset_1, \emptyset_2 ... \emptyset_i$ given by the microcomputer.
These times give certain information about the shape of the $\emptyset$
function. The microcomputer calculates these thresholds h_i by
successive approximations in such a way that the values of T_i
corresponds to a given percentatge of the period T, and can be
selfadjusted in a continous way. Taking as an example the
function $\emptyset(y)$ obtained from the scene of figure 4(b) the
thresholds h_i would befit to be located in the zones that
allow to detect local shape failures. In this case the program
would define the most adequate value of h_1, about an X percent
higher than the values v_o from the signal obtained when the
window moves over a model without failures.Fig.6. The margin X
is chosen as a compromise between sensitivity and the quantity
of false detections admited. The value h_2 is chosen as the
medium value between v_1 and v_2 observed during the previous
calibration phase.

The comparators used to detect variations of the function $\emptyset_{(y)}$
give the signals s_1 and s_2 which periodicity has to be
verified, as well as the variation of t_1, t_2, t_3,
corresponding to s_1 and t_5 and t_4 corresponding to s_2. Fig. 7.
The updating of the t_i values has to have a given time
constant in order to make them sensible to transistory
variations. At every period T of the $\emptyset(y)$ function, the
microcomputer receives an interruption in order to begin to
calculate the new values of the t_i times. Thus these
parameters are used to perform the tracking of the shape of
the function $\emptyset_i$ utilized.

331

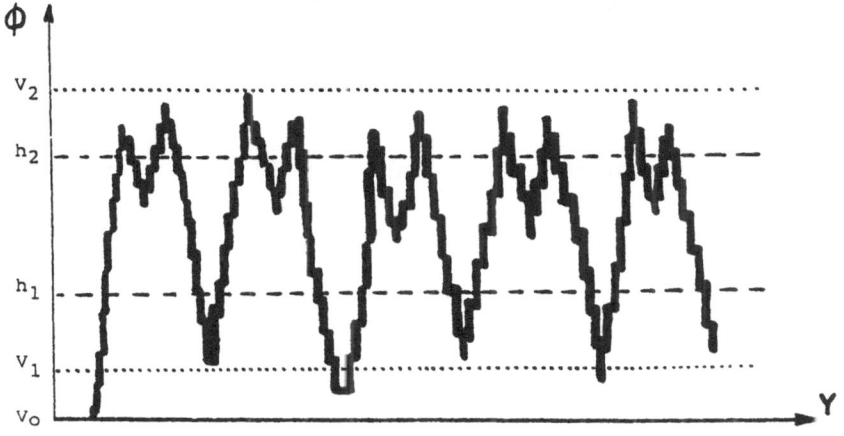

Fig.6   Definition of the h₁ thresholds

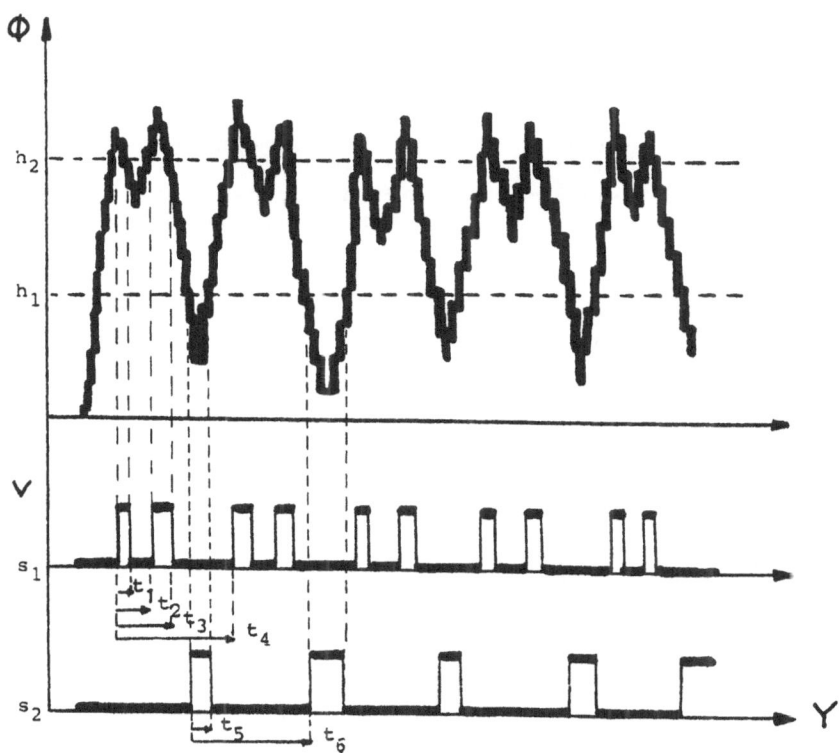

Fig.7   Periodicity of the function s₁ and s₂

3.2 Control Unit

The processor control can be assigned to a microprocessor since the functions to be performed don't need to be done at a high rate. This is due to the fact that the pixel to pixel operations to be done are implemented by the specially developed hardware. On the other hand, the use of a microprocessor instead of a specific control processor allows to modify the parameters given to the processor, such as window formats, thresholds etc., in a dynamical way. The calculus of these modifications are performed by means of an analysis of the obtained results and depending on some stablished criterious. The tasks realized by software can be carried out by standard microcomputers and even with a PC thus it is possible to take profit of their relatively high prestations, of their low cost and of the standardization and transportability of their software.The selection of the format to be utilized will depend on the kind of texture to be analysed. In this first processor designed and tested, only three different window formats to evaluate the degree of uniformity of the texture are available, due to the fact that this analysis is done by hardware.

The selection of the most appropriate window will be done by the microcomputer, testing the results obtained from the different lengths available formats during a previous testing phase, that can be considered as the learning one.

It will be considered as the most adequate format the one that supplies a "texture function" more uniform when it moves all over a given texture containing no failures.

In order to detect failures in textures it is hoped that when the window moves all over the image it is possible to detect local variations of the texture function greater than the usual boundaries. To define these boundaries in the previous calibration phase, the microcomputer calculates the values maximum and minimum of the function $\emptyset(y)$, fig.4, as well as the thresholds. These thresolds are obtained as the mean values between the relative maximums and minimums of the $\emptyset$ function therefore, the resulting signals S_1 and S_2 are periodic in spite of the fluctuations of the function $\emptyset$ along the image. The measure of the times "ti" will allow to determine any deviation of the periodicity of the signals S_1 generated, that are poduced by texture continuity failures.

4 Influence of the window format

In order to qualify a qualitative characteristic as it is the degree of uniformity of a texture, a function giving a measure of the density of "1ns" contained inside a given window and that could be called "texture function" has been used. The ideal texture function would be the one which shape was completely flat all along the image and producing a strong variation in a point of the image in the case that an anomaly appears in that point. Due to the difficulties of finding such an ideal function, the function chosen has been one suitable of being implementable in hardware, as it is the counting of 1ns.

The condition required to this function consists on having certain stability to mantain its average value. In order to assure the required degree of stability it is necessary to study the ripple of the function $\emptyset$ when it shifts along the line $\emptyset(x)$ and from line to line, $\emptyset(y)$.

Due to the fact that the texture function used consists on counting the number of "1"ns, what is really done is an averaging, while the used window moves along the image. In this case, the new texture function is a new signal multilevel and filtered. The filtering of this signal performed frcm a fixed number of samples is non recursive, and the level of atenuation attained is variable as can be seen in figure 8.

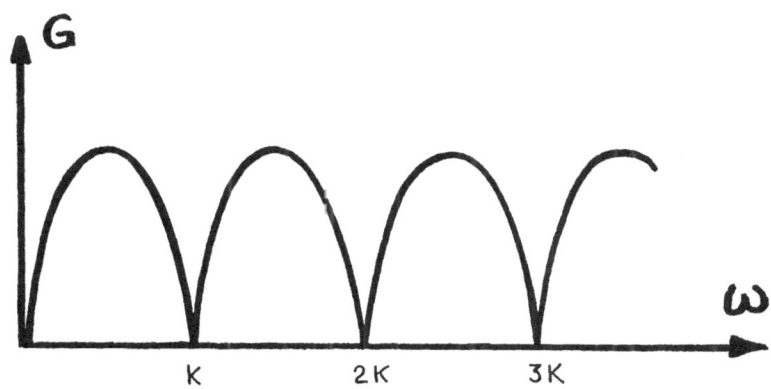

Fig.8 Non recursive filtering of the texture function

In this case, the points having a maximum attenuation nk, when shifting the window horizontally will be obtained only if the width of the used window coincide with the period of the repetitive texture of the inspected object or material. With such an operator when it is applied along a line, the magnitude of the signal $\emptyset(x)$ obtained will be less variable as it is nearer from the tuning point nk, thus, in the case that some imperfection modifying locally the texture exists, an appreciable variation of this function $\emptyset(y)$ will be produced.

By shifting the window line to line a fluctuation of the function $\emptyset$, respect to the variable y, will also be produced. This fluctuation can also be attenuated by averaging a given number of lines, being this number a multiple of the period of the analysed texture. Technological difficulties in implementing a so high window force to bound this highness to only 4 lines. Therefore, a function $\emptyset(x,y)$ corresponding to the points contained in a 4 lines and programmable width window, depending basically on the vertical shift is used.

As a result, the function $\emptyset(x,y)$ obtained, figure 4, is of relatively low frequency, (1 line = 64 µs), making it possible to analyse the obtained function and to detect the fluctuations that can appear as a consequence of their texture failures.

5 Conclusions

The use of computer vision in quality control applications requires solving the problem of attaining the high operation speed needed.In general, vision systems consist of a part of hardware and a part of software. The weight of each of these parts in the system determine the main characteristics of the whole equipment. The development of algorithms by means of software give great flexibility to the systems, moreover they are cheaper, but excessively slow for a great number of applications.

On the other hand, the use of hardwired processors provide a high operation speed, but at the expense of a higher cost and of less flexibility.

The vision system presented is based, in its first version, on a hardwired processor. The system has as main goal the obtention of a high operation speed giving at the same time a certain flexibility by doing a qualitative analysis of the image and suitable to be used in some different applications.

The system has been tested in inspection tasks in continous production systems having relatively high production speed. For instance, to inspect punched plate as is shown in Fig.9. Using this system, it is possible to locate punctual failures or installation adjustment failures almost instantaneously. In this way, it is possible to put in to action the corresponding correcting systems very quickly.Fig. 10

The use of high scale integration components as well as high speed microprocessors with outstanding capabilities, specially transputers will allow to build parallel architectures in order to obtain great advances in the field of visual inspection.

By complementing the quantitative vision systems with some qualitative evaluations many problems existing nowadays in the industry, specially in automatized inspection tasks could be solved.

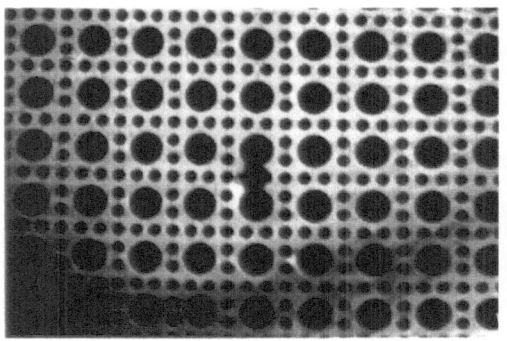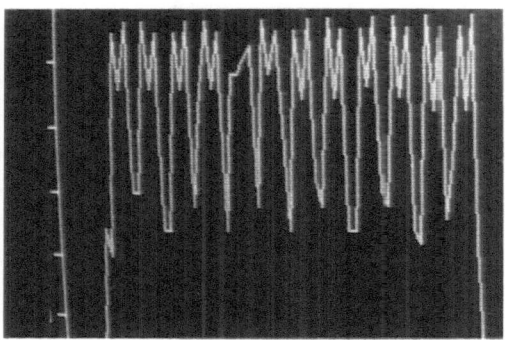

Fig.9   Detection of local failures on materials with
        repetitive texture

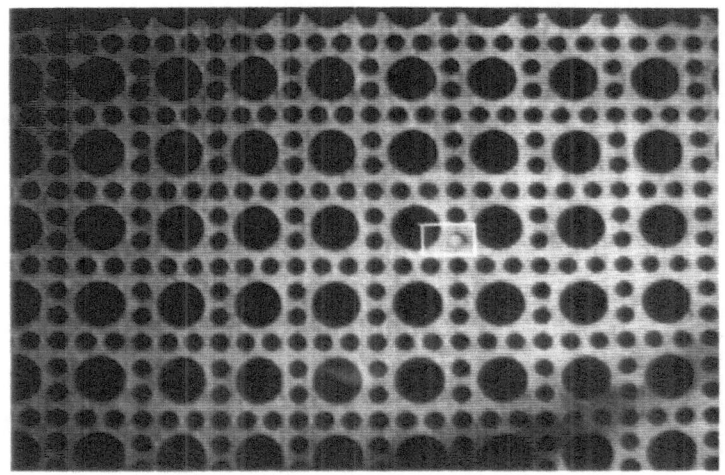

Fig. 10. Visualization of a failure by positioning in real
         time a mark on the monitor

BIBLIOGRAPHY

J. Bretschi
   "Automated Inspection Sytems for Industry"
   IFS Publications Ltd. 1981.

W.K. Pratt, O.D. Faugueras, A. Gagalowicz
   "Applications of Stochastic texture field models to image
   processing"
   Proc. IEEE vol.69 nº5, May 1981.

A.G. Makhlin, G.E. Tinsdale
   "Westinghouse grey scale vision system for real-time
   control and inspection"
   Robot vision. Ed. A. Pugh.
   IFS Publications Ltd. U.K. Springer-Verlag, 1983.

R. Freeling
   "The signifiance of lighting in industrial inspection
   tasks"
   1985 IEEE Conference on Robotics and Automation.

A. Casals
   "A simplyfied quality control system by means of visual
   perception"
   Regulación y Mando Automático. 1985 (in Spanish).

G. Nicolas, J.P. Hermann
   "Inspection of moulds by 3 dimensional vision".
   6th Inter. Conference on Robot Vision & Sensory Control
   1986.

A.L. Pain et.al
   "Automated visual inspection of aircraft engine combustor
   assemblies"
   1986 Int. Conference on Robotics & Automation.

J. Hollingum
   "Automated inspection for the gauge and panel lines"
   Sensor review, Vol.7 nª1, 1987.

R.S. Adrain et.al
   "Laser scanning cameras for in-reactor inspection"
   Sensor review, Vol.7 nº2, 1987.

C. Lee, Y.J. Chao
   "Surface texture dependence on surface roughness by
   computer vision"
   1987 IEEE Int. Conference on Robotics and Automation.

J. Amat, A. Casals
   "A high speed preprocessor for visual inspection
   applications"
   The International Society for Mini and Microcomputers.
   1988.

PREDICTIVE AND ESTIMATION SCHEMES
IN SENSOR-CONTROLLED TELEROBOTICS

G. Hirzinger
DFVLR Institute for Flight Systems Dynamics
D-8031 Wessling   (F.R. Germany)

Abstract

The paper discusses the problems that arise when sensorcontrolled robots in space are teleoperated from ground stations.
Predictive 3D-computer-graphics presently seems the only way
to successfully cope with the problem of transmission time
delays of several seconds. Appropriate estimation schemes are
outlined which include models of the delay lines the robot,
moving objects etc., and which derive the necessary corrections from sensory data as they are sent down from the spacecraft to earth. The space robot technology experiment ROTEX
scheduled for the next spacelab-mission D2 is taken as a basis
for the problem description.

Introduction

Among the many areas important in space technology automation
and robotics (A&R) will become one of the most attractive ones
for smaller countries like the Federal Republic of Germany; it
will allow experiment-handling, material processing, assembly
and servicing with a very limited amount of highly expensive
manned missions, and the expectation on an extensive technology transfer from space to earth seems to be much more
justified than in many other areas.

NATO ASI Series, Vol. F52
Sensor Devices and Systems for Robotics
Edited by A. Casals
© Springer-Verlag Berlin Heidelberg 1989

This is one of the main reasons why several activities towards space robotics have started now in Germany with the long-term goal of a major contribution to the space station, e.g. to the subsystem "man tended free flyer" (MTFF). In addition to study activities we made a proposal at the end of 1985 to fly a robot technology experiment ROTEX with the next "German" spacelab-mission D2. With the Challenger disaster the schedule is now shifted to 1990 or 91. Nevertheless ROTEX is kind of a starting shot for a German participation in space automation and robo-tics. In this paper - after a brief outline of the robot, its sensor-based gripper configuration and the tasks to be per-formed - the problem of teleoperation from ground is addressed, where long delay times have to be compensated by predictive estimation schemes using advanced 3D-computer graphics.

THE ROTEX PROPOSAL

It contains several items:

- A small, six-axis robot (working space < 1m) flies inside a space-lab rack (fig. 1). The robot will be built by DORNIER company. Its gripper will be provided with a number of sensors, especially a 6-axis force-torque wrist sensor, grasping force control, an array of 9 laser-range finders and a pair of optical fibers to provide a stereo image out of the gripper; in addition a fixed pair of cameras will provide a stereo image of the robot's working area.

- The robot is able to perform automatic, preprogrammed motions as well as teleoperated motions via an astronaut or an operator on ground (fig. 1), including experiment handling and servicing operations.

● The main goals of the experiment are:

- To verify joint control (including friction models)
  under zero gravity as well as μg-motion planning concepts based on the requirement that the robot's accelerations while moving must not disturb any μg-experi-
  ments nearby.

- To demonstrate and verify the use of DFVLR's senscr-
  based 6 dof-handcontrollers ("sensor balls") under zero
  gravity.

- To demonstrate the combination of a complex, multisensory robot system with powerful man-machine-interfaces (as
  are 3D-computergraphics, sensor-ball, stereo imaging,
  voice input-output), that allow teleoperation from
  ground, too.

In order to demonstrate servicing capabilities by teleoperation three basic tasks are envisioned:

a) assembling a mechanical grid structure
b) connecting/disconnecting an electrical plug
c) grasping a floating object.

For these tasks continuos or online sensory feedback is
involved.

SENSORS AND SENSORY FEEDBACK

The gripper sensors belong to the new generation of DFVLR
robot sensors with all analog preprocessing and digital computations performed inside the sensors or a least in the wrist
(fig. 2). Using a high speed serial bus only two signal wires
are comming out of the gripper (carrying forces-torques and

distances), augmented by two 20 kHz-power supply wires, from
which the sensors derive their DC-power supply voltages via
tiny transformers themselves. In particular we provide:

a) an array of 9 laser range finders based on triangulation,
   one "big" sensor (i.e. half the size of a match box) for
   the wider range of 3-50 cm, and 4 smaller ones in each
   finger for lower ranges of 0-3 cm. The range finders are
   the result of more than 5 years development aiming at a
   precise performance over a remakable range and independent
   of the slant angle and surface of the measured object.
   The "long range" finder will be provided with a scanning
   mechanism to aid the visual determination of position/orien-
   tation of a floating object to be grasped.

b) a "stiff" 6 axis force-torque sensor based on strain-gauge.
   measurements (fig. 3) or/and a compliant optical one (fig.
   4). Originally it seemed necessary to make a final decision
   between these two principles, but in fact at the moment in
   our lab a construction is under way that combines both
   principles in one compact sensor with the option to switch
   between them during an operation.

   The stiff sensor in fig. 3 resembles a double-maltese-
   cross, and uses 10 strain-gauge-half-bridges combined into
   7 full-bridges in a way that the sensor continues operating
   reliably with reduced accuracy if one of the strain-gauges
   is damaged. The sensor performs automatic temperature compensation based on the temperature characteristic as stored
   during the calibration process.

   The "compliant" optical force-torque sensor consists of an
   inner and an outer part (fig. 4). The basic measuring
   arrangement in the inner ring is composed of a LED, a slit
   and perpendicular to it a linear position sensitive detector (PSD) which is mobile against the remaining system.

Six of such systems (rotated by 60 degrees each) are mounted in a plane, whereby the slits alternatingly are vertical and parallel to the plane. The ring with PSD´s is
fixed inside the outer part and connected via springs with
the LED-slit-basis. The springs bring the outer part back
to neutral position when no forces-/torques are exerted.

c) the sensor or steering ball as a 6 dof-hand controller. For
a very natural 6 degree-of-freedom control of robots and
3D-computer graphic objects with only one human hand we
developed different kinds of plastic hollow balls with six-
axis force-torque sensors inside /3,4/. Our latest and pre-
ferred version uses the compliant sensor (fig. 4) inside
the ball, i.e. the only difference between wrist sensor and
ball is that the outer ring in fig. 4 is replaced by a
plastic hollow ball. Fig. 5 gives an idea of the hardware,
which serves not only as a 6-dimensional joystick but at
the same time issues forces and torques as exerted by the
human operator - an important feature in our teleoperation
concept.

TELEOPERATION WITH SENSORY FEEDBACK AND PREDICTIVE CONTROL

a) Fine-motion planning
Our sensor-based fine motion planning concept has been outlined in different papers (e.g. /7/). Its main features -
briefly recalled - are:

"Rudimentary" teach commands are derived either on-line
from a human teacher operating the sensor ball or from a
path-generator connecting preprogrammed points. They are
interpreted in a dual way as force/torque or positional/
orientational commands. Loosely speaking if the robot moves
in free space, the ball forces are transformed into trans-

lational commands, however if the robot senses contact
with the environment, it takes the ball inputs as nominal
force values and by closing the sensory loop at the robot's
site (see fig. 1) it always exerts only those forces which
are given by the human operator /8/. Thus although we are
not feeding the forces back to the human arm (as does "bi-
lateral" force control with the well-known stability problems in case of delays), the operator is sure that the robot is fully under his control and he easily may lock up
doors, assemble parts or plug in connectors. In other words,
the human operator (via stereovision and 3D-graphics) is
enclosed in the feedback loop on a very high level but low
band-width, while the low level sensory loops are closed
on-board at the robot directly with high bandwith. Thus we
try to prepare a supervisory control technique that will
shift more and more autonomy to the robot while always
offering real-time human interference.

Looking a little bit more into details (e.g. /7/) shows
that the above mentioned rudimentary teach commands $\Delta \underline{p}_T$
(i.e. the ball commands here) are in case of sensory con-
tact (fig. 6) projected into a position controlled subspace
to yield the position feedforward component $\Delta \underline{p}_{T,cp}$ and into
the orthogonal sensorcontrolled subspace $(\underline{f}_{T,cf})$; there
they are either neglected in case nominal sensor values
have to be kept constant autonomously or they are counter-
balanced with the wrist sensor data as mentioned above
yielding a robot that is fully under human control. We
treat the range finders as pseudo-force-sensors, too, thus
arriving at a unified treatment of completely different
sensors. For the generation of subspaces we use kind of a
simple knowledge-base, which depending on the present task
together with the actual sensor data generates these subspaces automatically.

b) Predictive control and estimation schemes in sensor-controlled telerobotics

When teleoperating a robot in spacecraft from ground or from another spacecraft so far away that a relay satellite is necessary for communication, the delay times are the crucial problem. At present, we think that predictive computer graphics is the only way to overcome the main problems. Fig. 7 is to outline that the human operator at the remote workstation handles the steering ball by looking at a "predicted" graphics (e.g. wire frame) model of the robot. The control commands issued to this instantaneously reacting robot simulator are sent to the remote robot as well using the time-delayed transmission links. Now the ground-station computer and simulation system contains a model of the uplink and downlink delay lines as well as a model of the actual states of the real robot and its environment (especially moving objects). Note that we have several alternatives to superimpose the predicted robot model (augmented by predictions of any other moving parts) with other information representations:

- the presently received (of source delayed) TV-stereo or mono image in case that the globally fixed camera pair is active

- the "delayed" graphics image derived from this delayed TV-image (including the case of hand-mounted cameras) and other sensory data

- the actual graphics image as derived from the state space model of robot and environment.

We do not have a final conclusion on what the most efficient way of superposition would be, but there is of course evidence that the most crucial problems lie in the derivation

of "output data" (e.g. positions/orientations) of moving
objects from say stereo images and range finders. But we
treat this as a special preprocessing problem not discussed
in more detail here.

Thus for all moving systems in the robot's working cell
we set up a linearized cartesian state space model (for
simplicity in single-input/single-output form)

$$\underline{x}(k+1) = \underline{A} \cdot \underline{x}(k) + \underline{b} \cdot u(k) \qquad (1)$$

$$y(k) = \underline{c}^T \underline{x}(k)$$

where k denotes the k-th sampling interval, $\underline{A}$ is the
systems dynamics matrix (containing robot, moving parts and
disturbances), $\underline{b}$ the input matrix (vector), $\underline{c}^T$ the measurement matrix (vector), which indicates e.g. whether a moving
object is observed out of the robot's hand, $\underline{x}$ is the state
vector, u the control input to the robot (e.g. cartesian
motion increments) and y contains the measured variables
(e.g. cartesian position/orientation of robot and moving
parts). Concerning disturbance models, assume that the robot just as in fig. 6 does not fully execute the rudimentary commands due to sensory contact with the environment;
the missing motion in the sensor-controlled subspace may be
interpreted as a constant disturbance which is very easy to
model. Of course if an environment model is contained in
the graphic simulation, this model knowledge can be used
to set initial estimates for these disturbances at the
approximate "first contact" instants. Let us now focus onto the down-link delay interpreted as delay-line in
the output. A formal way to take such delays into account
is to enlarge the system order by additional states.

$$z_1(k+1) = \underline{c}^T \underline{x}(k) = y(k+nd)$$

$$z_2(k+1) = \underline{c}^T \underline{x}(k-1) = z_1(k) \qquad (2)$$

$$\dot{z}_{nd}(k+1) = \underline{c}^T \underline{x}(k-nd+1) = z_{nd-1}(k)$$

where nd is the (presumably integer) number of sampling

intervals contained in the down-link delay time. Thus we arrive at an augmented system

$$
\underline{x}^*(k+1) = \left[\begin{array}{cc} \begin{array}{c} \underline{A} \\ \underline{c}^T \end{array} & \underline{O} \\ \hline \begin{array}{c} 1 \\ 1 \\ \\ \underline{O} \\ \\ 1 \end{array} & \begin{array}{c} \\ \ddots \\ O \\ O \end{array} \end{array} \right] \underline{x}^*(k) + \left[\begin{array}{c} \underline{b} \\ 0 \\ \vdots \\ \vdots \\ 0 \end{array} \right] u(k) \qquad (3)
$$

$$\underbrace{}_{\underline{A}^*} \qquad\qquad \underbrace{}_{\underline{b}^*}$$

$$y(k) = \underbrace{\left[\begin{array}{ccccccc} 0 & 0 & 0 & 0 & 0 & 0 & 1 \end{array} \right]}_{\underline{c}^T} \underline{x}(k)$$

This system may be of high order (e.g. if the down-link delay is 2 sec. and the sampling interval 40 msec., then we arrive at nd = 50), yet we are interested in simple procedures for estimating and predicting the system's state. Especially we are interested in an observer gain $\underline{K}$ for the general estimation equation

$$
\begin{aligned}
\hat{\underline{x}}(k+1) &= \underline{A}\, \hat{\underline{x}}(k) + \underline{b}\, u(k) + \underline{K}\, y(k) - \hat{y}(k) \qquad (4) \\
\hat{y}(k) &= \underline{c}^T \hat{\underline{x}}(k)
\end{aligned}
$$

and we know [11] that a characteristic polynomial $p(\lambda)$ for the estimation error dynamics is guaranteed by the gain

$$\underline{K} = -p(\underline{A})\underline{q}_n \qquad (5)$$

where $p(\underline{A})$ means that $\underline{A}$ has to be substituted into the characteristic polynomial and $\underline{q}_n$ is the last column of the inverse observability matrix $\underline{Q}^{-1}$. Let us request a time-optimal estimation, i.e.

$$p(\lambda) = \lambda^{n+nd} \qquad (6)$$

where n is the original system's order, then with the well-known definition of the observability matrix we arrive at

$$p(\underline{A}^*) = \underline{A}^{*^{n+n_T}} = \begin{bmatrix} \underline{A}^{n+n_T} & \\ \underline{c}^T\underline{A}^{n+n_T-1} & \\ \vdots & \underline{0} \\ \vdots & \\ \underline{c}^T\underline{A}^n & \end{bmatrix} \qquad (7)$$

$$\underline{Q}^* = \begin{bmatrix} \underline{c}^{*^T} \\ \underline{c}^{*^T}\underline{A} \\ \vdots \\ \underline{c}^{T}\underline{A}^{n+n_T-1} \end{bmatrix} = \begin{bmatrix} 0 & \cdot & \cdot & \vline & \cdot & \cdot & 0 & 1 \\ 0 & \cdot & \cdot & \vline & \cdot & \cdot & 1 & 0 \\ 0 & \cdot & \cdot & \vline & \cdot & 1 & 0 & 0 \\ & & & \vline & & & & \\ 0 & 0 & \cdot & 1 & 0 & 0 & 0 & 0 \\ \hline \underline{c}^T & & & & & & & \\ \underline{c}^T\underline{A} & & & \underline{0} & & & & \\ \underline{c}^T\underline{A}^{n-1} & & & & & & & \end{bmatrix} = \begin{bmatrix} & & & & & 1 \\ \underline{0} & & & & 1 & \\ & & & 1 & & \\ & & 1 & & & \\ \hline \underline{0} & & & \underline{0} & & \end{bmatrix} \quad (8)$$

$$\underline{Q}^{*^{-1}} = \begin{bmatrix} \underline{0} & \vline & \underline{0}^{-1} \\ & 1 & \vline \\ & 1 & \vline & \underline{0} \\ & 1 & \vline \\ 1 & & \vline \end{bmatrix} = \begin{bmatrix} \underline{0} & \vline & \underline{q}_1, \cdots \underline{q}_n \\ & 1 & \vline \\ & 1 & \vline & \underline{0} \\ & 1 & \vline \\ 1 & & \vline \end{bmatrix} \qquad (9)$$

$$\underline{q}_n^* = \begin{bmatrix} \underline{q}_n \\ 0 \\ 0 \\ \vdots \\ 0 \end{bmatrix} \qquad (10)$$

Thus the gain $\underline{K}^*$ to be applied to the delay-line system becomes

$$\underline{K}^* = -p(\underline{A}^*)\underline{q}_n^* = -\begin{bmatrix} \underline{A}^{n+n_T} \\ \underline{c}^T\underline{A}^{n+n_T-1}\underline{q}_n \\ \cdot \\ \cdot \\ \cdot \\ \cdot \\ \underline{c}^T\underline{A}^n\underline{q}_n \end{bmatrix} = -\begin{bmatrix} \underline{A}^{n_T} \\ \underline{c}^T\underline{A}^{n_T-1} \\ \cdot \\ \cdot \\ \cdot \\ \underline{c}^T \end{bmatrix} \cdot \underbrace{\begin{bmatrix} \underline{A}^n\underline{q} \end{bmatrix}}_{\underline{K}} \qquad (11)$$

In other words, in order to design the time-optimal estimator for this delay-line system, we compute the gain $\underline{K} = \underline{A}^n\underline{q}$ for the lower order system without delay and then set up the simulation structure implying software registers and correcting inputs as shown in the right half of fig. 8. Without proof we emphasize [9] that this structure is applicable to non-time-optimal gains, i.e. "slower" observer or Kalman Filter gains as well; furthermore, it can be shown, that the system works in a way so that after a disturbance has occured, of course first nd sampling intervals have to pass until the disturbance is measurable, but then the estimation error settles down as if the dead time would not be present at all. Clearly, the design of the observer matrix $\underline{K}$ has to take into account the relative precision of the different sensory informations.

The left or predictive part of fig. 8 is easier to interpret, as it just takes the best estimate of the system state and (together with the control inputs still in the up-link delay "pipe-line") performs a precomputation by solving the state-space equation (1) iteratively. Thus we arrive at an optimal prediction of robot and environment as a basis for presently issued ball (or otherwise generated) commands.

CONCLUSION

The ROTEX proposal as a first step of Germany´s engagement in space robotics aims at the demonstration of a fairly complex system with a multisensory robot on board and human telerobotic interference that makes use of sensor-based 6 dof handcontrollers, new concepts for predictive 3D-computergraphics and stereo display. For us teleoperation from ground is a very challenging technique that forces us to move even stronger into on-onboard autonomy. The control structures are thus that the human operator may step more and more towards supervisory control without changing the control loop structures.
However we feel that for a number of years remotely operating robots will show up limited intelligence only, so that human "anytime" interference will remain important for a long-time.. Advanced real-time 3D-graphics seem to become one of the most powerful man-machine interfaces in the attempts to overcome large transmission delay times by appropriate estimation schemes as outlined in the paper.

REFERENCES

1. Bejczy, A.K.: Smart sensors for smart hands. AIAA paper
    No. 78-1714, AIAA-NASA Conf. on "Smart" Remote Sensors,
    Hampton, Virginia. 14-16, Nov. 1978.

2. Lee, S., Bekey, G., Bejczy, A.K.: Computer control of
    space-borne teleoperators with sensory Feedback, IEEE Int.
    Conference on Robotics and Automation, St. Louis, Mis-
    souri, March 25-28, 1985, Proceedings S. 205-214.

3. Heindl, J., Hirzinger, G.: Device for programming move-
    ments of a robot. US-Patent: No. 4,589,810, May 20, 1986.

4. Hirzinger, G., Dietrich, J.: Multisensory robots and sen-
    sorbased path generation. IEEE Int. Conference on Robotics
    and Automation, San Francisco, April 7-10, 1986.

5. Craig, J.J., Raibert, M.H.: Hybrid Position/Force Control
    of Manipulators. Transaction of the ASME, Journal of Dyna-
    mic Systems, Measurements and Control, Vol 102, (6/1982)
    126-132.

6. Mason, M.T.: Compliance and force control for computer
    Controlled Manipulators, IEEE Trans. on Systems, Man and
    Cybernetics, Vol. SMC-11, No. 6, (1981) 418-432.

7. Hirzinger, G., Landzettel, K.: Sensory feedback structrur-
    es for robots with supervised Learning. IEEE Ing. Conf. on
    Robotics and Automation, St. Louis, Missouri, March 1985.

8. Hirzinger, G., Heindl, J.: Sensor programming, a new way
    for teaching a robot paths and forces torques simultaneous-
    ly. 3rd Int. Conf. on Robot Vision and Sensory Controls,
    Cambridge, Massachusetts/USA, Nov. 7-10, 1983.

9. Hirzinger, G.: On the control of sampled-data dead-time
    systems (in German) Internel Report, No. 552-77/38.

10. Sheridan, T.B.: Human supervisory control of robot
    systems. IEEE Int. Conf. on Robotics and Automation, San
    Francisco, April 7-10, 1986.

11. Ackermann, J.: Sampled-Data Control Systems, Springer-Verlag
    Berlin, Heidelberg, New York, 1985

350

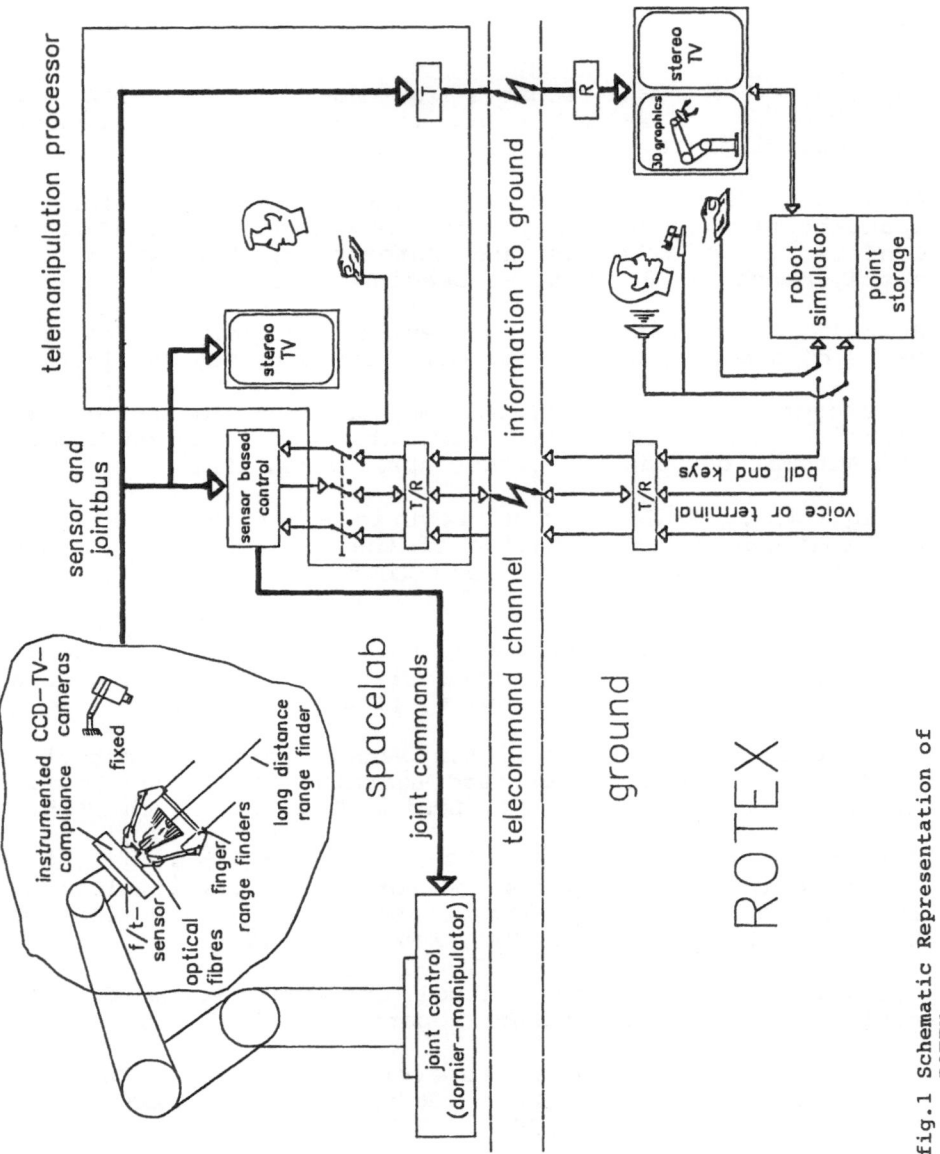

fig.1 Schematic Representation of
ROTEX

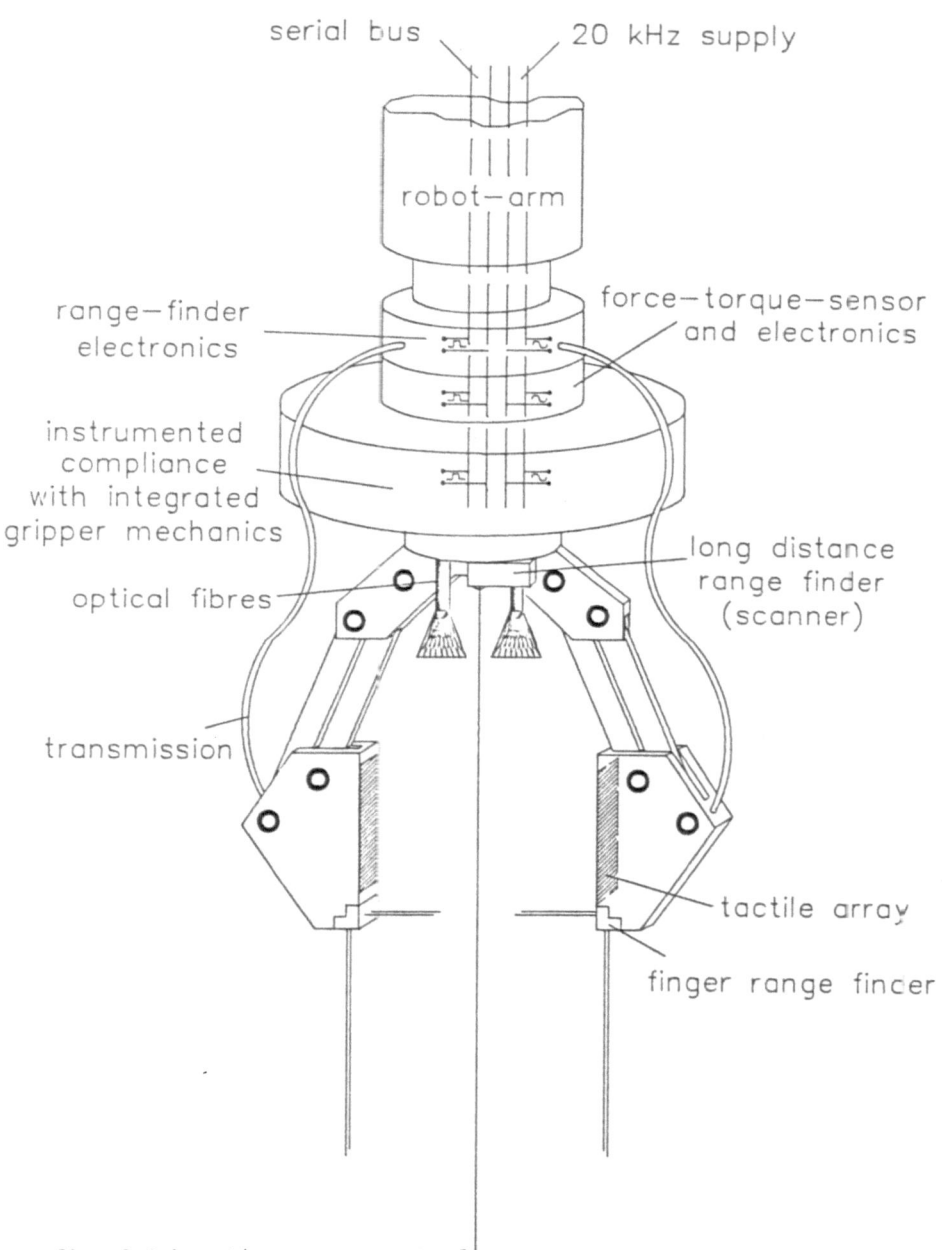

fig. 2 Schematic arrangement of sensors
        in a prototype gripper

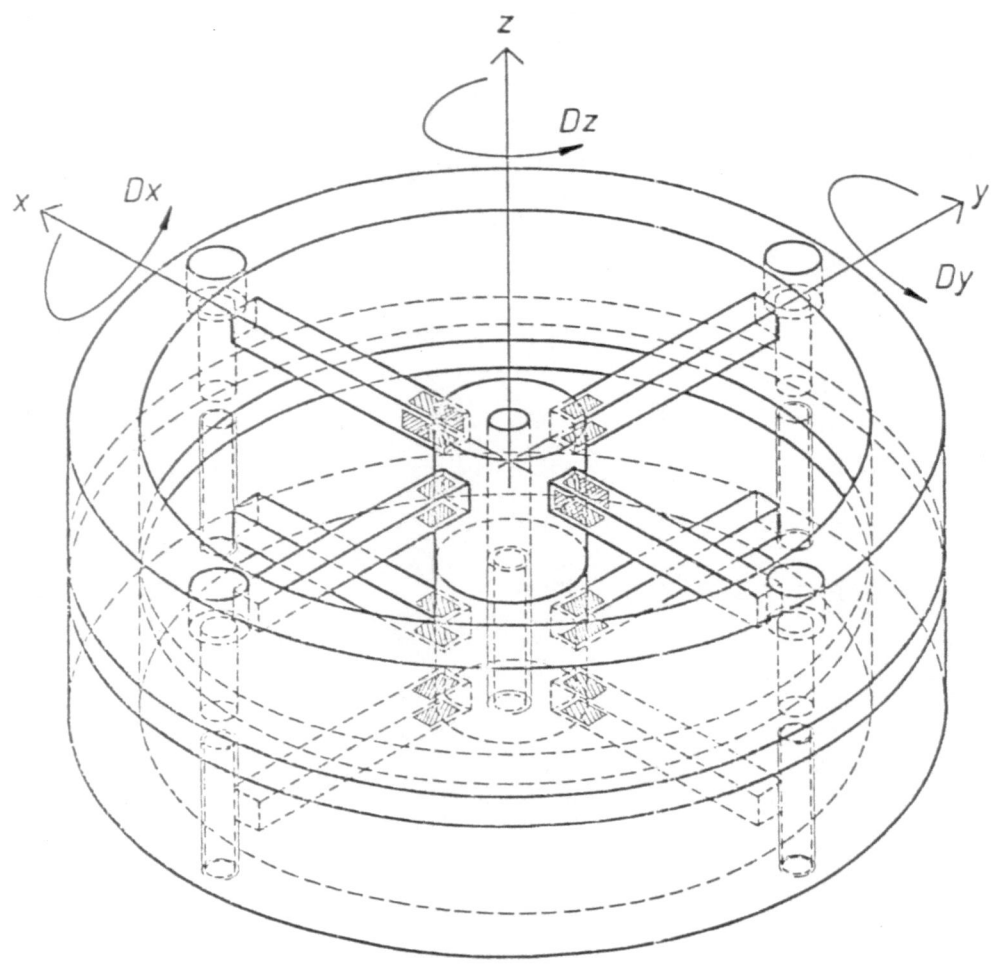

fig. 3 Stiff strain-gauge force/torque sensor

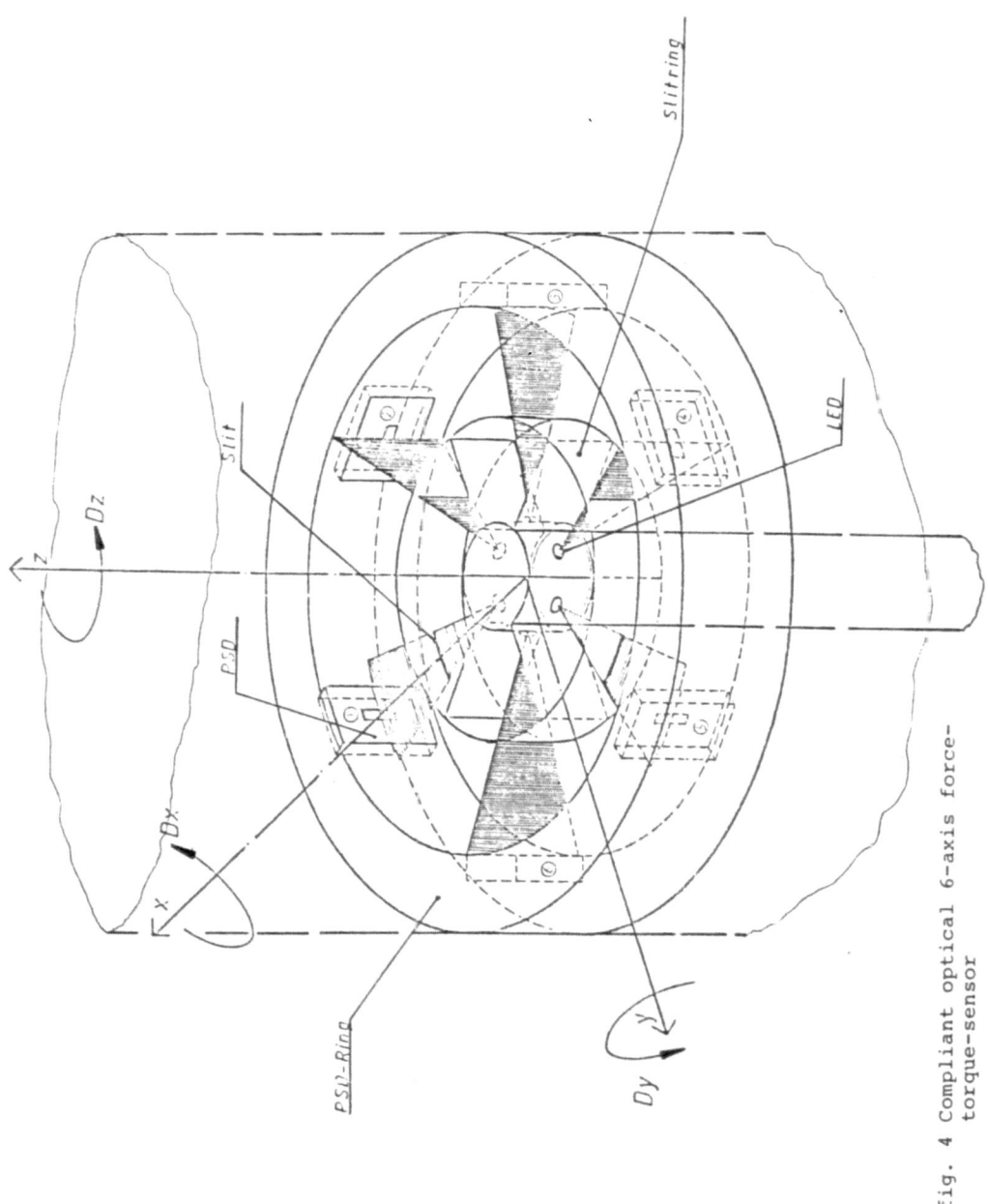

fig. 4 Compliant optical 6-axis force-
torque-sensor

fig. 5 The DFVLR sensor ball as 6 dof-handcontroller

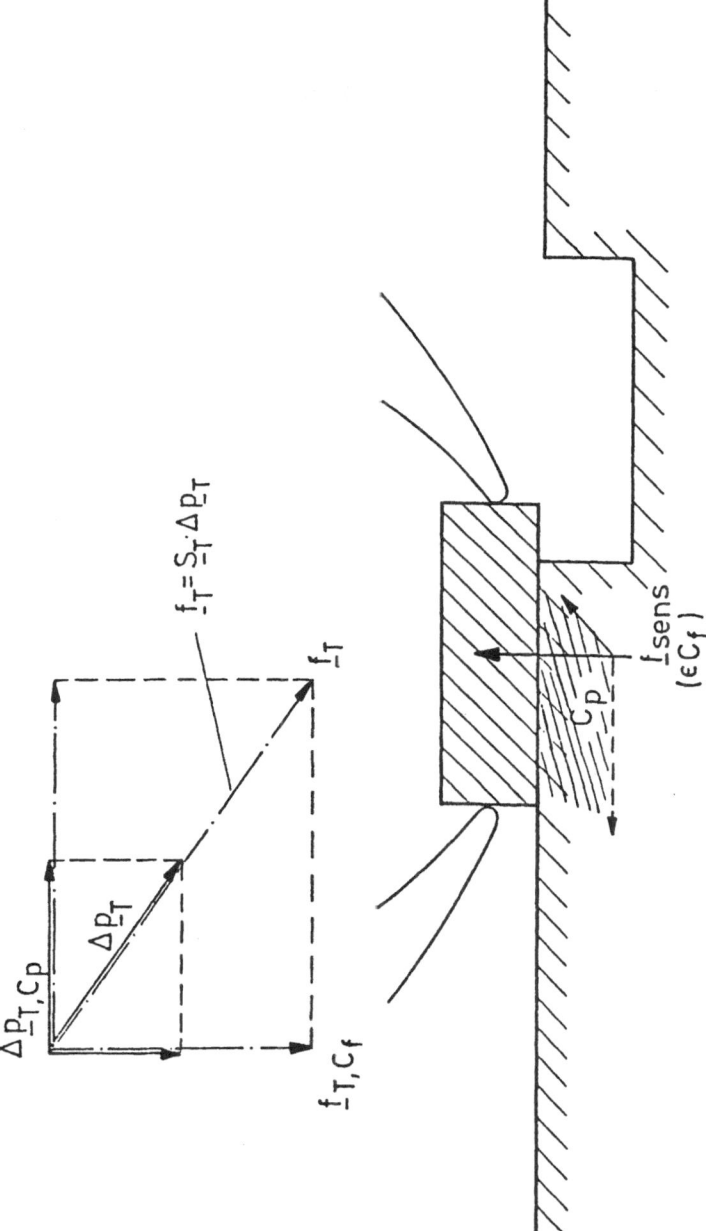

fig. 6 Projection of rudimentary commands
       into position and sensorcontrolled
       subspaces.

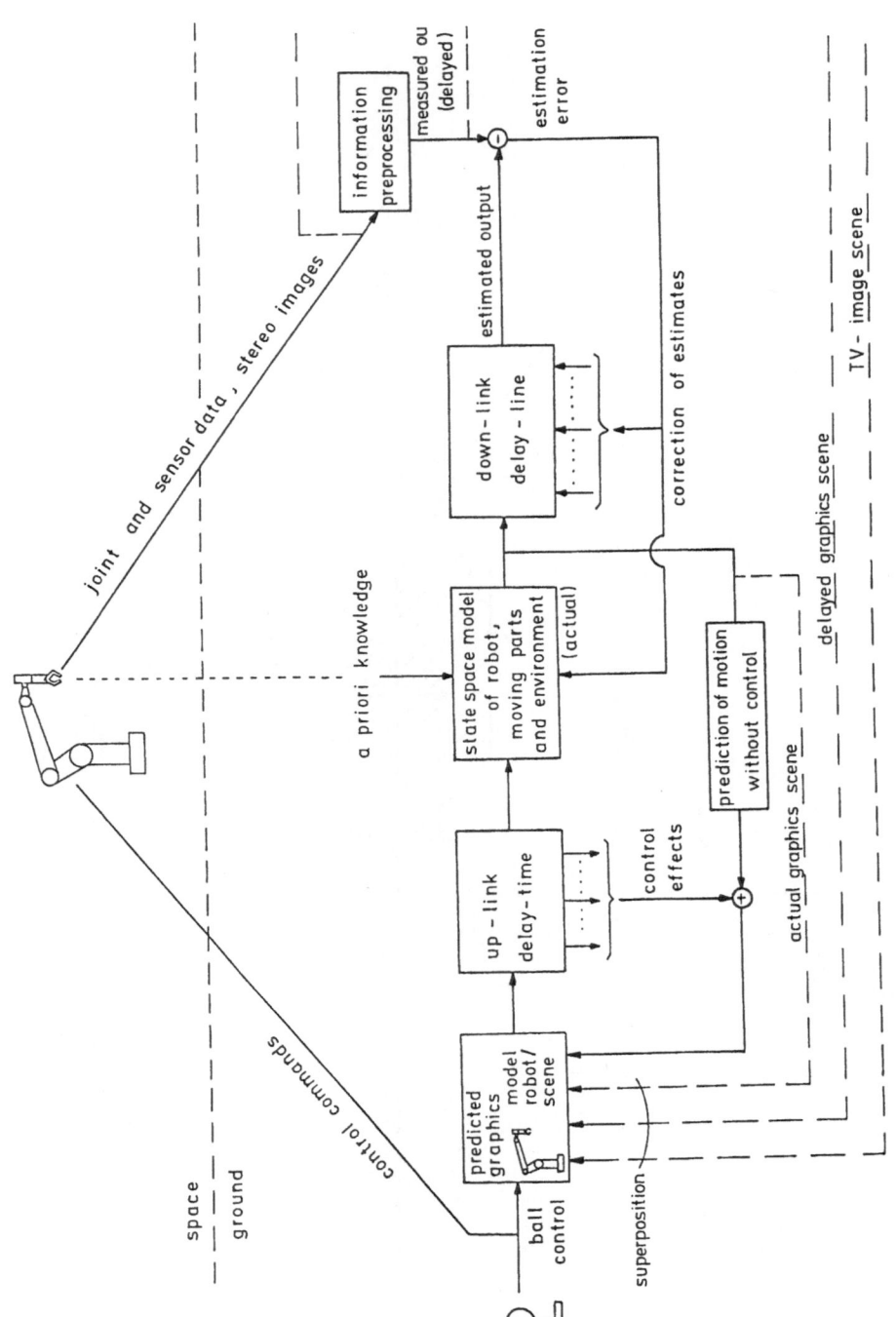

fig. 7 Block structure of predictive estimation scheme

357

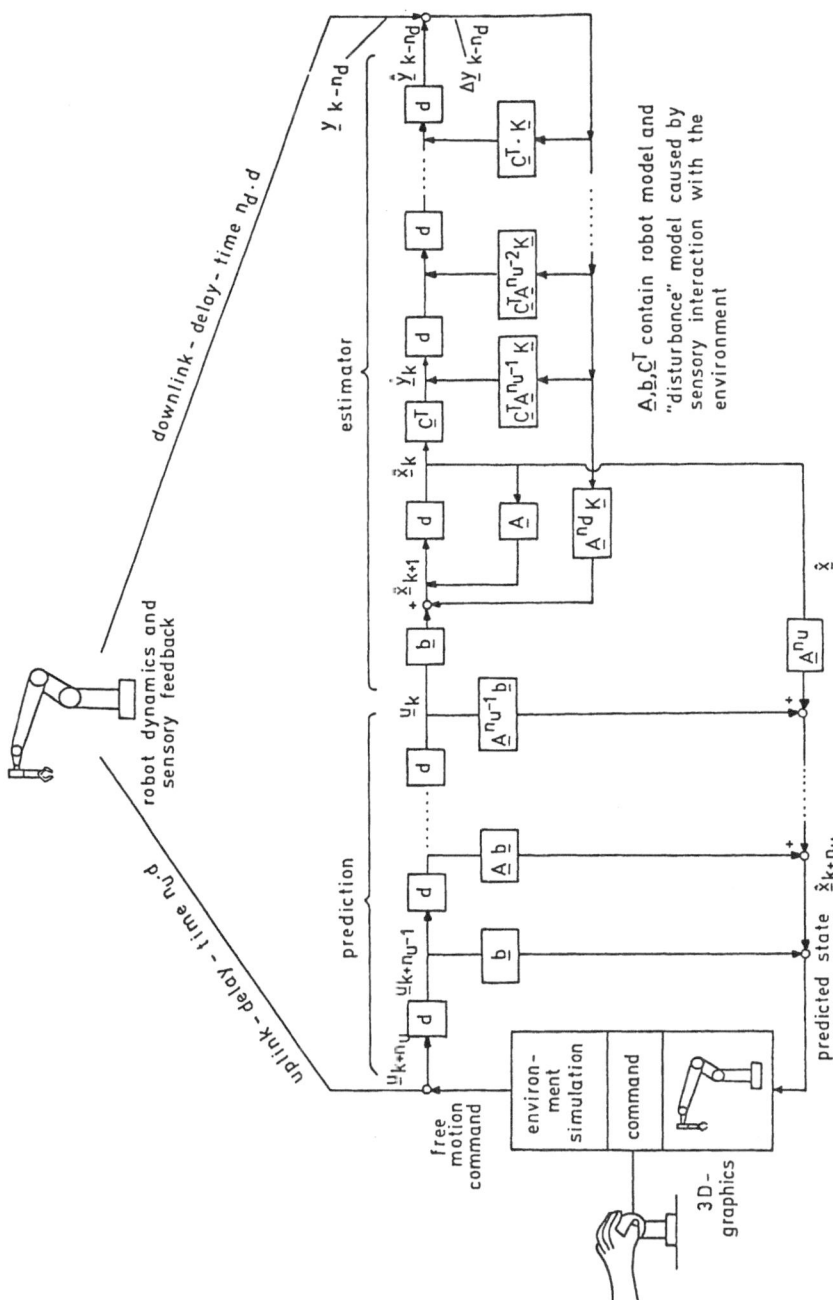

fig. 8 Detailed predictive estimation scheme for the graphic simulation of a teleoperated robot that has sensory feedback on board and a large time delay in the transmission links.

LIST OF SPEAKERS

J. Aguilar
LAAS-CNRS
7, Avenue du Colonel-Roche
31077 TOULOUSE
France

H. Van Brussel
Dept. of Mechanical Engineering
Katholieke Universiteit Leuven
Celestijnenlaan 300 B
3030 LEUVEN
Belgium

A. Cameron
Dept. of Engineering Science
University of Oxford
Parks Road
OXFORD OX1 3PJ
U.K.

A. Casals
Dept. E.S.A.I.I.
Facultat d'Informàtica Barcelona
Univ. Politècnica Catalunya
Pau Gargallo 5
08028 BARCELONA
Spain

R. Ceres
Instituto de Automática Industrial
Desvio Km 22.800, N-III, La Poveda
28500 ARGANDA DEL REY / Madrid
Spain

J.B.C. Davies
Dept. of Mechanical Engineering
Heriot-Watt University
EDINBURGH EH14 4AS
U.K.

F. Dessen
Norw. Inst. of Technology
Div. of Engineering Cybernetics
O.S. Bragstads plass 8
7034 TRONDHEIM
Norway

O.D. Faugueras
INRIA
Domaine Voluceau-Rocquencourt
BP 105, 78153 LE CHESNAY Cédex
France

A.A. Goldenberg
Dept. of Mechanical Engineering
University of Toronto
5 King's College Road
TORONTO, Ontario M5S 1A4
Canada

D. Herault
Thomson-CSF, Div. Tubes Electr.
38 rue Vauthier, BP 305
92102 BOULOGNE-BILLANCOURT Cédex
France

G. Hirzinger
DFVLR Oberpfaffenhofen
8031 WESSLING
FRG

P. Levi
Institut für Informatik
Technische Universität München
Arcisstr. 21
8000 MÜNCHEN 2
FRG

V. Llario
Dept. E.S.A.I.I.
Facultat d'Informàtica Barcelona
Univ. Politècnica Catalunya
Pau Gargallo 5
08028 BARCELONA
Spain

S. Monchaud
INSA - LATEA
35043 RENNES Cédex
France

O. Monga
INRIA
Domaine Voluceau-Rocquencourt
BP 105, 78153 LE CHESNEY Cédex
France

H. Moravec
Robotic Institute
Carnegie-Mellon University
PITTSBURGH, PA 15213
USA

A. Mukerjee
Dept. of Computer Science
Texas A&M University
COLLEGE STATION, TX 77843-3112
USA

A. Oosterlinck
Director E.S.A.T. Division
University of Leuven
de Croylaan 52B
3030 HEVERLEE
Belgium

A. Romiti
Dip. di Meccanica
Politecnico di Torino
Corso Duca degli Abruzzi 24
TORINO 10129
Italy

Y. Shirai
Dept. of Mechanical Engineering
Fac. of Engineering
Osaka University
2-1- YAMAKAOKA, SUITASI
Japan, 565

J. Schoenwald
Science Center
Rockwell Intern. Corporation
1049 Camino Dos Rios
THOUSAND OAKS, CA 91360
USA

H. Urban
Krupp Atlas Elektronik GmbH
Sebaldsbrücker Heerstr. 235
2800 BREMEN 44
FRG

P.W. Verbeek
Dept. of Applied Physics
Technical University of Delft
Lorentzweg 1
2628 CJ DELFT
Netherlands

D.G. Whitehead
Dept. of Electronic Engineering
University of Hull
HULL, N. Humberside HU6 7RX
U.K.

LIST OF PARTICIPANTS

J. Amat
Dept. E.S.A.I.I.
Facultat d'Informàtica Barcelona
Univ. Politècnica Catalunya
Pau Gargallo 5 - 08028 BARCELONA
    (Spain)

J.M. Balchen
The Norwegian Inst. of Technol.
Div. Engineering Cybernetics
O.S. Bragstads plass 8
N-7034 THRONDHEIM-NTH
    (Norway)

L.Basañez
Inst. de Cibernètica
Diagonal 647. 080028 BARCELONA
    (Spain)

M. Bergamasso
University Pisa
Via Diotesalvi 2
46100 PISA
    (Italy)

F. Bertran
E.T.S. Ing. Ind
Maria Zambrano, 50
50015 ZARAGOZA
    (Spain)

A. Català
Dept. E.S.A.I.I.
Facultat d'Informàtica Barcelona
Univ. Politècnica Catalunya
Pau Gargallo 5 - 08028 BARCELONA
    (Spain)

A. Civit
Facultad de Física
Reina Mercedes s/n
41012 SEVILLA
    (Spain)

E. Díaz Delgado
Facultad de Física
Universidad de Sevilla
Reina Mercedes s/n.
41012 SEVILLA
    (Spain)

A.B. Martínez
Dept. E.S.A.I.I.
Facultat d'Informàtica Barcelona
Univ.Politècnica Catalunya
Pau Gargallo 5 - 08028 BARCELONA
    (Spain)

Mr. Meléndez
Thomson Tubos Electrónicos
c/ Albacete 5.
28027 MADRID
    (Spain)

S. Mimenza
Ikerlan
Apartado de Correos 146
20500 MONDRAGON- (Guipúzcoa)
    (Spain)

E. Montseny
Dept. E.S.A.I.I.
Facultat d'Informàtica Barcelona
Univ. Politècnica Catalunya
Pau Gargallo 5 -08028 BARCELONA
    (Spain)

J. No
Inst. de Automática Industrial
Ctra. Madrid-Valencia, Km.22,800
ARGANDA DEL REY. (Madrid)
    (Spain)

G. Ormazabal
Ikerlan
Apartado de Correos 146
20500 MONDRAGON. (Guipúzcoa)
    (Spain)

M. Papadopoulos
Applied Electronics Lab.
School of Engineering
University of Patras
PATRAS
(Greece)

F. Puig
Automática e Instrumentación
c/ Concepción Arenal 5-7
08027 BARCELONA
(Spain)

H.Sandkuehler
Krupp Atlas Electronik GmbH
Sebaldsbrüker Heerst. 235
D 2800 BREMEN 44
(RFA)

A. Sanfeliu
Inst. de Cibernètica (UPC-CSIC)
Diagonal 647. 08028 BARCELONA
(Spain)

L.Tremosa
Revista de Robótica
Avda. de Madrid 95,Atc.1ª
08028 BARCELONA
(Spain)

NATO ASI Series F

Including Special Programme on Sensory Systems for Robotic Control (ROB)

NATO ASI Series F

Vol. 22: Software System Design Methods. The Challenge of Advanced Computing Technology. Edited by J. K. Skwirzynski. XIII, 747 pages. 1986.

Vol. 23: Designing Computer-Based Learning Materials. Edited by H. Weinstock and A. Bork.
IX, 285 pages. 1986.

Vol. 24: Database Machines. Modern Trends and Applications. Edited by A. K. Sood and
A. H. Qureshi. VIII, 570 pages. 1986.

Vol. 25: Pyramidal Systems for Computer Vision. Edited by V. Cantoni and S. Levialdi. VIII,
392 pages. 1986. *(ROB)*

Vol. 26: Modelling and Analysis in Arms Control. Edited by R. Avenhaus, R. K. Huber and
J. D. Kettelle. VIII, 488 pages. 1986.

Vol. 27: Computer Aided Optimal Design: Structural and Mechanical Systems. Edited by
C. A. Mota Soares. XIII, 1029 pages. 1987.

Vol. 28: Distributed Operating Systems. Theory und Practice. Edited by Y. Paker, J.-P. Banatre
and M. Bozyiğit. X, 379 pages. 1987.

Vol. 29: Languages for Sensor-Based Control in Robotics. Edited by U. Rembold and
K. Hörmann. IX, 625 pages. 1987. *(ROB)*

Vol. 30: Pattern Recognition Theory and Applications. Edited by P. A. Devijver and J. Kittler. XI,
543 pages. 1987.

Vol. 31: Decision Support Systems: Theory and Application. Edited by C. W. Holsapple and
A. B. Whinston. X, 500 pages. 1987.

Vol. 32: Information Systems: Failure Analysis. Edited by J. A. Wise and A. Debons. XV, 338
pages. 1987.

Vol. 33: Machine Intelligence and Knowledge Engineering for Robotic Applications. Edited by
A. K. C. Wong and A. Pugh. XIV, 486 pages. 1987. *(ROB)*

Vol. 34: Modelling, Robustness and Sensitivity Reduction in Control Systems. Edited by
R. F. Curtain. IX, 492 pages. 1987.

Vol. 35: Expert Judgment and Expert Systems. Edited by J. L. Mumpower, L. D. Phillips, O. Renn
and V. R. R. Uppuluri. VIII, 361 pages. 1987.

Vol. 36: Logic of Programming and Calculi of Discrete Design. Edited by M. Broy. VII, 415
pages. 1987.

Vol. 37: Dynamics of Infinite Dimensional Systems. Edited by S.-N. Chow and J. K. Hale. IX, 514
pages. 1987.

Vol. 38: Flow Control of Congested Networks. Edited by A. R. Odoni, L. Bianco and G. Szegö.
XII, 355 pages. 1987.

Vol. 39: Mathematics and Computer Science in Medical Imaging. Edited by M. A. Viergever and
A. Todd-Pokropek. VIII, 546 pages. 1988.

Vol. 40: Theoretical Foundations of Computer Graphics and CAD. Edited by R. A. Earnshaw.
XX, 1246 pages. 1988.

Vol. 41: Neural Computers. Edited by R. Eckmiller and Ch. v. d. Malsburg. XIII, 566 pages.
1988.

NATO ASI Series F